Networking and Computation

Thomas G. Robertazzi • Li Shi

Networking and Computation

Technology, Modeling and Performance

Second Edition

 Springer

Thomas G. Robertazzi
Stony Brook University
Stony Brook
New York, USA

Li Shi
Stony Brook University
Stony Brook
New York, USA

Electronic Supplementary Material The online version of this book (https://doi.org/10.1007/978-3-030-36704-6)
contains supplementary material, which is available to authorized users.

ISBN 978-3-030-36706-0   ISBN 978-3-030-36704-6   (eBook)
https://doi.org/10.1007/978-3-030-36704-6

This Springer imprint is published by the registered company Springer Nature Switzerland AG.
The registered company address is: Gewerbestrasse 11, 6330 Cham, Switzerland

To the next generation
—Thomas G. Robertazzi

To my wife, Pengmei, and my son, Brayden
—Li Shi

Preface

We are living in a world of terabit networking and exaflops computing. Both networking and computing are equally important and more and more are used in an integrated manner. Networking and computing courses in particular are popular with students who, even in college, sense the field's importance and excitement.

The purpose of this book is to provide an upper-level undergraduate/first-year graduate text covering both networking and computing, largely with a mathematical and algorithmic flavor. While there are many existing books on networking and computing, finding both topics treated in the same text is less common. This book should appeal to electrical and computer engineers, computer and information scientists, and applied mathematicians who are either students or practitioners.

To some extent this book is based on undergraduate and graduate courses the first author has taught since 1983. The book starts with an introductory chapter on networking and computing technology. Chapter 2 covers fundamental stochastic (i.e., random) models for networking. Chapter 3 provides an introduction to queueing theory, a widely used tool for modeling and predicting the performance of networked systems. In Chap. 4, fundamental deterministic algorithms for networking are studied. Chapter 5 provides an extensive and updated tutorial on divisible load scheduling theory, a relatively new performance evaluation methodology that has applications to parallel processing. Amdahl's and other related laws are examined in Chap. 6. Finally Chap. 7, by the second author, reviews machine learning for networking. This is a topic of much recent interest.

New material has been added to the original 2007 work *Networks and Grids: Technology and Theory* to create this text.

At the undergraduate level, one can teach the quantitative material (say Chaps. 2 and 4 and parts of Chaps. 3 and 5) first while students are fresh and more receptive and save the qualitative technology description for the second half of the course using a text such as *Introduction to Computer Networking* by the first author. At the graduate level, one can focus on the quantitative and algorithmic material in depth, allowing students more independence in learning the technology.

In terms of acknowledgements, we would like to acknowledge the assistance of Profs. Scott Smolka and Wendy Tang in reading certain sections of the manuscript. Thank you to Tricia Chigan, Victor Frost, Shan Lin, and Fan Ye for making suggestions for this edition, not all of which I followed. This work has benefited from the manuscript's use by the first author's students at Stony Brook. Thanks go to Carlos Gamboa, Jui Tsun Hung, Shevata Kaul, Mequanint Moges, and Harpreet Pruthi for creating some of the figures. Thank you to Guanchao Feng, Yang Liu, Li Shi, and Hechuan Wang for assistance with manuscript preparation. The first author has benefited from the IT assistance of John Joseph and Anthony Olivo. Parts of this manuscript were typed by Sandy Pike and Rachel Swensen to whom the first author is thankful for a great job. The first author's work at Stony Brook has been made easier over the years by the administrative and secretarial skills of Judy Eimer, Carolyn Huggins, Debbie Kloppenburg, Maria Krause, Rachel Ingrassia, Susan Nastro, Angela Scauso, Cathy Mooney, and Nancy Davies.

The authors are grateful for the excellent editorial efforts of Mary James and Zoe Kennedy and the excellent production effort of Brian Halm, Maria David and Kala Palinisamy.

Finally, the first author dedicates the book to the next generation. The second author dedicates the book to his wife, Pengmei, and his son, Brayden.

Stony Brook, NY, USA   Thomas G. Robertazzi
Santa Monica, CA, USA   Li Shi

Contents

Introduction

Abstract

In this introductory chapter basic transmission and networking technologies, parallel processing, and machine learning for networking are discussed. This includes the topics of coaxial cables, twisted pair wiring, fiber optics, microwave line of sight, satellites in various orbits, cellular systems, ad hoc networking, and wireless sensor networks. Also discussed are multiplexing (frequency division multiplexing (FDM), time division multiplexing (TDM), frequency hopping, and direct sequence spread spectrum), circuit switching versus packet switching and layered protocols. The chapter concludes by introducing parallel processing, divisible load scheduling, Amdahl's law, and also machine learning for networking.

1.1 Networking Overview

There is something about technology that allows people and their computers to communicate with each other that makes networking a fascinating field, both technically and intellectually.

What is a network? It is a collection of computers (nodes) and transmission channels (links) that allow people to communicate over distances, large and small. A Bluetooth personal area network may simply connect your home PC with its peripherals. An undersea fiber optic cable may traverse an ocean. The Internet and telephone networks span the globe. Networks range in size from networks on chips to deep space networks.

The Internet has been developed over the past 50 years or so. The 1980s and 1990s saw the introduction and growth of local area networks and SONET fiber networks. The 1990s and the early years of the new century have seen the development and expansion of WDM fiber multiplexing. New wireless standards continue to appear. Cloud computing and data centers are increasingly becoming a foundation of today's networking/computing world. Networking is also becoming more software oriented.

This chapter now starts with an introduction to the applied foundation of networking.

1.2 Achieving Network Connectivity

A variety of transmission methods, both wired and wireless, are available today to provide connectivity between computers, networks, and people. Wired transmission media include coaxial cable, twisted pair wiring, and fiber optics. Wireless technology includes microwave line of sight, satellites, cellular systems, ad hoc networks, and wireless sensor networks. We now review these media and technologies.

1.2.1 Coaxial Cable

This is the thick cable you may have in your house to connect your cable TV set up box to the outside wiring plant. This type of cable has been around for many years and is a mature technology (Wikipedia). The coaxial cable was invented by Oliver Heaviside in 1880 and operates as a transmission line. The first transatlantic coaxial cable was deployed in 1956. While still popular for cable TV systems today, it was also a popular choice for wiring local area networks in the 1980s. It was used in the wiring of the original 10 Mbps Ethernet.

A coaxial cable has four parts: a copper inner core, surrounded by insulating material, surrounded by a metallic outer conductor; finally surrounded by a plastic outer cover. Essentially in a coaxial cable, there are two wires (copper inner core and outer conductor) with one geometrically inside the other. This configuration reduces interference to/from the coaxial cable with respect to other nearby wires.

© Springer Nature Switzerland AG 2020
T. G. Robertazzi, L. Shi, *Networking and Computation*, https://doi.org/10.1007/978-3-030-36704-6_1

The bandwidth of a coaxial cable is on the order of 1 GHz. How many bits per second can it carry? Modulation is used to match a digital stream to the spectrum carrying ability of the cable. Depending on the efficiency of the modulation scheme used, 1 bps requires anywhere from a fraction of a hertz to several hertz. For short distances, a coaxial cable may use 8 bits/Hz or carry 8 Gbps.

There are also different types of coaxial cable. One with a $50\,\Omega$ termination is used for digital transmissions. One with a $75\,\Omega$ termination is used for analog transmissions or cable TV systems. The most frequently used coaxial cable for home television systems is RG-6.

A word is in order on cable TV systems. Such networks are locally wired as tree networks with the root node called the head end. At the head end, programming is brought in by fiber or satellite. From the head end cables (and possibly fiber) radiate out to homes. Amplifiers may be placed in this network when distances are large.

For many years, cable TV companies were interested in providing two-way service. While early limited trials were generally not successful (except for Video on Demand), in recent years cable TV has winners in broadband access to the Internet and in carrying telephone traffic.

1.2.2   Twisted Pair Wiring

Coaxial cable is generally no longer used for wiring local area networks. One type of replacement wiring has been twisted pair. Twisted pair wiring typically had been previously used to wire phones to the telephone network. A twisted pair consists of two wires twisted together over their length. The twisted geometry reduces electromagnetic leakage (i.e., cross talk) with nearby wires. Twisted pairs can run several kilometers without the need for amplifiers. The quality of a twisted pair (carrying capacity) depends on the number of twists per inch.

About 1990, it became possible to send 10 Mbps (for Ethernet) over unshielded twisted pair (UTP). Higher speeds are also possible if the cable and connector parameters are carefully implemented.

One type of unshielded twisted pair is category 3 UTP. It consists of four pairs of twisted pair surrounded by a sheath. It has a bandwidth of 16 MHz. Many offices used to be wired with category 3 wiring.

Category 5 UTP has more twists per inch. Thus, it has a higher bandwidth (100 MHz). Newer standards include category 6 versions (250 MHz or more) and category 7 versions (600 MHz or more). Category 8 is at 2000 MHz bandwidth for 40 Gbps Ethernet (Wikipedia).

The fact that twisted pair is lighter and thinner than coaxial cable has speeded its widespread acceptance.

1.2.3   Fiber Optics

Fiber optic cable consists of a silicon glass core that conducts light, rather than electricity as in coaxial cables and twisted pair wiring. The core is surrounded by cladding and then a plastic jacket.

Fiber optic cables have the highest data carrying capacity of any wired medium. Fiber has a capacity of approximately 100 Tbps (terabits per second or 100×10^{12} bits per second). In fact, this data rate for years has been much higher than the speed at which standard electronics could load the fiber. This mismatch between fiber speed and nodal electronics speed has been called the "electronic bottleneck." Decades ago the situation was reversed, links were slow and nodes were relatively fast. This paradigm shift has led to a redesign of protocols.

There are two major types of fiber: multi-mode and single mode. Pulse shapes are more accurately preserved in single mode fiber, lending to a higher potential data rate. One of the reasons multi-mode fibers have a lower performance is dispersion. Under dispersion, square digital pulses tend to spread out in time, thus lowering the potential data rate. Special pulse shapes (such as hyperbolic cosines) called solitons, that dispersion is minimized for, have been the subject of research.

Mechanical fiber connectors to connect two fibers can lose 10% of the light that the fiber carries. Fusing two ends of the fiber results in a smaller attenuation.

Fiber optic cables today span continents and are laid across the bottom of oceans between continents. Repeater distances are 70–150 km (Wikipedia). Fiber optics is also used by organizations to internally carry telephone, data, and video traffic.

Wavelength division multiplexing (WDM) systems multiplex multiple optical signals, each at a different optical frequency (or color), on a single fiber to boost the data per time a fiber carries. For instance, a fiber with 100 channels, each of 10 Gbps, carries a total of 1000 Gbps or 1 Tbps. In 2018 researchers implemented a system carrying 159 Tbps over a single multi-mode fiber.

1.2.4   Microwave Line of Sight

Microwave radio energy travels largely in straight lines. Thus, some network operators construct networks of tall towers kilometers apart and place microwave antennas at different heights on each tower. While the advantage is that there is no need to dig trenches for cables, the expense of tower construction and maintenance must be taken into account. Miniature units to connect two nearby buildings have been developed. It should be noted that transmissions at microwave frequencies are also used in other applications such as cellular phones and space communications (including satellite communications).

1.2.5 Satellites

There are about 3700 satellites of all types in orbit, of which about 1100 are operational (Wikipedia). Satellites are deployed for such functions as communications, navigation, weather, earth sensing, and research. Satellites are extensively used for communication purposes. They fill certain technological niches very well: providing connectivity to mobile users, for large area broadcasts and for communications for areas with poor infrastructure. Two major communication satellite architectures are geostationary satellites and low earth orbit satellites (LEOS). Both are now discussed.

Geostationary Satellites

You may recall from a physics course that a satellite in a low orbit (hundreds of kilometers) around the equator seems to move against the sky. As its orbital altitude increases, its apparent movement slows. At a certain altitude of approximately 36,000 km from the earth's surface, it appears to stay in one spot in the sky, over the equator, 24 h a day. In reality, the satellite is moving around the earth but at the same angular speed that the earth is rotating, giving the illusion that it is hovering in the sky.

This is very useful. For instance, a satellite TV service can install home antennas that simply point to the spot in the sky where the satellite is located. Alternatively, a geostationary satellite can broadcast a signal to a large area (its "footprint") 24 h a day. Some geostationary satellite orbital locations are more economically preferable than others, depending on which regions of the earth are under the location.

A typical geostationary satellite will have several dozen transponders (relay amplifiers), each with a bandwidth of tens of MHz (Tanenbaum 02). Such a satellite may weigh several thousand kilograms and consume several kilowatts using solar panels.

The number of microwave frequency bands used has increased over the years as the lower bands have become crowded and technology has improved. Frequency bands include L (1.5/1.6 GHz), S (1.9/2.2 GHz), C (4/6 GHz), Ku (11/14 GHz), and Ka (20/30 GHz) bands. Here the first number is the downlink band and the second number is the uplink band. The actual bandwidth of a signal may vary from about 15 MHz in the L band to several GHz in the Ka band (Tanenbaum 02). However there are higher bands extending up to about 170 GHz used for satellite communication and broadcasting, military communications, and also radio astronomy and automotive radar.

It should be noted that extensive studies of satellite signal propagation under different weather and atmospheric conditions have been conducted. Excess power for overcoming rain attenuation is often budgeted above 11 GHz.

Low Earth Orbit Satellites

Low earth orbit (LEO) is an orbit around the earth of 1200 miles or less of altitude. Most spacecraft and satellites that have been sent into space have been in a LEO. The most famous such early communication system was Iridium from Motorola. It received its name because the original proposed 77 satellite network has the same number of satellites as the atomic number of the element Iridium. In fact, the actual system orbited had 66 satellites but the system name Iridium was kept.

The purpose of Iridium was to provide a global cell phone service. One would be able to use an Iridium phone anywhere in the world (even on the ocean or in the Arctic). Unfortunately, after spending five billion dollars to deploy the system, talking on Iridium cost a dollar or more a minute while local terrestrial cell phone service was under 25 cents a minute. While an effort was made to appeal to business travelers, the system was not profitable and was sold and is now operated by a private company.

Technologically though, the Iridium system is interesting. There are 11 satellites in each of 6 polar orbits (passing over the North Pole, south to the South Pole, and back up to the North Pole, see Fig. 1.1).

At any given time, several satellites are moving across the sky over any location on earth. Using several dozen spot beams, the system can support almost a quarter of a million conversations. Calls can be relayed from satellite to satellite.

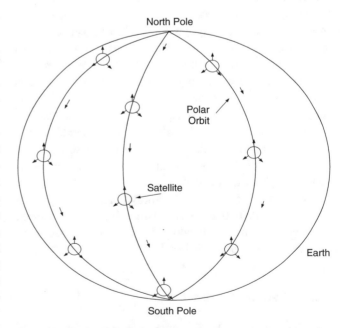

Fig. 1.1 Low earth orbit satellites (LEOS) in polar orbits

It should be noted that when Iridium was hot, several competitors were proposed but not built. One used a "bent pipe" architecture where a call to a satellite would be beamed down from the same satellite to a ground station and then sent over the terrestrial phone network rather than being relayed from satellite to satellite. This was done in an effort to lower costs and simplify the design. During 2017–2019 launches took place of a second generation of 66 Iridium satellites called Iridium NEXT. Iridium NEXT has data transmission capabilities that were not emphasized in the original design. More recently SpaceX has plans to orbit 12,000 LEO satellites to create a network for worldwide broadband coverage.

Other Orbits: MEO and Molniya

There are other possible orbits for communications satellites besides the geosynchronous and low earth orbits (Wikipedia).

A satellite in *medium earth orbit (MEO)* is between 2000 and 35,786 km above the earth. Satellites in MEO operate in a fashion similar to LEO satellites but are above a spot on the earth's surface for longer periods than LEO satellites (normally from 2 to 8 h for MEOS). Because of the larger staying power and larger footprint (i.e., the area of earth coverage under a satellite) of MEO satellites, fewer are needed to form a global network. However signal delay between the earth's surface and a MEO satellite is larger and signals are weaker (unless more power is used) than for a LEO satellite.

As one moves further north (or south), geostationary satellites which are over the equator appear lower in the sky (that is near the horizon). This creates problems of multi-path interference as signals bounce off the ground so that multiple time shifted signals reach ground receivers. This creates a need for larger signal power. These signal power problems can be mitigated by launching satellites in *Molniya orbit* for countries such as Russia. Such orbits have the satellites spend a good portion of time over far northern latitudes with the satellite signal footprint moving minimally. A Molniya orbit satellite orbits the earth twice a day, being accessible over a spot in the far northern latitudes for 6–9 h every second orbit. Thus, three Molniya orbit satellites can provide continuous connectivity for the intended region.

There are other issues with Molniya orbit satellites. Less energy is needed to place a satellite in a Molniya orbit than a geostationary orbit. Molniya satellites will also move through the Van Allen radiation belt four times a day.

1.2.6   Cellular Systems

Starting around the early 1980s, cellular telephone systems which provide connectivity between mobile phones

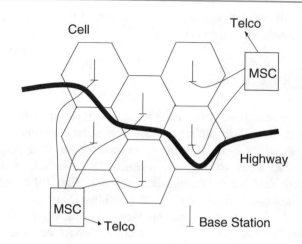

Fig. 1.2 Part of a cellular network

and the public switched telephone network were deployed. In such systems, signals go from/to a cell phone to/from a local "base station" antenna which is hard wired into the public switched telephone network. Figure 1.2 illustrates such a system. A geographic region such as a city or suburb is divided into geographic sub-regions called "cells."

In Fig. 1.2 base stations are shown at the center of cells. Nearby base stations are wired into a switching computer (the mobile switching center or MSC) that provides a path to the telephone network.

A cell phone making a call connects to the nearest base station (i.e., the base station with the strongest signal). Base stations and cell phones measure and communicate received power levels. If one is driving and one approaches a new base station, its signal will at some point become stronger than that of the original base station one is connected to and the system will then perform a "handoff." In a handoff, connectivity is changed from one base station to an adjacent one. Handoffs are transparent, the talking user is not aware when one occurs.

Calls to a cell phone involve a paging like mechanism that activates (rings) the called user's phone.

The first cellular system was deployed in 1979 in Japan by NTT. The first US cellular system was AMPS (Advanced Mobile Phone System) from AT&T. It was first deployed in 1983. These were first generation analog systems. Second generation systems (first deployed in 1991) were digital. Second generation systems made texting ubiquitous. The most popular was the European originated GSM (Global System for Mobile), which was installed all over the world. Third and fourth generation cellular systems provide increased data rates for such applications as Internet browsing, picture transmission, and streaming. Third generation cellular systems enabled the widespread adoption of smart phones and started with the WCDMA standard in Japan in 2001. Fourth generation cellular systems promised increases in data rate up to a factor of ten. The first fourth generation systems were WIMAX in North America and LTE, first offered in Scandinavia. Fifth generation systems will promote even higher data rates (Wikipedia).

1.2.7 Ad Hoc Networks

Ad hoc networks [172, 179] are radio networks where (often mobile) nodes can come together, transparently form a network without any user interaction and maintain the network as long as the nodes are in range of each other and energy supplies last [167, 187]. In an ad hoc network messages hop from node to node to reach an ultimate destination. For this reason ad hoc networks used to be called multi-hop radio networks. In fact, because of the nonlinear dependence of energy on transmission distance, the use of several small hops uses much less energy than a single large hop, often by orders of magnitude.

Ad hoc network characteristics include multi-hop transmission, possibly mobility and possibly limited energy to power the network nodes. Applications include mobile networks, emergency networks, wireless sensor networks, and ad hoc gatherings of people, as at a convention center.

Routing is an important issue for ad hoc networks. Two major categories of routing algorithms are topology based routing and position based routing. Topology based routing uses information on current links to perform the routing. Position based routing makes use of a knowledge of the geographic location of each node to route. The position information may be acquired from a service such as the Global Positioning System (GPS).

Topology based algorithms may be further divided into proactive and reactive algorithms. Proactive algorithms use information on current paths as inputs to classical routing algorithms. However to keep this information current a large amount of control message traffic is needed, even if a path is unused. This overhead problem is exacerbated if there are many topology changes (say due to movement of the nodes).

On the other hand, reactive algorithms such as DSR, TORA, and AODV maintain routes only for paths currently in use to keep the amount of information and control overhead more manageable. Still, more control traffic is generated if there are many topology changes.

Position based routing does not require maintenance of routes, routing tables, or generation of large amounts of control traffic other than information regarding positions. "Geocasting" to a specific area can be simply implemented. A number of heuristics can be used in implementing position based routing.

1.2.8 Wireless Sensor Networks

The integration of wireless, computer, and sensor technology has the potential to make possible networks of miniature elements that can acquire sensor data and transmit the data to a human observer. Wireless sensor networks (WSN) are also known as wireless sensor and actuator networks (WSAN) when there is control of the actuators through bi-directional links (Wikipedia). It is assumed that such wireless sensor networks will use ad hoc radio networks to forward data in a multi-hop mode of operation.

Typical parameters for a wireless sensor unit (including computation and networking circuitry) include a size from $1\,mm$ to $1\,cm$, a weight less than $100\,g$, cost less than one dollar, and power consumption less than $100\,\mu W$ [210]. By way of contrast, a wireless personal area network Bluetooth transceiver consumes more than a $1000\,\mu W$. A cubic millimeter wireless sensor can store, with battery technology, at least $1\,J$ allowing at least a $10\,\mu W$ energy consumption for 1 day [128]. Thus, energy scavenging from light or vibration has been proposed. Note also that data rates are often relatively low for sensor data (100's bps to $100\,Kbps$).

Naturally, with these parameters, minimizing energy usage in wireless sensor networks becomes important. While in some applications wireless sensor networks may be needed for a day or less, there are many applications where a continuous source of power is necessary. Moreover, communication is much more energy expensive than computation. The energy cost of transmitting 1000 bits for $100\,m$ equals the energy cost of processing three million instructions on a 100 million instructions per second/W processor (Wikipedia).

While military applications of wireless sensor networks are fairly obvious, there are many potential scientific and civilian applications of wireless sensor networks. Scientific applications include geophysical, environmental, and planetary exploration. One can imagine wireless sensor networks being used to locate forest fires, investigate volcanoes, measure weather, check water quality, monitor beach pollution, or record planetary surface conditions.

Biomedical applications include applications such as glucose level monitoring and retinal prosthesis [209]. Such applications are particularly demanding in terms of manufacturing sensors that can survive in and not affect the human body.

Sensors can be placed in machines (where vibration can sometimes supply energy) such as rotating machines, semiconductor processing chambers, robots, and engines. Wireless sensors in engines could be used for pollution control telemetry.

Finally, among many potential applications, wireless sensors could be placed in homes and buildings for climate control. Note that wiring a single sensor in a building can cost several hundred dollars. Ultimately, wireless sensors could be embedded in building materials.

1.3　Multiplexing

Multiplexing involves sending multiple signals over a single medium. Thomas Edison invented a four to one telegraph multiplexer that allowed four telegraph signals to be sent over one wire. The major forms of multiplexing for networking today are frequency division multiplexing (FDM), time division multiplexing (TDM), and spread spectrum. Each is now reviewed.

1.3.1　Frequency Division Multiplexing (FDM)

Here a portion of spectrum (i.e., band of frequencies) is reserved for each channel (Fig. 1.3a). All channels are transmitted simultaneously but a tunable filter at the receiver only allows one channel at a time to be received. This is how AM, FM, and analog television signals are transmitted. Moreover, it is how distinct optical signals are transmitted over a single fiber using wavelength division multiplexing (WDM) technology.

1.3.2　Time Division Multiplexing (TDM)

Time division multiplexing is a digital technology that, on a serial link, breaks time into equi-duration slots (Fig. 1.3b). A slot may hold a voice sample in a telephone system or

a packet in a packet switching system. A frame consists of N slots. Frames, and thus slots, repeat. A telephone channel might use slot 14 of 24 slots in a frame during the duration of a call, for instance.

Time division multiplexing was used in the second generation cellular system, GSM. It is also used in digital telephone switches. Such switches in fact can use electronic devices called time slot interchangers that transfer voice samples from one slot to another to accomplish switching.

1.3.3　Frequency Hopping

Frequency hopping is one form of spread spectrum technology and is typically used on radio channels. The carrier (center) frequency of a transmission is pseudo-randomly hopped among a number of frequencies (Fig. 1.4a). The hopping is done in a deterministic, but random looking pattern that is known to both transmitter and receiver (i.e., "pseudo-random sequence"). If the hopping pattern is known only to the transmitter and receiver, one has good security. Frequency hopping also provides good interference rejection. Multiple transmissions can be multiplexed in the same local region if each uses a sufficiently different hopping pattern. Frequency hopping dates back to the era of World War II.

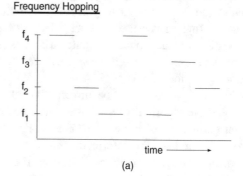

Frequency Hopping

(a)

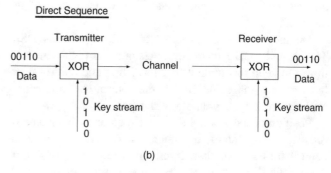

Direct Sequence

(b)

Fig. 1.4 (**a**) Frequency hopping spread spectrum, (**b**) direct sequence spread spectrum

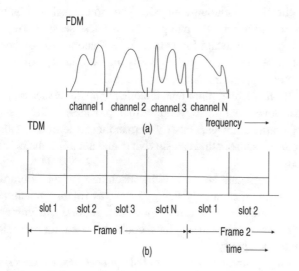

Fig. 1.3 (**a**) Frequency division multiplexing, (**b**) time division multiplexing

Table 1.1 XOR truth table

Key	Data	Output
0	0	0
0	1	1
1	0	1
1	1	0

1.3.4 Direct Sequence Spread Spectrum

This alternative spread spectrum technology uses exclusive or (xor) gates as scramblers and de-scramblers (Fig. 1.4b). At the transmitter data is fed into one input of an xor gate and a pseudo-random key stream into the other input.

From the xor truth table (Table 1.1), one can see that if the key bit is a zero, the output bit equals the data bit. If the key bit is a one, the output bit is the complement of the data bit (0 becomes 1, 1 becomes 0). This scrambling action is quite strong under the proper conditions. Unscrambling can be performed by an xor gate at the receiver. The transmitter and receiver must use the same (synchronized) key stream for this to work. Again, multiple transmissions can be multiplexed in a local region if the key streams used for each transmission are sufficiently different.

1.4 Circuit Switching Versus Packet Switching

Two major architectures for networking and telecommunications are circuit switching and packet switching. Circuit switching is the older technology, going back to the years following the invention of the telephone in the late 1800s. As illustrated in Fig. 1.5a, for a telephone network, when a call has to be made from node A to node Z, a physical path with appropriate resources called a "circuit" is established. Resources include link bandwidth and switching resources. Establishing a circuit requires some setup time before actual communication commences. Even if one momentarily stops talking, the circuit is still in operation. When the call is finished, link and switching resources are released for use by other calls. If insufficient resources are available to set up a call, the call is said to be blocked.

Packet switching was created during the 1960s. A packet is a group of bits consisting of header bits and payload bits. The header contains the source and destination address, priority levels, error check bits, and any other information that is needed. The payload is the actual information (data) to be transported. However, many packet switching systems have a maximum packet size. Thus, larger transmissions are split into many packets and the transmission is reconstituted at the receiver.

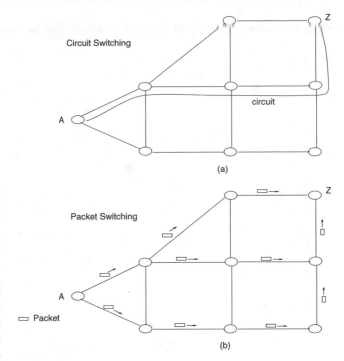

Fig. 1.5 (a) Circuit switching, (b) packet switching

The diagram of Fig. 1.5b shows packets, possibly from the same transmission, taking multiple routes from node A to node Z. This is called datagram or connectionless oriented service. Packets may indeed take different routes in this type of service as nodal routing tables are updated periodically in the middle of a transmission.

A hybrid type of service is the use of "virtual circuits" or connection oriented service. Here packets belonging to the same transmission are forced to take the same serial path through the network. A virtual circuit has an identification number which is placed in packet headers to be used by nodes to continue forwarding the packet along its preset (circuit) path. As in circuit switching, a virtual circuit needs to be set up prior to its use for communication. That is, entries need to be made in routing tables implementing the virtual circuit.

An advantage of virtual circuit usage is that packets arrive at the destination in the same order that they were sent. This avoids the need for buffers for reassembling transmissions (reassembly buffers) that are needed when packets arriving at the destination are not in order, as in datagram service.

Packet switching is advantageous when traffic is bursty (occurs at irregular intervals) and individual transmissions are short. It is a very efficient way of sharing network resources when there are many such transmissions. Circuit switching is not well suited for bursty and short transmissions. It is more efficacious when transmissions are relatively long (to minimize setup time overhead) and provide a constant traffic rate (to well utilize the dedicated circuit resource).

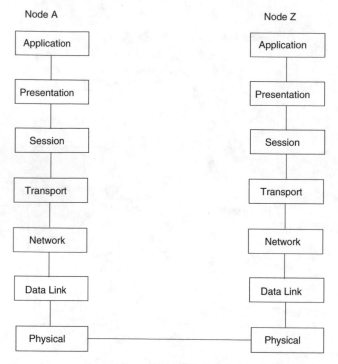

Node A

Application

Presentation

Session

Transport

Network

Data Link

Physical

Node Z

Application

Presentation

Session

Transport

Network

Data Link

Physical

Fig. 1.6 OSI protocol stack for a communicating source and destination

1.5 Layered Protocols

Protocols are the rules of operation of a network. A common way to engineer a complex system is to break it into more manageable and coherent components. Network protocols are often divided into layers in the layered protocol approach. Figure 1.6 illustrates the generic OSI (open systems interconnection) protocol stack. Proprietary protocols may have different names for the layers and/or a different layer organization but pretty much all networking protocols have the same functionality.

Transmissions in a layered architecture (see Fig. 1.6) move from the source's top layer (application), down the stack to the physical layer, through a physical channel in a network, to the destination's physical layer, up the destination stack to the destination application layer. Note that any communication between peer layers must move down one stack, across, and up the receiver's stack. It should also be noted that if a transmission passes through an intermediate node, only some lower layers (e.g., network, data link, and physical) may be used at the intermediate nodes.

It is interesting that a packet moving down the source's stack may have its header grow as each layer may append information to the header. At the destination, each layer may remove information from the packet header, causing it to decrease in size as it moves up the stack.

In a particular implementation, some layers may be larger and more complex, while others are relatively simple.

In the following, we outline the functionality of each layer.

1.5.1 Application Layer

Applications for networking include email, remote login, file transfer, and the worldwide web.

1.5.2 Presentation Layer

This layer controls how information is formatted, such as on a screen (number of lines, number of characters across).

1.5.3 Session Layer

This layer is important for managing a session, as in remote logins. In other cases, this is not a concern.

1.5.4 Transport Layer

This layer can be thought of as an interface between the upper and lower layers. More importantly, it is designed to give the impression to the layers above that they are dealing with a reliable network, even though the layers below the transport layer may not be perfectly reliable. For this reason, some think of the transport layer as the most important layer.

1.5.5 Network Layer

The network layer manages multiple links. Its most important function is to do routing. Routing involves selecting the best path for a circuit or packet stream.

1.5.6 Data Link Layer

Whereas the network layer manages multiple link functions, a data link protocol manages a single link. One of its potential functions is encryption, which can either be done on a link by link basis (i.e., at the data link layer) or on an end to end basis (i.e., at the transport layer) or both. End to end encryption is a more conservative choice as one is never sure what type of sub-network a transmission may pass through and what its level of encryption, if any, is.

1.5.7 Physical Layer

The physical layer is concerned with the raw transmission of bits. Thus, it includes engineering physical transmission media, modulation and de-modulation, and radio technology. Many communication engineers work on physical layer aspects of networks. Again, the physical layer of a protocol stack is the only layer that provides actual direct connectivity to peer layers.

Introductory texts on networking usually discuss layered protocols in detail.

1.6 Parallel Processing

Parallel processing, the topic of Chaps. 5 and 6 in this book, involves the simultaneous execution of computations on different processors. The processors may be physically distinct or perhaps "cores" within a chip. Such parallel processing reduces the total processing time compared to doing all of the computation on a single serial processor. In fact, an important performance measures, *speedup*, or S is the ratio of the time to solve a problems on one processor, $T(1)$, to the time to solve it on n processors, $T(n)$

$$S = \frac{T(1)}{T(n)} \tag{1.1}$$

Note that speedup is greater than one if there is an improvement in processing time through the use of parallel processing.

There are different types of "parallelism" and their hardware implementations (Wikipedia).

There is bit level parallelism where one boosts computer performance by increasing the word size (going from 32 to 64 bit words for instance). There is instruction level parallelism (for instance using internal pipelining). In task scheduling, different tasks are processed in different processors or cores. In indivisible task scheduling, a single task can be performed on at most one processor. Such optimization leads to NP complete style computational issues. In divisible load scheduling (DLT), the subject of Chap. 5, a computation/communication "load" (think data) can be arbitrarily partitioned among processors and links.

An example of a divisible load is a trillion numbers that have to be summed. One can send different groups of numbers to different processors for intermediate summing, collect partial results at one processor, and sum those for the final answer.

Since 1988 a very elegant theory on divisible load scheduling, transmission, and processing has been developed. Given processor and links speeds, interconnection network topology and the amount of computation intensity versus communication intensity one can with minimal computation determine how much load and when to allocate load to processors and links for a minimal processing time solution (i.e., minimum finish time or makespan). Chapter 5 covers this basic theory and elaborations.

Chapter 6 discuss Amdahl's law and related relationships. Amdahl's law is a relationship that allows one to determine the boost in speedup one gets using a parallel system versus a single serial processor. Amdahl's law has been a feature of parallel processing performance evaluation since it was first proposed by Gene Amdahl in 1967. In this book we look at Amdahl's and related laws and their application to today's parallel systems.

1.7 Machine Learning

After decades of development, networking systems have evolved from small local area networks interconnecting a limited number of computers to very complicated and varied systems. While the performance of modern networking systems has been phenomenal and has enabled a great many novel applications that have largely changed the world, the scale and complexity of the problems to solve in such a system has dramatically increased. For example, while naive traffic routing focuses on finding paths with a single optimization criterion like minimizing distance or maximizing the minimum bandwidth, more advanced traffic routing may need to consider future traffic demand on critical links for time-sensitive network traffic and consider link failure probabilities for latency-sensitive network traffic. Conventional methods, such as heuristics and linear programming technique, have appeared to have more and more limitations in solving these networking problems.

On the other hand, machine learning techniques-techniques that can automatically learn the solution space from historical data and perform a new solution search based on the learned experience-have shown a promising ability to solve many complex networking problems, yet their applications to the networking area are still at an early stage.

Chapter 7 in this book, beginning with an overview of machine learning, reviews state-of-the-art machine learning applications in the network area with the purpose of providing some insights on existing solutions and future opportunities. Specifically, we look at the applications of machine learning techniques to traffic classification, traffic routing, and resource management, as these three domains of the network problems are the most closely related to the previous chapters of this book.

Abstract

A number of basic probability models and solutions are discussed in this chapter. These include basic distributions and the Poisson process. A large number of probability problems with a networking orientation are presented. Also examined are multiple access models, switching elements, tree networks, some interconnection networks and a multiple bus system.

2.1 Introduction

A stochastic model is a mathematical model involving random quantities. Stochastic models are very useful for modeling networking and computation.

A "switch" (or "hub") is a computerized device that provides electronic connectivity between PCs, workstations, wireless devices, and other computers connected to it. This naturally means a switch will have a number of input and output links (connections) to these various user devices (Fig. 2.1).

A packet switch accepts packets of data on incoming links and routes them on outgoing links towards their destination. Note that a packet (i.e., bundle of bits) typically consists of a header and the payload. The header holds control information such as source and destination addresses, packet priority level, and error checking codes. The payload is the actual information to be transmitted. Packet switching service has been of much interest since the implementation of early wide area packet switching networks or datagram service, starting in the 1960s.

An older technology compared to packet switching is circuit switching. Here traditionally a physical path is reserved from one local switch to a (possibly distant) switch.

Thus in the classical telephone network the path extends from one phone to another passing through a number of intermediate (hierarchical) switches on the way. The use of circuit switching sometimes is said to provide connection oriented service.

In both cases the fundamental design problem is how does one design and build the highest performance switch. Performance typically involves such parameters as the best values of throughput (capacity) and time delay (to transmit a packet or setup a call) for the money invested. Typical approaches for designing the internal architecture of a high speed packet switch could include the use of a computer bus, shared memory, or VLSI switch designs based on space division switching.

A major goal of a performance evaluation study is to determine the performance of various architectural alternatives using statistical and mathematical models. While experimental work is crucial in engineering a system that is reliable and has high performance, performance evaluation allows a cost-effective, preliminary consideration of technological alternatives prior to implementation. Furthermore, analytical performance evaluation results can give an intuitive understanding of design tradeoffs. The rest of this chapter will discuss two paradigms for modeling arriving streams of traffic. These are the continuous time and discrete time paradigms. Probabilistic models are emphasized in this chapter because of their simplicity and utility.

In Sect. 2.2 the fundamental Bernoulli and Poisson processes are discussed. Bernoulli process related distributions are presented in Sect. 2.3. Many worked out probability problems appear in Sect. 2.4. Multiple access performance is covered in Sect. 2.5. This includes discussions of Ethernet models and Aloha communication. Finally switching elements and networks are examined in Sect. 2.6

© Springer Nature Switzerland AG 2020
T. G. Robertazzi, L. Shi, *Networking and Computation*, https://doi.org/10.1007/978-3-030-36704-6_2

Fig. 2.1 A switching hub

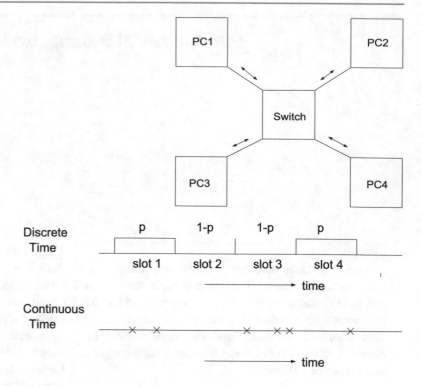

Fig. 2.2 Bernoulli and Poisson processes

2.2 Bernoulli and Poisson Processes

2.2.1 The Discrete Time Bernoulli Process

Discrete time systems typically involve the use of equi-length time slots. That is, if one plots the signal flow on a single input link versus time, one might have the situation shown in Fig. 2.2a.

This figure is based on the Bernoulli process assumption that the probability of a (single) packet being present in a particular time slot is p. The probability of no packet being present in a particular slot is $1 - p$. Here each event in a slot (a packet or no packet) is independent of all others (which will simplify calculations below).

There are possible variations of the Bernoulli process including allowing 0 to N arrivals, per slot. This will occur, for instance, in modeling a switching element (module) in a VLSI switching array with N inputs that are each mathematically modeled as a Bernoulli process input. Naturally we are assuming here that time slot boundaries are aligned so that in each slot 0 to N packets arrive simultaneously to the switching element.

In some packet based systems all packets are of the same length. Also in some systems a number of time slots comprising a "frame" (say slot 1 through 100) repeat in a periodic fashion so that each user gets one slot of transmission time every time the frame repeats. One packet may fit into one slot. Time division multiplexing, for instance, works in this manner.

In order to examine the Bernoulli process, let

$$Prob(1\ arrival/slot) = P_1 = p \qquad (2.1)$$

$$Prob(no\ arrival/slot) = P_0 = 1 - p \qquad (2.2)$$

$$Prob(2\ or\ more\ arrivals/slot) = 0.0 \qquad (2.3)$$

One can then compute statistics for the Bernoulli process such as the mean (average) number of customers per slot or the variance in the number of customers per slot.

So, from first principles, the mean or expected number of packets, $\bar{n}$, is

$$E[n] = \bar{n} = \sum_{i=0}^{1} i P_i = 0 \times P_0 + 1 \times P_1 \qquad (2.4)$$

$$\boxed{\bar{n} = P_1 = p} \qquad (2.5)$$

This result is quite intuitive. If there were always no arrivals, one would anticipate $E[n] = 0$. If each and every slot carries a packet, one would expect $E[n] = 1$. In fact the probability p is between 0 and 1. One might consider p as a performance measure called "throughput." Throughput is the (possibly normalized) amount of useful information carried per unit time by a network element, be it a link, buffer, or switch.

Next, from first principles one can compute the variance of the number of packets on a link as

Fig. 2.3 Poisson process state transition diagram

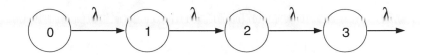

$$\sigma^2 = \sum_{i=0}^{1} (i - E[n])^2 P_i \qquad (2.6)$$

$$\sigma^2 = (E[n])^2 P_0 + (1 - E[n])^2 P_1 \qquad (2.7)$$

$$\sigma^2 = p^2(1 - p) + (1 - p)^2 p \qquad (2.8)$$

$$\sigma^2 = p(1 - p)[p + 1 - p] \qquad (2.9)$$

$$\boxed{\sigma^2 = p(1 - p)} \qquad (2.10)$$

Recall that variance is the sum of the squared differences between the values of a random variable and the random variable's mean which are weighted by the probability of the random variable taking on specific values. Note that the variance in the number of customers per slot (above) is maximized when p is 0.5.

At this point, it should be noted that popular transform techniques (such as Laplace and Fourier transforms) allow one to treat a time based signal in the frequency domain and for certain types of work this simplifies calculations. In fact a random process can often be transformed into a frequency like domain using the moment generating function (really a z transform using z^n rather than signal processing's z transform's usual z^{-n}). Here z is a complex number. From the moment generating function definition one has for the Bernoulli process

$$P(z) = \sum_{i=0}^{1} P_i z^i \qquad (2.11)$$

$$P(z) = P_0 z^0 + P_1 z^1 \qquad (2.12)$$

$$P(z) = (1 - p)z^0 + pz \qquad (2.13)$$

$$\boxed{P(z) = (1 - p) + pz} \qquad (2.14)$$

While no further use of the moment generating function appears in this chapter, it is a useful analytical tool for such purposes as finding transient and advanced queue discipline statistics.

2.2.2 The Continuous Time Poisson Process

The continuous time analog to the discrete time Bernoulli process is the Poisson process. With continuous time model-

ing an event, such as a switch arrival, may occur at any time instant (e.g., $t = 2.49346$ s, rather than at integer slot times (slot 1, slot 2...). It can be seen from Fig. 2.2b that arrivals are random in placement. If one takes a certain interval of a homogeneous (i.e., constant arrival rate) Poisson process with M arrivals, the M events are in fact randomly placed in a uniform manner over the interval. Although the time invariant or homogeneous Poisson process is quite tractable, time varying arrival rates ($\lambda(t)$) can also be modeled.

To find the basic equations governing a Poisson process, let $P_n(t)$ be the probability of n arrivals in a time interval $[0, t]$.

As will be discussed below, a state machine like diagram called a state transition diagram can be drawn. The circles (Fig. 2.3) are states (0, 1, 2, .. customers having arrived up to this time) and transitions between them are labeled with the rate of making a transition from state i to state $i + 1$.

In fact we will initially use a time slotted model to characterize the Poisson process. However these are not the fixed engineered macroscopic packet transmission times of the Bernoulli process. Instead the "slots" are the usual mathematically arbitrarily small intervals from $t \rightarrow t + \Delta t$ where we'll eventually let $\Delta t \rightarrow 0.0$ to create a continuous time model.

Going through the steps, from first principles one has the difference equations $n = 1, 2 \ldots$

$$P_n(t + \Delta t) = P_n(t)P_{n,n}(\Delta t) + P_{n-1}(t)P_{n-1,n}(\Delta t) \qquad (2.15)$$

$$P_0(t + \Delta t) = P_0(t)P_{0,0}(\Delta t) \qquad (2.16)$$

Here $P_{n,n}(\Delta t)$ is the probability of going from n arrivals at a point in time $[t]$ to n arrivals at time $[t, t + \Delta t]$. Also $P_{n-1,n}(\Delta t)$ is the probability of going from $n - 1$ arrivals to n arrivals in time Δt by means of a single arrival.

So Eq. (2.15) says that the probability one has n arrivals by time $t + \Delta t$ is the sum of two terms. One term is the probability that one has n arrivals by time t and the probability that there are no arrivals in slot $[t, t + \Delta t]$. The second term corresponds to the probability of having $n - 1$ arrivals by time t and the probability that there is one arrival in slot $[t, t + \Delta t]$ bringing the number of arrivals from $n - 1$ to n by time $t + \Delta t$.

Equation (2.16) says that the probability of no (zero) arrivals by time $t + \Delta t$ is equal to the probability that there are no arrivals by time t and the probability that there are no arrivals in the slot $[t, t + \Delta t]$.

Intuitively, $P_{n-1,n}(\Delta t)$ should be proportional to the arrival rate and the time slot width. Thus as Δt becomes small

$$P_{n-1,n}(\Delta t) = \lambda \Delta t \qquad (2.17)$$

$$P_{n,n}(\Delta t) = 1 - \lambda \Delta t \qquad (2.18)$$

Substituting, using algebra and letting $\Delta t \to 0$, one arrives at

$$\frac{dP_n(t)}{dt} = -\lambda P_n(t) + \lambda P_{n-1}(t) \quad n = 1, 2 \ldots \qquad (2.19)$$

$$\frac{dP_0(t)}{dt} = -\lambda P_0(t) \qquad (2.20)$$

Note that for n arrivals we have a family of n linear differential equations. The second equation electrical and computer engineers will recognize as being similar to the differential equation for capacitive voltage discharge through a resistor.

It is well known that the solution to this equation is

$$P_o(t) = e^{-\lambda t} \qquad (2.21)$$

Substituting this solution into the $n = 1$ equation and solving will yield

$$\frac{dP_1(t)}{dt} = -\lambda P_1(t) + \lambda e^{-\lambda t} \qquad (2.22)$$

$$P_1(t) = \lambda t e^{-\lambda t} \qquad (2.23)$$

Continuing to substitute the current solutions into the next differential equation yields

$$P_2(t) = \frac{\lambda^2 t^2}{2} e^{-\lambda t} \qquad (2.24)$$

$$P_3(t) = \frac{\lambda^3 t^3}{6} e^{-\lambda t} \qquad (2.25)$$

Or simply

$$P_n(t) = \frac{(\lambda t)^n}{n!} e^{-\lambda t} \quad n = 0, 1, 2 \ldots \qquad (2.26)$$

This is the Poisson distribution. Given the Poisson arrival rate λ and time interval length t, one can use the Poisson distribution to easily find the probability of there being n arrivals in the interval. Figure 2.4 illustrates the Poisson distribution for $n = 0$ to $n = 5$.

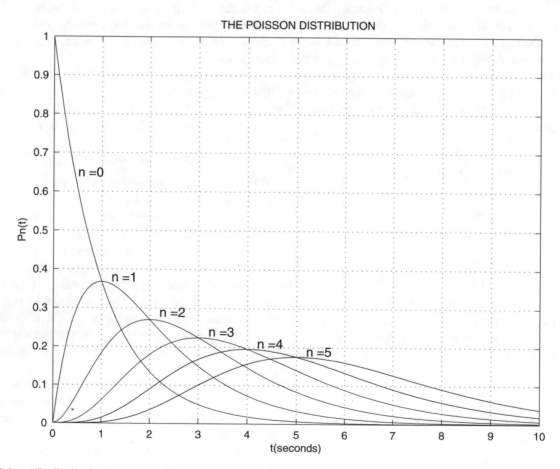

Fig. 2.4 Poisson distribution family

In fact it is straightforward to find the average or mean number of arrivals in a time interval of length t. From first principles (see Appendix A) one has

$$\bar{n} = \sum_{n=0}^{\infty} n P_n(t) \tag{2.27}$$

Substituting the Poisson distribution one has

$$\bar{n} = \sum_{n=0}^{\infty} n \frac{(\lambda t)^n}{n!} e^{-\lambda t} \tag{2.28}$$

As $e^{-\lambda t}$ is not affected by the summation index, n, one can move $e^{-\lambda t}$ past the summation. Also clearly the $n = 0$ term is zero as n is a factor. Thus

$$\bar{n} = e^{-\lambda t} \sum_{n=1}^{\infty} \frac{(\lambda t)^n}{(n-1)!} \tag{2.29}$$

With a change of variables $(n-1) \rightarrow n$ one has

$$\bar{n} = e^{-\lambda t} \sum_{n=0}^{\infty} \frac{(\lambda t)^{n+1}}{n!} \tag{2.30}$$

Again a term, this time a factor of λt, does not depend on n and so the factor can be brought outside the summation

$$\bar{n} = e^{-\lambda t} \lambda t \sum_{n=0}^{\infty} \frac{(\lambda t)^n}{n!} \tag{2.31}$$

Finally using a summation formula from the Appendix A has

$$\boxed{\bar{n} = e^{-\lambda t} (\lambda t) e^{\lambda t} = \lambda t} \tag{2.32}$$

The fact that the average number of customers arriving in an interval is λt is quite intuitive. For instance, if 100 calls arrive per minute to a telephone switch, the average number arriving in $10\,\text{min}$ is $10 \times 100 = 1000$ calls. In fact it can be seen in Fig. 2.4 where $\lambda = 1$ that the distributions $n = 1, 2, \ldots$ peak at λt.

It is useful to know that if N independent Poisson processes, each with Poisson arrival rate λ_i, are merged, the merged process is Poisson with mean arrival rate $\sum_{i=1}^{N} \lambda_i$. Also, suppose a Poisson process of rate λ is split randomly into N processes. That is, an arrival is sent to split process i with independent probability p_i. Naturally

$$p_1 + p_2 + p_3 \ldots + p_N = 1$$

Then each split process is an independent Poisson process with rate $p_i \lambda$. Naturally merged processes arise in multiplexing and split processes arise in demultiplexing.

Interestingly, the homogeneous Poisson process can be used as a model of the spatial (geographic) distribution of station location in two dimensions [73]. In this case λ_A is the average number of stations located in a unit area. Within an actual area, A, of any shape, the actual number of stations placed is uniformly (perfectly randomly) distributed.

Using the one-dimensional Poisson distribution as a starting point, one has the probability that there are n stations in area A, with spatial location density λ_A as $P_n(A)$

$$P_n(A) = \frac{(\lambda_A A)^n}{n!} e^{-\lambda_A A} \tag{2.33}$$

Intuitively the average number of stations placed in area A is

$$\bar{n} = \lambda_A A \tag{2.34}$$

Other models of spatial location are possible [67, 169].

2.3 Bernoulli Process Statistics

Given a discrete time Bernoulli process, there are a number of standard distributions associated with natural questions about the process. Four such distributions are discussed below

Question 1 (Geometric Distribution) How many slots does it take on average, for the first arrival?

To answer this question use can be made of the geometric distribution. Consider a data record of $i-1$ idle slots followed by the ith slot containing a packet. Naturally p here is the probability of an arrival in a slot. The probability of no arrival in a slot is $1 - p$. It is assumed that arrivals/non-arrivals in each slot are independent of each other. Then $P(i)$, the probability that the first packet arrival occurs in the ith slot is

$$P(i) = \underbrace{(1-p)(1-p)(1-p)}_{i-1\ times} \ldots p \tag{2.35}$$

$$\boxed{P(i) = (1-p)^{i-1} p \quad i = 1, 2 \ldots} \tag{2.36}$$

This is the geometric distribution.

Let us find the mean number of slots till the first arrival. Using first principles

$$\bar{n} = \sum_{i=1}^{\infty} i P(i) \tag{2.37}$$

$$\bar{n} = \sum_{i=1}^{\infty} i p (1-p)^{i-1} \tag{2.38}$$

As p is not a function at the summation index, i, one can bring p outside of the summation

$$\bar{n} = p \sum_{i=1}^{\infty} i(1-p)^{i-1} \qquad (2.39)$$

Multiply and dividing by $1 - p$ yields

$$\bar{n} = \frac{p}{1-p} \sum_{i=1}^{\infty} i(1-p)^{i} \qquad (2.40)$$

$$\bar{n} = \frac{p}{1-p} \sum_{i=0}^{\infty} i(1-p)^{i} \qquad (2.41)$$

Here $i = 1$ becoming $i = 0$ is allowed as the value of the zeroth term is zero. Using a summation from the Appendix A

$$\bar{n} = \frac{p}{1-p} \frac{1-p}{p^2} \qquad (2.42)$$

$$\boxed{\bar{n} = 1/p} \qquad (2.43)$$

Thus one can tabulate the average number of slots until the first arrival

p	$\mu = 1/p$
0.1	10
0.2	5
0.5	2
0.9	1.1
1.0	1.0

Intuitively these values make sense. For instance, if an arrival occurs with probability $p = 0.2$, it takes five slots on average until the first arrival.

To find the variance in the number of arrivals, from first principles

$$\sigma^2 = \sum_{i=1}^{\infty} (i - \mu)^2 P(i) \qquad (2.44)$$

Here $P(i)$ is the geometric distribution.
One can show after a few steps

$$\boxed{\sigma^2 = (1-p)/p^2} \qquad (2.45)$$

Example Let us consider a web server example. A major web site has a load balancing computer that feeds requests for web pages to one of 12 computers that store the web pages. "Load balancing" is done so that no single computer is overloaded. Let q be the independent probability that a computer can accept a web page request (the computer is not overloaded). Let $1 - q$ be the independent probability a computer cannot accept a web page request.

When it gets a web page request the load balancing computer checks with each of 12 computers in turn (sequentially) to see whether or not is overloaded. Find an expression for the probability that the ith computer it checks can accept the job (is not overloaded).

Solution The situation is clearly modeled by a geometric distribution. The probability $P(i)$ that the ith computer accepts the request is

$$P(i) = (1-q)^{i-1}q$$

Here $i - 1$ computers reject the request (with probability $(1-q)^{i-1}$ and the ith computer accepts it (with probability q)). Note that there is a finite probability that none of the computers accepts the request with probability

$$P(reject) = 1 - \sum_{i=1}^{12} (1-q)^{i-1}q$$

Or

$$P(reject) = \sum_{i=13}^{\infty} (1-q)^{i-1}q$$

Question 2 (Binomial Distribution) What is the probability of n arrivals in N slots?

This question is answered by using a binomial distribution. Let us say that in N slots there are n arrivals. The arrivals are assumed to be independent of one another. The probability of n arrivals in N slots in a specific pattern of placement is

$$p^n(1-p)^{N-n} \qquad (2.46)$$

The probability of n arrivals in *any* pattern of placement in N slots is the probability of a single pattern occurring, times the number of patterns. Thus we need to multiply the probability expression above by the number of patterns involving n arrivals distributed among N slots (of course $n \leq N$). The number of patterns is equal to the binomial coefficient

$$\binom{N}{n} \qquad (2.47)$$

One then has the binomial distribution for the probability of n arrivals in N slots

$$\boxed{\binom{N}{n} p^n(1-p)^{N-n}} \qquad (2.48)$$

A bit of thought will show that for a given p, the average number of arrivals in N slots is Np.

Example A Consider the previous web server example. Find an expression for the probability that six or more of the computers are overloaded. You may use a summation.

Solution to Example A The probability, $P_{\geq 6/12}$ that six or more of the twelve computers are overloaded is a sum of binomial probabilities

$$P_{\geq 6/12} = P(6 \ overloaded) + P(7 \ overloaded) + \cdots$$
$$+ P(12 \ overloaded)$$

$$P_{\geq 6/12} = \sum_{i=6}^{12} \binom{12}{i} (1 - q)^i q^{12-i}$$

Alternately

$$P_{\geq 6/12} = 1 - P(0 \ overloaded) - P(1 \ overloaded) + \cdots$$
$$+ P(5 \ overloaded)$$

$$P_{\geq 6/12} = 1 - \sum_{i=0}^{5} \binom{12}{i} (1 - q)^i q^{12-i}$$

Example B A fault tolerant computer system in a Jupiter space probe uses majority logic and three independent computers. "Votes" are taken and if at least two computers agree to activate thrusters, the thrusters are activated. Let p be the probability that a single computer makes the wrong decision. Find the probability that a vote is incorrect.

Solution to Example B The probability that a vote is incorrect is the probability that two computers are wrong plus the probability that three computers are wrong. Using the binomial distribution

$$P(vote \ incorrect) = P(2 \ computers \ wrong)$$
$$+ P(3 \ computers \ wrong)$$
$$= \binom{3}{2} p^2 (1 - p) + \binom{3}{3} p^3 (1 - p)^0$$
$$= 3p^2(1 - p) + p^3$$

If $p = 0.01$, the probability of an incorrect vote is

$$P(vote \ incorrect) = 0.0000297 + 10^{-6} = 0.000298$$

This is a 30 times improvement in the reliability of the system. Naturally, a complete reliability analysis will consider such additional factors as the reliability of the voting

logic and the degree to which the computers are independent (e.g., overheating in the space probe may cause simultaneous failures).

Question 3 (Pascal Distribution) What is the distribution of the time until the kth arrival where $k \geq 1$?

The distribution in question is the Pascal distribution [104].

To find the Pascal distribution, with some thought one can see that the probability of k arrivals, in N slots, $P_N(k)$, is

$$P_N(k) = P[A]P[B] \tag{2.49}$$

Here $P[A]$ is the probability of $k - 1$ arrivals, in $N - 1$ slots. Also $P[B]$ is the probability of an arrival in the Nth slot.

Thus, using binomial style statistics (see above)

$$P[A] = \binom{N-1}{k-1} p^{k-1}(1 - p)^{N-1-(k-1)} \tag{2.50}$$

$$P[B] = p \tag{2.51}$$

Since $P[AB] = P[A]P[B]$, then

$$\boxed{P_N(k) = \binom{N-1}{k-1} p^k(1 - p)^{N-k} \quad k = 1, 2, 3 \ldots}$$

$$\tag{2.52}$$

Naturally the mean number of slots holding k arrivals is $\mu = k/p$ (see Eq. (2.43)). Again, the expression makes intuitive sense.

Example A buffer is fed by a packet stream that can be modeled as a Bernoulli process. The buffer dumps its contents onto a network when the tenth packet arrives. What is the probability that the tenth packet arrives in the 30th slot?

Solution Using the Pascal distribution with these parameters

$$P_{30}(10) = \binom{29}{9} p^{10}(1 - p)^{20}$$

The mean number of slots holding ten packets is $10/p$.

Question 4 (Multinomial Distribution) What is the distribution of n_i packet arrivals of type i (where the number of types is 3 or greater) in N slots in any pattern?

Let p_i be the probability of arrival of type i. Then the probability of a specific pattern of packet arrivals where there are four types of packet (say voice, data, video, and empty slots) is

$$Prob(specific\ pattern) = p_v^{n_v} p_d^{n_d} p_{tv}^{n_{tv}} p_{idle}^{n_{idle}} \quad (2.53)$$

Any pattern with the same number of each arrival type has the same probability expression and probability. To find the probability of the specific number of arrivals in any pattern, just multiply the above probability by the number of such patterns. The number of such patterns is the multinomial coefficient

$$\binom{N}{n_v, n_d, n_{tv}, n_{idle}} = \frac{N!}{n_v!, n_d!, n_{tv}!, n_{idle}!} \quad (2.54)$$

So one has the multinomial distribution, a generalization of the binomial distribution

$$Prob(any\,pattern) = \frac{N!}{n_v!, n_d!, n_{tv}!, n_{idle}!} p_v^{n_v} p_d^{n_d} p_{tv}^{n_{tv}} p_{idle}^{n_{idle}} \quad (2.55)$$

If one has only two packet types, then the multinomial distribution reduces to the binomial distribution.

If one knows the probabilities of arrival of each packet type as well as knows the number of each packet type one is interested in, then the probability of that number of each packet type arriving in any pattern can be calculated. Additional terms can be added in the natural way to change the number of packet types.

A related example of this distribution appears in Sect. 2.4.1 (A Stream of Different Packet Types).

2.4 Probability Problems

In this section a large number of probability problems with a networking flavor will be solved. More such problems appear as end of chapter exercises.

2.4.1 Packet Streams

In these type of problems a stream(s) of packets is modeled as a Bernoulli process. That is one might assume a sequence of time slots of equal width. Here each time slot has a single packet with independent probability p and a time slot has no packet (is empty or idle) with independent probability $1 - p$. That is, what happens in each slot is independent of what happens in other slots. Again, mathematically this is a Bernoulli process. There are many real-world applications that can be modeled as a Bernoulli process including coin flipping.

Single Packet Stream

What are typical probability questions one can ask about a single Bernoulli stream of packets (see Fig. 2.5)? Here are some

(a) Write an expression for the probability of the exact sequence in Fig. 2.5 occurring.

One has the probability of an empty slot 1 is $(1 - p)$, the probability of a packet in slot 2 is p, and so on.... So

$$Prob = (1-p)p(1-p)pp(1-p) = p^3(1-p)^3 \quad (2.56)$$

The overall probability result here is called the "joint" probability. It is the probability of no packet in slot 1 and a packet in slot 2 and no packet in slot 3, and so on... The overall probability involves simply multiplying the event probabilities. This can be done as the event probabilities (i.e., what happens in a time slot) are independent (see any introductory text on probability).

(b) Write an expression of exactly three packets occurring in six slots in any pattern.

Any sequence of three packets in six slots occurs with the same probability. So one just has to multiply that probability (really the answer of part (a)) by the number of possible sequences (i.e., the binomial coefficient) consisting of three packets in six slots, $\binom{6}{3}$, or

$$\binom{6}{3} p^3(1-p)^3 \quad (2.57)$$

This is a binomial distribution. The "bi" in binomial means two (as in biplane or bifocals) and there are two options in each slot: packet or no packet. So the binomial distribution is the appropriate choice of distribution. If there were more than two choices (say a variety of packet types), we would use the multinomial distribution for this type of problem.

(c) Write an expression for the probability that the first packet appears in slot 10.

The question is answered by the geometric distribution. One has nine empty time slots (each occurring with independent probability $(1 - p)$) followed by an arrival in the tenth slot occurring with probability p.

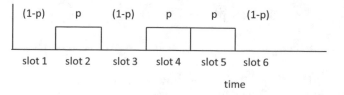

Fig. 2.5 Single stream of packets and empty slots

$$Prob(10) = \underbrace{(1-p)(1-p)\dots(1-p)}_{nine\ empty\ slots}$$

$$\times \underbrace{p}_{packet\ in\ 10th\ slot} = (1-p)^9 p \quad (2.58)$$

The individual probabilities can be simply multiplied to find the joint probability since the events in each slot are independent of any event in other slots.

(d) Write an expression for the probability of the first arrival being in slot 10 followed by arrivals in slot 11 and 12.

Since each time slot is independent of every other time slot, the probabilities simply multiply to produce to the joint probability.

$$Prob(10, 11, 12) = ((1-p)^9 \times p) \times p \times p = (1-p)^9 p^3 \quad (2.59)$$

(e) What is the probability that the fifth packet arrives in the tenth slot?

The Pascal (negative binomial) distribution can be used to answer this question. One has

$$Prob(kth\ packet\ in\ Nth\ slot) = \binom{N-1}{k-1} p^k (1-p)^{N-k} \quad (2.60)$$

$$Prob(5th\ packet\ in\ 10th\ slot) = \binom{10-1}{5-1} p^5 (1-p)^{10-5}$$

$$= \binom{9}{4} p^5 (1-p)^5 \quad (2.61)$$

The Pascal distribution is a generalization of the geometric distribution. See the previous Sect. 2.3, Question 3, for information on the Pascal distribution.

Dual Packet Streams

Suppose now that a switching element has two inputs lines, each of which can be modeled as a Bernoulli process (Fig. 2.6). It is assumed that the two Bernoulli streams are independent of each other and the individual streams consist of independent events.

The following questions are considered.

(a) Write an expression for the probability of no arrivals in both streams in ten consecutive time slots.

Then all 20 (i.e., 2×10) slots are empty and one has the following:

$$Prob = \underbrace{(1-p)(1-p)\dots(1-p)}_{twenty\ times} = (1-p)^{20} \quad (2.62)$$

The individual event probabilities can be multiplied to find the joint probability because of the independence of events (i.e., arrivals and non-arrivals).

(b) Write an expression for the probability of the specific sequences shown in the figure occurring.

There are seven packets and seven empty slots. Thus

$$Prob = \underbrace{ppppppp}_{7\ packets} \underbrace{(1-p)(1-p)(1-p)(1-p)(1-p)(1-p)(1-p)}_{7\ empty\ slots} = p^7 (1-p)^7 \quad (2.63)$$

(c) Write an expression for the probability of one or two packets arriving in a single time slot (considering both streams).

$$Prob = \underbrace{2p(1-p)}_{1\ arrival} + \underbrace{p^2}_{2\ arrivals} \quad (2.64)$$

The first term represents one packet being on one line (with probability p) and no packet arriving on the other line (with probability $1-p$). The single arriving packet can be on either line so we multiply by a factor of 2 to get the overall probability. The second term is the probability of one arrival for each stream for a total of two arrivals.

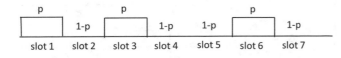

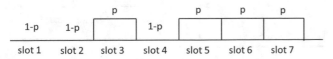

Fig. 2.6 Dual streams of packets and empty slots

(d) Write an expression for the probability of the first arrival of one or two packets in a single time slot occurring in slot 11.

This is just the probability of no packets in the first ten individual time slots multiplied by the result of part (c) (the probability of one or two packets arriving in a single time slot).

$$Prob = (1 - p)^{20}\left(2p(1 - p) + p^2\right) \qquad (2.65)$$

A Stream of Different Packet Types

When there are different types of packets the multinomial distribution is often useful. Consider Fig. 2.7 showing a sequence with two voice packets (packets carrying voice information), one data packet (a packet carrying data), and three idle time slots with no packets, all in six time slots. It is assumed that the probability of a voice packet arrival is p_v, the probability of a data packet arrival is p_d, and the probability of no packet in a slot is p_{idle}. We assume all arrivals are independent of each other. Naturally $p_v + p_d + p_{idle} = 1$. That is, in a time slot one of the three events occur with probability of 1. Also let n be the total number of arrivals of all types ($n = n_v + n_d + n_{idle}$).

(a) Suppose now we wish to determine the probability of two voice packets, one data packet, and three idle slots occurring in six slots *in any order*. The probability of such a particular sequence is $(p_v^2)(p_d^1)(p_{idle}^3)$. Multiplying by the number of such sequences/patterns, which is the multinomial coefficient, yields the multinomial distribution. So for any such order/pattern

$$Prob(n_v, n_d, n_{idle})$$

$$= \frac{n!}{n_v! n_d! n_{idle}!}(p_v^{n_v})(p_d^{n_d})(p_{idle}^{n_{idle}}) \qquad (2.66)$$

For the parameters being considered one has

$$Prob(n_v = 2, n_d = 1, n_{idle} = 3)$$

$$= \frac{6!}{2!1!3!}(p_v^2)(p_d^1)(p_{idle}^3) \qquad (2.67)$$

The individual probabilities can be substituted on the right side of the equations to find the joint probability at the left side of the equation.

(b) Another question for this situation is what is the average number of voice packets in six slots? Of data packets? Of voice or data packets?

The average number of voice packets is $np_v = 6p_v$.
The average number of data packets is $np_d = 6p_d$.
The average number of voice or data packets is $(6p_v + 6p_d) = 6(p_v + p_d)$.

Packet Generators

Finally, consider multiple Bernoulli streams from the outputs of a packet generator (see Fig. 2.8). Here a "packet generator" is a device creating packet streams, perhaps for testing purposes. It is assumed that both the packets within a specific stream and the streams with respect to each other are independent of each other.

Possible problems involving a packet generator providing four streams of packets include the following:

(a) Write the probability in one slot of a packet on both outputs 1 and 2 and no packets on outputs 3 and 4.
    Since the outputs are independent of each other the probability of this particular output pattern is:

$$p^2(1 - p)^2 \qquad (2.68)$$

(b) What is the probability in a time slot of exactly two packets at the outputs in any pattern?
    Each output pattern with exactly two output packets and two idle outputs has identical probability $p^2(1-p)^2$. This is simply multiplied by the number of such patterns (i.e., the binomial coefficient).

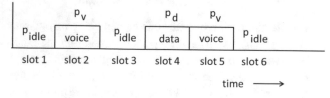

Fig. 2.7 A packet stream with packets of different types

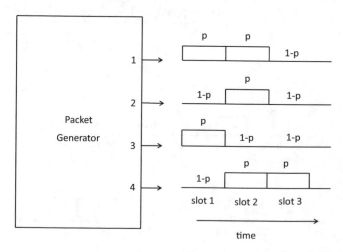

Fig. 2.8 Bernoulli packet generator with four output streams

$$Prob(exactly\ 2\ packets\ in\ any\ pattern)$$

$$= \binom{4}{2} p^2 (1-p)^2 \qquad (2.69)$$

This is the binomial distribution.

$$\overline{number\ of\ output\ packets} = \sum_n np(n) = \sum_{n=1}^{4} n \binom{4}{n} p^n (1-p)^{4-n} \qquad (2.70)$$

Here the over lined quantity is simply the average of that quantity. The second, simpler answer, is just to recognize that the binomial mean is simply Np where N is the number of output streams or just $4p$. Since p is a number between 0 and 1, $4p$ is a number between 0 and

There are two terms above, one for exactly three packets at the outputs in any pattern and one for exactly four packets at the outputs (there is only one pattern for this second case).

(e) Write an expression in terms of q of part (d) for the probability that the ith time slot is the first slot to have three or more packets appearing across all the outputs.

This is just a geometric distribution with parameter q.

$$Prob = (1-q)^{i-1}q \qquad (2.72)$$

2.4.2 Switching Elements

"Switching elements" are devices that connect signal inputs to signal outputs according to some rules in a structured manner. That is, packets or calls are routed from input to outputs. Interconnection networks are patterned connections of relatively simple switching elements into an overall switching network. Such interconnection networks are often implemented on chips [191, 198].

Switching Element Inputs
Consider a switching element in Fig. 2.9 that has four inputs and two outputs.

Time is slotted into slots of equal width. In a time slot a packet arrives at each input with independent probability p and a packet does not arrive at each input in a time slot with independent probability $1 - p$. In the following consider only the inputs.

(c) Write an expression for the average number of packets in a slot across all four outputs.

There are two possible answers. We are really asking for the mean of the binomial distribution. One answer is simply the weighted average definition of the mean.

4 which makes sense since there are four outputs. That is there are most four output packets in a slot.

(d) Write an expression for the probability of three or more packets appearing in a slot across all four outputs. Call the answer q.

$$Prob(3\ or\ more\ output\ packets) = q = \underbrace{\binom{4}{3} p^3 (1-p)^1}_{3\ packets} + \underbrace{\binom{4}{4} p^4 (1-p)^0}_{4\ packets} \qquad (2.71)$$

(a) What is the probability of exactly two incoming packets in a slot across all four inputs in any pattern?

There are two options at each input in each slot, packet or idle slot (i.e., no packet). This is a binomial distribution problem.

$$Prob(exactly\ 2\ incoming\ packets) = \binom{4}{2} p^2 (1-p)^2 \qquad (2.73)$$

That is, there are two inputs with packets (occurring with probability p^2), two inputs without packets (occurring with probability $(1-p)^2$). The number of ways this can happen is $\binom{4}{2} = 6$ so the probability of two arriving

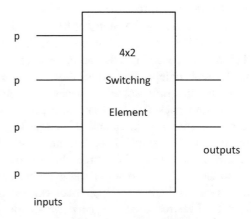

Fig. 2.9 A four-input, two-output, switching element

packets and two idle slots in some specific pattern is multiplied by this amount.

(b) What is the probability of at least two incoming packets in a slot across all inputs?

One can write this as

$$Prob(at\ least\ 2\ incoming\ packets)$$

$$= Prob(2\ packets) + Prob(3\ packets)$$

$$+ Prob(4\ packets) \tag{2.74}$$

$$= \binom{4}{2} p^2 (1-p)^2 + \binom{4}{3} p^3 (1-p)^1$$

$$+ \binom{4}{4} p^4 (1-p)^0 \tag{2.75}$$

Or

$$Prob(at\ least\ 2\ incoming\ packets)$$

$$= 1 - Prob(0\ packets) - Prob(1\ packets) \tag{2.76}$$

$$= 1 - \binom{4}{0} p^0 (1-p)^4 - \binom{4}{1} p^1 (1-p)^3 \tag{2.77}$$

$$= 1 - (1-p)^4 - 4p(1-p)^3 \tag{2.78}$$

(c) What is the average number of packets at the inputs in a time slot?

This is the mean value of the binomial distribution or $4p$. More formally

$$\overline{number\ of\ packets} = \sum_{n=1}^{4} n \binom{4}{n} p^n (1-p)^{4-n} = 4p$$

$$\tag{2.79}$$

In the above, the over lined quantity means the average of that quantity.

2 × 1 Switching Element

Now consider a switching element that has two inputs and one output. Time is slotted again. In each time slot either 0 or 1 packets arrive at each input. A packet arrives to an input in a time slot with independent probability p and does not arrive to an input in a time slot with independent probability $1 - p$. An output packet is produced in a time slot if at least one packet arrives to the inputs. If one packet arrives to the inputs in a time slot, it is sent to an output. If two packets arrive to the inputs in a time slot, one packet is forwarded to the output and other packet is dropped/cleared from the system. A higher level protocol can resend the dropped packet if desired. Switching networks are sometimes designed to handle traffic on a statistical basis in this manner.

(a) Write an expression for the probability of a packet at the output in a time slot. Call this answer q.

The probability that there is a packet at the output is equal to the probability that there is at least one packet at the inputs. This is equal to one minus the probability there are no packets at the inputs

$$Prob(one\ output\ packet)$$

$$= Prob(at\ least\ one\ input\ packet) \tag{2.80}$$

$$Prob(at\ least\ one\ input\ packet)$$

$$= 1 - Prob(no\ input\ packets) \tag{2.81}$$

$$Prob(one\ output\ packet) = 1 - (1-p)^2 = q \tag{2.82}$$

(b) Suppose that the process starts with slot 1. What is an expression for the probability that the first output packet appears in slot i?

The answer to this question is a geometric distribution. There is no packet for $i - 1$ slots with probability $(1-q) \times (1-q) \times \ldots (1-q) = (1-q)^{i-1}$ and in the ith slot there is an output packet with probability q. To find the overall joint probability these probabilities multiply since events in each slot are independent

$$Prob(1st\ output\ packet\ in\ slot\ i) = (1-q)^{i-1} q$$

$$\tag{2.83}$$

(c) What is the probability that i consecutive slots all have output packets?

This is just the probability of there being a packet at the output in a slot, q, multiplied i times or simply q^i.

4 × 2 Switching Element

Consider a four-input, two-output (4 × 2) switching element as in Fig. 2.9.

There is at most one packet per input per time slot. If one or two packets arrive at the inputs in a time slot, then the packets go to the outputs. If three or four packets arrive, two packets are randomly chosen to go to the outputs and the remaining packet(s) is erased/dropped. Let p be the independent probability of a packet arrival on an input in a time slot and let $1 - p$ be the independent probability of no arriving packet at an input in a time slot.

(a) What is the probability that exactly one packet arrives in one slot across all inputs and exactly two packets arrive in the next slot across all inputs?

Arrivals from slot to slot are independent. So the requested probability is the product of two binomial probabilities

$$Prob(1, \; then \; 2 \; arrivals)$$
$$= P(1 \; arrival) Prob(2 \; arrivals) \qquad (2.84)$$

$$Prob(1, \; then \; 2 \; arrivals)$$
$$= \left[\binom{4}{1} p^1 (1-p)^3 \right] \left[\binom{4}{2} p^2 (1-p)^2 \right] \qquad (2.85)$$

(b) What is the probability that there are two packets at the outputs?

The probability of two packets at the outputs equals the probability of 2–4 arrivals at the inputs

$$Prob(2 \; output \; packets) = \sum_{n=2}^{4} \binom{4}{n} p^n (1-p)^{4-n} \qquad (2.86)$$

(c) What is the mean thruput (the average number of packets at the outputs in a time slot)?

$$\overline{Thruput} = 1 \cdot Prob(1 \; output \; packet)$$
$$+ 2 \cdot Prob(2 \; output \; packets) \qquad (2.87)$$

The average thruput is a weighted sum. It is one times the probability of one packet at the inputs (leading to a single output packet) plus two times the probability of two to four packets at the inputs (which leads to two packets at the outputs). Thus

$$\overline{Thruput} = 1 \cdot \binom{4}{1} p^1 (1-p)^3 + 2 \cdot \sum_{n=2}^{4} \binom{4}{n} p^n (1-p)^{4-n} \qquad (2.88)$$

Concatenator

Consider a switching element serving as a "concatenator." There are four inputs and four outputs. Time is slotted. The independent probability of there being a packet at each input in a time slot is p. The independent probability of there being no packet at an input in a time slot is $1 - p$.

Referring to Fig. 2.10, if one input packet arrives across all inputs in a time slot, it is sent to the top output. If two packets arrive across all the inputs in a time slot, the packets are sent to the top two outputs. If three packets arrive across all the inputs in a time slot, the packets are sent to the top three outputs. If four packets arrive at the inputs in a time slot, each output gets a packet.

(a) What is the probability that there is a packet at the top output in a time slot?

One must be careful—the answer is not 1 as no packet may arrive to the inputs in a time slot. The probability in question is equal to one minus the probability of no arrivals or the probability of at least one arrival.

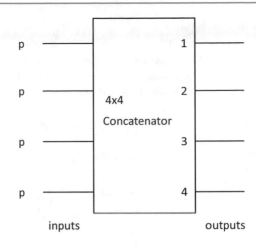

Fig. 2.10 A four-input, four-output concatenator

$$Prob(packet \; at \; top \; output) = 1 - (1-p)^4 \qquad (2.89)$$

Or

$$Prob(packet \; at \; top \; output) = \sum_{n=1}^{4} \binom{4}{n} p^n (1-p)^{4-n} \qquad (2.90)$$

(b) What is the probability that in a time slot, output 3 has no packet?

This probability is just the probability that two or less packets arrive to the inputs in a time slot. One has

$$Prob(no \; packet \; at \; output \; 3)$$
$$= Prob(0 \; input \; packets) + Prob(1 \; input \; packet)$$
$$+ Prob(2 \; input \; packets) \qquad (2.91)$$
$$= (1-p)^4 + \binom{4}{1} p(1-p)^3 + \binom{4}{2} p^2 (1-p)^2 \qquad (2.92)$$

(c) What is the probability that output 4 has a packet in a time slot?

This is simply the probability of four input packets.

$$Prob(packet \; at \; output \; 4) = \binom{4}{4} p^4 (1-p)^0 = p^4 \qquad (2.93)$$

2.4.3 Clusters of Computers

Clusters (i.e., groups) of computers provide parallel processing power to solve problems in scientific, engineering and business computing and in data centers.

A Cluster in a Rack

Consider an equipment rack in a data center housing 96 computers. Let p be the independent probability that a computer is down (i.e., not working). Let $1 - p$ be the independent probability that a computer is up (i.e., working).

(a) Write an expression for the probability that exactly one server is down.

This is just a binomial distribution

$$Prob(exactly\ 1\ server\ down) = \binom{96}{1} p(1-p)^{96-1}$$
$$= 96p(1-p)^{95} \tag{2.94}$$

(b) Write an expression for the probability that two or less computers are down.

This is

$$Prob(2\ or\ less\ down)$$
$$= Prob(0\ down) + Prob(1\ down) + Prob(2\ down) \tag{2.95}$$

$$= \binom{96}{0} p^0 (1-p)^{96} + \binom{96}{1} p(1-p)^{95}$$
$$+ \binom{96}{2} p^2 (1-p)^{94} \tag{2.96}$$

Simplifying

$$= (1-p)^{96} + 96p(1-p)^{95} + 4560p^2(1-p)^{94} \tag{2.97}$$

(c) As the number of computers is increased, does the answer of (b) increase or decrease? Why?

The more computers, the more likely that there are failed computers. Thus the probability of part (b) decreases.

Clusters

Let there be three computers in each of four clusters (see Fig. 2.11).

Let p be the independent probability that a computer is up (i.e., working) and $1 - p$ be the independent probability

that a computer is down (i.e., not working). Computers and clusters are independent with respect to each other.

(a) A cluster is "functioning" if at least one computer in it is working. Find an expression for the probability that all four clusters are functioning.

The probability that at least one computer is up in a specific cluster is

$$Prob(at\ least\ 1\ computer\ up\ in\ a\ cluster)$$
$$= r = 1 - (1-p)^3 \tag{2.98}$$

This is one minus the probability that all three computers in a specific cluster are down (or overall the probability that at least one computer in a specific cluster is up).

Now, across all clusters the probability that each cluster is functioning is:

$$Prob(every\ cluster\ is\ up) = (1 - (1-p)^3)^4 = r^4 \tag{2.99}$$

The individual cluster probabilities multiply since the clusters are independent with respect to each other.

(b) Find an expression for the expected (i.e., average) number of clusters that are functioning.

Using r from part (a) one has the binomial mean which can be written in two ways.

$$\overline{number\ of\ up\ clusters} = 4r = \sum_{i=1}^{4} i \binom{4}{i} r^i (1-r)^{4-i} \tag{2.100}$$

Here r is a probability between 0 and 1 so that the expected number of clusters that are functioning, $4r$, is a number between 0 and 4. This makes intuitive sense.

2.4.4 Linear Networks

Linear networks have a simple topological structure that allows for interesting problems.

Idle Paths and Blocking

Consider the linear network in Fig. 2.12.

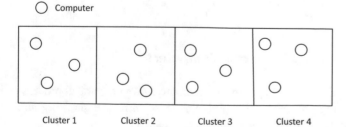

Fig. 2.11 Four clusters consisting of three computers each

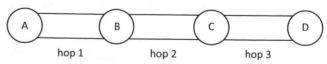

Fig. 2.12 A linear network with two links between each pair of consecutive nodes

In this academic exercise each link can support at most one call (actual telecommunication links support many calls). A new call can only use a link if it is "idle" (not busy with an existing call). The independent probability that a link is busy with an existing call is p. The independent probability that a link is idle and can accommodate a new call is $1 - p$.

(a) Write an expression for the probability that there is at least one idle path for a call from A to D (see the figure).

$$Prob(1 \ or \ 2 \ idle \ paths \ from \ A \ to \ D) = \quad (2.101)$$

$$= (Prob(1 \ or \ 2 \ idle \ links \ from \ A \ to \ B))^3 = q^3 = \quad (2.102)$$

Here q is the probability of there being at least one idle link from A to B.

$$q^3 = \left(\underbrace{\binom{2}{1}(1-p)p}_{1 \ idle \ link \ A \ to \ B} + \underbrace{\binom{2}{2}(1-p)^2 p^0}_{2 \ idle \ links \ A \ to \ B} \right)^3$$

$$= \left(2(1-p)p + (1-p)^2 \right)^3 \quad (2.103)$$

Alternately the probability that at least one of the two AB links is idle is the same as one minus the probability both AB links are busy. So

$$= (1 - Prob(2 \ AB \ links \ busy))^3 = q^3 = \quad (2.104)$$

$$= \left(1 - \binom{2}{2} p^2 (1-p)^0 \right)^3 = (1 - p^2)^3 \quad (2.105)$$

(b) Write an expression for the probability a path from A to D being blocked between nodes B and C.

This is simply the probability both links between B and C are busy or p^2.

(c) Let the linear network in the figure go on to the right forever. Write an expression for the probability that a new call originating at node A is blocked at hop i.

This is an application of the geometric distribution. Each hop is successfully transited with probability q. It is blocked eventually at hop i with probability $(1 - q)$. Thus

$$Prob(call \ is \ blocked \ at \ i) = q^{i-1}(1 - q) \quad (2.106)$$

(d) What is the average number of hops transited by a new call for part (c)?

number of successful hops

$$-\sum_i (i - 1)q^{i-1}(1 - q) \quad (2.107)$$

$$= \sum_i i q^{i-1}(1 - q) - \sum_i q^{i-1}(1 - q) \quad (2.108)$$

$$= \frac{1}{1-q} - 1 = \frac{1 - (1-q)}{1-q} = \frac{q}{1-q} \quad (2.109)$$

If the call is stopped at the ith hop, it has successfully transited $i - 1$ hops. This accounts for the $(i - 1)$ term in the summation of Eq. (2.107). The last equation is found by some algebraic manipulation to use an infinite summation formula one can look up in the Appendix A or in a mathematical handbook. See Question 1 in Sect. 2.3 for a similar derivation.

2.4.5 Now What?

We will look at more complex switching problems in the following sections. But this section has given us a good foundation of probability problem solution techniques. In practicing such problems, say for example for an exam, the authors recommend solving each problem to completion and only then looking at the answers. Similar problems as in this section also appear in the end of the chapter exercises.

2.5 Multiple Access Performance

2.5.1 Introduction

An early problem in computer networking was the multiple access problem. That is, how does one share a common medium (i.e., channel) among a number of users in a decentralized manner. One important early (circle 1980) application was the first version of Ethernet, the popular local area network protocol. Here the channel was supported on a coaxial cable. A second earlier application was the Aloha packet radio network interconnecting the University of Hawaii. In the following we analyze the performance of two discrete time tractable Ethernet models and a continuous time Aloha model. In both cases we will find that there is an intermediate value of the traffic load "offered" to the network (called offered load) that maximizes throughput.

2.5.2 Discrete Time Ethernet Model

The first implementation of Ethernet involved stringing a coaxial cable within an office in a linear fashion. Computers tap into the cable which serves as a "private" radio channels

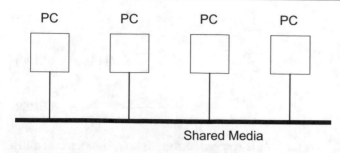

Fig. 2.13 Shared media Ethernet

in the sense that transmissions are confined (electromagnetically) to the cable (see Fig. 2.13).

The basic Ethernet protocol is called CSMA/CD. This stands for carrier sense multiple access with collision detection. It functions as follows. Any station sensing an idle channel will attempt to transmit as soon as it has a message (or packet). While all stations will be able to read the header of a packet only the station whose address matches the destination address will pay attention to it.

In this sense the original Ethernet cable served as a broadcast medium. The basic shortcoming of the CSMA/CD protocol is that if two or more stations sense the cable to be idle at about the same time, the two or more stations may launch packets onto the cable at the same time, the transmissions will overlap and this will be unintelligible to the stations. This overlapping of transmissions is called a "collision."

Stations in the early 1980s implementations of Ethernet can detect a collision situation. The multiple stations involved in a collision will each reschedule their transmissions into the future hoping to transmit them without a collision.

It is possible to create an approximate discrete time model of Ethernet transmission using the binomial distribution. That we will do in this subsection. A more accurate model is discussed in the next subsection.

Let time be divided into equi-spaced slots. Each slot can hold a fixed length packet. Each of N stations will attempt to transmit a single packet in a specific slot with independent probability p (and not transmit with independent probability $1 - p$). Thus each slot holds 0 to N packets. However useful information is conveyed only if there is a single transmission (packet). A "collision" occurs if two or more stations attempt to transmit in the same time slot. The channel is idle if no station transmits in a specific slot. Thus there are three possible mutually exclusive events (idleness, useful transmission, collision) that can occur in a specific slot. Using the binomial distribution one has

$$P[0 \; xmssns] = (1 - p)^N \qquad (2.110)$$

$$P[1 \; xmssn] = Np(1 - p)^{N-1} \qquad (2.111)$$

To calculate the probability of 2 or more transmissions in a slot, one could sum the probabilities of each of n (i.e., $2 \le n \le N$) stations transmitting. A more clever approach is to realize that

$$P[collision] = 1 - P[0 \; xmssns] - P[1 \; xmssn] \quad (2.112)$$

$$P[collision] = 1 - (1 - p)^N - Np(1 - p)^{N-1} \quad (2.113)$$

Naturally we can plot each of these three probabilities versus p, the "offered load" to the network. Most interesting is the plot of useful throughput ($P[1 \; xmssn]$) versus p. As p increases the throughput is initially linear in p (light load) and then saturates and decreases (heavy load). Thus a heavily loaded 10 Mbps Ethernet may only carry, say, 4 or 5 Mbps of actual traffic. The decrease in throughput is due to the increasing fraction of time wasted in collisions as p is increased. This is the price paid for a completely decentralized system with no centralized scheduling.

We can find the value of p that maximizes throughput. Using our calculus knowledge, one can set the derivative of throughput ($P[1 \; xmssn]$) with respect to p equal to zero, solve and obtain

$$p_{optimal} = 1/N \qquad (2.114)$$

This makes intuitive sense. If N stations attempt to access the cable, each with probability $1/N$, the offered load is simply $N \times (1/N) = 1.0$, or one packet per slot on average. That is, beyond this match between offered load and normalized network capacity, more and more collisions result.

It was mentioned above that the binomial model is approximate. This is because of several implicit assumptions made for the binomial model that do not perfectly represent actual Ethernet operation. For instance, in reality packets/frames are of variable length and can be transmitted at any (continuous) time instant. Secondly, the propagation delay of the coaxial cable should be modeled to adequately represent the collision process. Also in reality the probability a specific attempt to transmit in a slot is correlated with transmission failures in the recent past. Finally a station usually has buffers to hold packet waiting to be transmitted. However our binomial based model does portray the key feature of CSMA/CD performance, the drop in throughput as load is increased.

The CSMA/CD protocol used in Ethernet is said to be 1 persistent. That is if the channel is sensed as idle by a station with a packet to send, the station transmits instantly with probability 100% (i.e., 1 persistent). In a p persistent protocol the probability that a station with a packet transmits on an idle channel is $p \times 100\%$. With such less greedy stations, throughout actually does improve, though with a corresponding increase in delay. That is the delay experienced by each packet increases with the less greedy access policy.

A word is in order on the differences between the Ethernet cable (bus) and the typical "bus" inside a computer. Computer buses typically have a high throughput achieved by using a number of wires (say 32 or 64 wires) in parallel. That is for a 32 bit "wide" bus each wire may operate at 100 Mbps so that the total capacity is 32×100 Mbps or 3.2 Gbps of capacity.

Local area network (e.g., Ethernet) "buses," on the other hand, are essentially supported on a single wire. Given the large distances (10 to 100s of meters) involved, this makes economic sense.

Because computer buses have a small physical size, their use can be governed by a bus scheduler. A bus scheduler determines, in some fair manner, how much bus time to grant to each system (CPU, I/O...) requesting service. There are no collisions. An early Ethernet cable, on the other hand, has significant propagation delay due to its multimeter size. Thus at the time it was reasonable to trade off wasted capacity due to collisions for the simplicity of no central scheduler.

In fact Ethernet performance degrades as physical network size is increased. This has led to some interesting tradeoffs as Ethernet speed pushed above 10 Mbps, which are discussed below. The trend in the more recent versions of both Ethernet and other local area networks is to make use of a hub/star architecture.

In a hub architecture stations are wired directly to a hub (switch), which can be the size of a cigar box or smaller, in a star type pattern. A number of stations wired to a hub may make use of a shared media card which mimics the larger coaxial Ethernet connections including the presence of collisions. In a switched Ethernet card, on the other hand, buffers hold transmissions until they can be scheduled on a bus in the hub. In this case there are no collisions. It should be noted that the trend towards hub architecture includes such older local area network protocols as fast Ethernet (100 Mbps) and Gigabit Ethernet (1000 Mbps).

2.5.3 Ethernet Design Equation

With some more elaboration we can develop the basic Ethernet equation that predicts channel efficiency (i.e., utilization) as a function of data rate, minimum frame length, and network size.

Assume a heavy load of k stations attempting to access the shared media (cable). Following [227], let p be the independent probability that a single station attempts to access the media. Then the probability only a single station acquires the channel, A, is given by a binomial distribution

$$A = kp(1 - p)^{k-1} \qquad (2.115)$$

Now from the previous section we know A is maximized when $p = 1/k$. The probability that a "contention interval" has j slots is given by a geometric distribution

$$P(j) = P[j \ slot \ contention \ interval] = (1 - A)^{j-1} A \qquad (2.116)$$

Here the contention interval is the interval during which stations contend unsuccessfully for access to the channel until one station is successful. Then the mean (average) number of slots per contention interval is

$$\sum_{j=0}^{\infty} j P(j) = \sum_{j=0}^{\infty} j(1 - A)^{j-1} A \qquad (2.117)$$

$$= A \sum_{j=0}^{\infty} j(1 - A)^{j-1} \qquad (2.118)$$

$$= \frac{A}{1 - A} \sum_{j=0}^{\infty} j(1 - A)^{j} \qquad (2.119)$$

$$= \frac{A}{1 - A} \times \frac{1 - A}{A^2} \qquad (2.120)$$

$$= \frac{1}{A} \qquad (2.121)$$

Here we have used the same procedure used in simplifying a previous summation in this chapter (see Sect. 2.3, Question 1). Again the slotted model we use is an approximation to what is really a continuous time system.

Let the propagation delay from one end of the cable to the other end be τ. The worst case contention interval duration occurs when a station at one end of the cable transmits onto an idle channel, only to have the station at the opposite end of the cable transmit just before the first station's signal reaches it. In this case the time between the first station beginning transmission and finding out a collision occurs is 2τ (the round trip propagation delay). In the worst case, then, each contention interval slot is 2τ seconds long.

The mean contention interval (consisting of $1/A$ contention interval slots) is thus $2\tau/A$ seconds. However with an optimal (throughput maximizing) choice of p, $1/A = e$ as $k \to \infty$. So $2\tau/A = 5.4\tau$.

Next an expression for channel efficiency (utilization) is needed. This is the ratio of the time useful information is transmitted to the total time it takes to transmit the information. If P is the frame (packet) length in seconds

$$U = \frac{useful \ time}{total \ time} = \frac{P}{P + \frac{2\tau}{A}} \qquad (2.122)$$

But

$$P = \frac{F}{B} \qquad (2.123)$$

Here F is minimum frame size (in bits) and B is the data rate (in bits per second). Then

$$U = \frac{\frac{F}{B}}{\frac{F}{B} + \frac{2\tau}{A}} \qquad (2.124)$$

But from just before, $2\tau/A = 2\tau e$, so

$$U = \frac{\frac{F}{B}}{\frac{F}{B} + 2\tau e} \qquad (2.125)$$

$$U = \frac{1}{1 + \frac{2B\tau e}{F}} \qquad (2.126)$$

However the one way propagation delay is L/c where L is the cable length (in meters) and c is the speed of light or electromagnetic radiation (in meters per second). Thus

$$\boxed{U = 1 / \left(1 + \frac{2BLe}{cF}\right)} \qquad (2.127)$$

This is the basic Ethernet design equation. The original Ethernet standard (IEEE 802.3) produced reasonable utilizations with a 10 Mbps data rate, 512 bit minimum frame size, and a maximum cable size of 500–1000 m.

The challenge to the Ethernet community over the years was how to boost data rate while maintaining utilization. Clearly if B is simply increased, U will drop. Thus in producing Fast Ethernet with a 100 Mbps data rate in the early to mid 1990s B was increased by a factor of ten and the maximum network size, L, was reduced by a factor of ten (to about 50 m) so the product, BL, in the equation is constant.

In designing gigabit (1000 Mbps) Ethernet during the late 1990s, this trick could not be repeated since network size, L, would be an unrealistic 5 m. Instead B was increased by a factor of ten and F, the minimum frame size, was increased

by a factor of eight (from 512 bits to 512 bytes). Thus the ratio B/F is approximately constant and utilization levels are largely maintained.

Finally, note that utilization increases as frame size increases. This is because for longer frames, collisions are a smaller portion of the transmission time of a packet.

2.5.4 Aloha Multiple Access Throughput Analysis

In the second look at sharing a communication medium, the Aloha packet radio network developed circa 1970 for the University of Hawaii is examined. The idea at the time was to connect the Hawaiian island campuses with a distributed radio network. The basic layout appears in Fig. 2.14. The central station monitors a single incoming channel of packets from the satellite islands. What the central station "hears" on the incoming channel is broadcast back to all satellite islands on a second channel.

There is no central control in this system. A station on a satellite island simply broadcasts into the main island in the hopes it is the only incoming transmission at that time. That is, if two or more stations transmit towards the central repeater at about the same time, the messages will overlap and be unintelligible. This is the Aloha equivalent of an Ethernet collision. Individual satellite stations monitor the outbound channel to hear what the central repeater hears (an intelligible transmission, unintelligible transmission or idleness).

However without collision detection and with the relatively larger transmission distances the performance of Aloha is lower than that of Ethernet for the same channel speed. Intuitively though, one would expect the type of performance of Aloha to be similar to Ethernet. That is, as offered load is increased, a linear growth in throughput first saturates and then decreases due to collision-like behavior.

The by now standard analysis [3, 4, 199, 204] to be used to measure this performance is a bit different in the details compared to the Ethernet like analysis of the previous section.

Fig. 2.14 ALOHA network geography (topology)

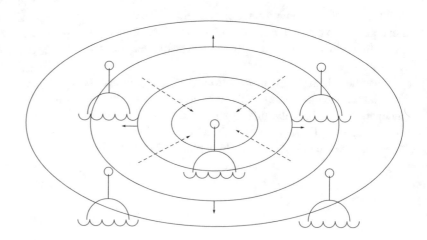

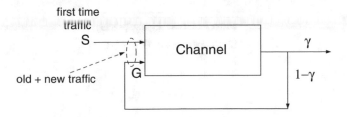

Fig. 2.15 ALOHA channel diagram

Now consider Fig. 2.15. In the channel system of Fig. 2.15 there are two streams of traffic attempting to enter the channel, S packets per unit time T of new, first time traffic and G packets per unit time T of new traffic and repeated transmissions which did not get through on earlier attempts. Let the probability of a successful transmission be γ. Since the throughput in a stable channel equals the outflow (or inflow)

$$S = G\gamma \qquad (2.128)$$

But what is γ? There are two related scenarios here. In the first scenario, known as slotted Aloha, time is broken into equi-spaced slots of duration T seconds per slot. One packet fits in one slot. Collision like behavior occurs when more than one packet attempts to be transmitted in the same slot. We assume that the probability $P(k)$, that there are k packets transmitted in a specific slot, is Poisson. That is, the arrivals in one slot time follow a Poisson distribution. This is reasonable as the transmissions from the satellite islands should be independent of each other, at least to a first approximation.

$$P(k) = \frac{(Gt/T)^k}{k!} e^{-(G/T)t} \qquad (2.129)$$

Here G/T is the equivalent of the packet arrival rate, with overall units of the number of packets in an interval t. Here also we use a "tagged" packet approach to estimate throughput. That is, we observe one given (tagged) packet and observe its chance of being successfully transmitted. For our tagged packet to be successful, there must be no other new or old transmissions ($k = 0$). Thus

$$\gamma = e^{-(G/T)t} \qquad (2.130)$$

Here small t is the window of vulnerability in seconds. That is, it is the time period when a packet(s) other than the tagged packet may attempt to transmit, causing a collision. For slotted Aloha, clearly $t = T$ (the slot width and packet transmission time). Thus

$$\gamma = e^{-(G/T)t} \qquad (2.131)$$

And

$$\gamma = e^{-G} \qquad (2.132)$$

$$\boxed{S = Ge^{-G}} \qquad (2.133)$$

Plotting throughput (S) versus offered traffic (G), one finds (Fig. 2.16) throughput is maximized at a value of 36.8% at $G = 1.0$.

The second possible scenario is called "pure" Aloha. Here packets may arrive at any time instant. If one considered partially overlapping packets leading to collision, then the window of vulnerability during which a tagged packet is susceptible to an overlapped transmission is, with some thought, $t = 2T$. Then

$$\boxed{S = Ge^{-2G}} \qquad (2.134)$$

Now throughput is maximized at a value of 18.4% at $G = 0.5$ (Fig. 2.16). While this is significantly lower than the 36.8% of slotted Aloha, there is no need for slot timing boundaries. A key implicit assumption in our look at Aloha is that G, the arrival stream of new and old traffic, is Poisson. It is actually clearer that this would be true for the new distributed traffic stream, S, than for G. This is because, over relatively short periods of time, new incoming transmissions are independent of each other. Actually G is correlated with S. That is, a failed transmission attempt increases the rate of future attempts. This correlation aside, the performance evaluation we have, is reasonable for a first look at Aloha. See [201] for a more detailed treatment.

2.5.5 Aloha Multiple Access Delay Analysis

With some work we can determine the average delay experienced by packets attempting to transit an Aloha channel. Let us look first at slotted Aloha. If a packet is not successful on an attempt through the channel, it is randomly rescheduled some time into the future. Of course it may take a number of attempts to get a packet through the channel, particularly if the load is heavy.

Let us say a packet that needs to be rescheduled is uniformly likely to be transmitted over K slots (in each with probability $1/K$). Then following Saadawi, the average number of slots a packet waits before transmitting is $\bar{i}$

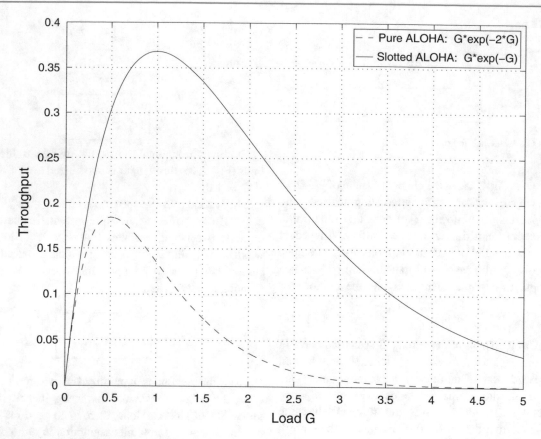

Fig. 2.16 ALOHA system throughput operating curves

$$\bar{i} = \sum_{i=0}^{K-1} i \frac{1}{K} \tag{2.135}$$

$$\bar{i} = \frac{1}{K} \sum_{i=0}^{K-1} i \tag{2.136}$$

$$\bar{i} = \frac{1}{K} \frac{K(K-1)}{2} \tag{2.137}$$

$$\bar{i} = \frac{K-1}{2} \tag{2.138}$$

Here we use $K - 1$ in the summation rather than K as we are interested in the time a packet waits before transmitting (0 to $K - 1$ slots at most).

If T is the duration of a slot in this "backoff" algorithm, then the average backoff time is $\bar{i}T$.

Next, the time cycle duration of an unsuccessful attempt consists of three components

$$T_u = 1 + R + \frac{K-1}{2} \tag{2.139}$$

One can see this includes the packet transmission time (1), propagation delay (R), and the average backoff time (($K - 1$)/2). With algebra

$$T_u = R + \frac{K+1}{2} \tag{2.140}$$

Here the units are in slots. Multiplying by the number of seconds/slot, T, one has

$$T_u = T \left[R + \frac{K+1}{2} \right] \tag{2.141}$$

Here T_u, the average unsuccessful cycle time, is in seconds.

Now let E be the average number of retransmissions. Then the average delay (in slots) to get a packet through the channel, D, has three components

$$D = \frac{1}{2} + \frac{T_u}{T} E + (1 + R) \tag{2.142}$$

Here 1/2 is the average slot time difference between a packet becoming ready to send and actually be sent (uniformly on slots of normalized duration 1). The second term is the time for unsuccessful transmissions (measured in slots). The last term is the time to successfully transmit the packet plus propagation delay (R). Here also propagation delay is the time for the radio signal to physically transit the channel.

Naturally we need to calculate E. From the previous section we know the probability a packet successfully transits the Aloha channel is e^{-G}. Using the geometric distribution, the probability a packet transits the channel in n attempts, $P(n)$, is

$$P(n) = (1 - e^{-G})^{n-1} e^{-G} \qquad (2.143)$$

The average number of attempts, including the successful one, is then (see Sect. 2.3)

$$\bar{n} = \sum_{n=1}^{\infty} n P(n) \qquad (2.144)$$

$$\bar{n} = \sum_{n=0}^{\infty} n (1 - e^{-G})^{n-1} e^{-G} \qquad (2.145)$$

$$\bar{n} = e^{G} \qquad (2.146)$$

Then substituting

$$E = e^{G} - 1 \qquad (2.147)$$

Here we subtract the successful last attempt from the total number of attempts to find E, the average number of unsuccessful attempts, E. We can also say (as $S = Ge^{-G}$ from the previous subsection)

$$E = \frac{G}{S} - 1 \qquad (2.148)$$

If one plots D (average delay) versus G (total throughput) for slotted Aloha, one obtains an exponential curve as in Fig. 2.17. For low or medium load average delay is small. However as one approaches a fully loaded channel delay increases exponentially. This type of behavior is similar in spirit to multiple access resource sharing systems such as web servers and Markovian queues (see the next chapter).

In any case, putting everything together one has

$$\boxed{D = \frac{1}{2} + \left[R + \frac{K+1}{2} \right] \left[e^{G} - 1 \right] + [1 + R]}$$

$$(2.149)$$

For pure Aloha, the delay equation, using a similar approach as that for slotted Aloha, is in slots

$$D = \frac{T_u}{T} E + (1 + R) \qquad (2.150)$$

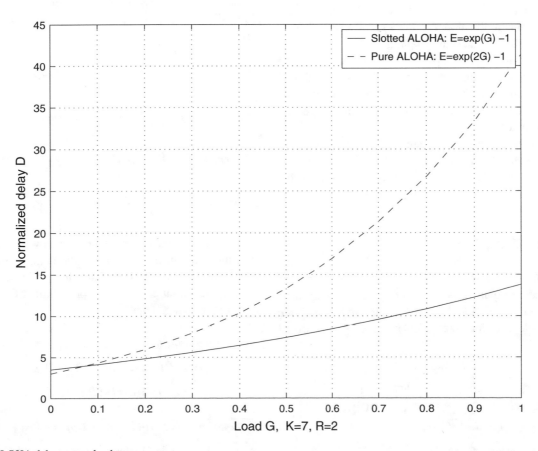

Fig. 2.17 ALOHA delay versus load curves

Since a packet is transmitted as soon as it is ready in pure Aloha, there is no average submission time delay "1/2" term. Note here $E = e^{2G} - 1$. Thus

$$D = \left[R + \frac{K+1}{2} \right] \left[e^{2G} - 1 \right] + [1 + R] \qquad (2.151)$$

For the parameters chosen in Fig. 2.17, it can be seen that pure Aloha has significantly more delay at heavy loads than slotted Aloha. This is a consequence of pure Aloha's e^{2G} term, versus slotted Aloha's e^G term.

2.6   Switching Elements and Networks

2.6.1   Introduction

How does one design a high speed packet switch? The switching architecture with the best performance is space division switching. That is, one uses a multiplicity of relatively simple "switching elements" tied together by a structured interconnection network. The term "space division" comes from the fact that individual switching elements and paths are spatially separated. The term also allows a contrast to the older technology of time division switching [204].

An important motivation for the use of space division switching is the potential to implement this architecture in VLSI. The ability to simply copy (replicate) the same switching element many times on a chip speeds implementation. Also specialized VLSI chips have the potential to process many more packets per second than a general purpose computer.

In the following sections fundamental and representative statistical models of switching elements are considered. Tree networks and some interconnection networks are next discussed. Finally, a multiple bus system is studied.

2.6.2   Switching Elements

Consider the $m \times n$ switching elements of Fig. 2.18. There are m inputs and n outputs. Input and output processes are time slotted in nature. All slots have the same duration and one packet fits into one slot exactly. Each input's packet arrival

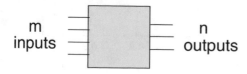

Fig. 2.18 $m \times n$ switching element

process is a Bernoulli process. The independent probability of a packet arriving in a given slot at a given input is p. The independent probability of no packet being in a slot at an input is $1 - p$. Arrivals are independent of each other from slot to slot and input to input.

Let us assume that time slot boundaries are synchronized. That is, the time slot boundaries for all inputs occur at the same time instants. The same is true for the outputs.

A switching element may implement one of a number of routing policies (for transferring incoming packets to output ports). Let us consider two representative policies

Policy A Say one has a 3×2 switching element. If 1 or 2 packets arrive in a slot on the inputs, the packets are sent to the output ports. It is not important here which of two arriving packets goes to which output port as long as only one of the two packets goes to each output port. If three packets arrive, one packet is randomly and fairly selected to be dropped (erased) and the other two packets proceed to the output ports.

Let us now look at some performance measures. A performance measure is a quantity that represents the performance of the system under consideration. Throughput, delay, blocking probability, and loss probability are common network performance measures. From the binomial distribution the probability of 0, 1, 2, or 3 arrivals across all inputs in a slot is

$$P(0 \; arrivals) = \binom{3}{0} p^0 (1-p)^3 = (1-p)^3 \qquad (2.152)$$

$$P(1 \; arrival) = \binom{3}{1} p^1 (1-p)^2 = 3p(1-p)^2 \qquad (2.153)$$

$$P(2 \; arrivals) = \binom{3}{2} p^2 (1-p) = 3p^2(1-p) \qquad (2.154)$$

$$P(3 \; arrivals) = \binom{3}{3} p^3 = p^3 \qquad (2.155)$$

The average (mean) throughput or flow per slot for the switching element can now be simply found. It is a weighted combination of the number of arriving packets successfully going to the output ports and the probability of each number of packets arriving. Thus

$$\begin{aligned} \overline{Throughput} \\ = 1 \cdot P(1 \; arrival) + 2 \cdot P(2 \; arrivals) \\ + 2 \cdot P(3 \; arrivals) \qquad (2.156) \end{aligned}$$

$$\overline{Throughput} = 1 \cdot 3p(1-p)^2 + 2 \cdot 3p^2(1-p) + 2p^3 \qquad (2.157)$$

$$\overline{Throughput} = 3p(1-p)^2 + 6p^2(1-p) + 2p^3 \qquad (2.158)$$

Here the weight multiplying $P(3\ arrivals)$ is 2, and not 3, as if there are three arrivals only two packets proceed to the two output ports.

Finally consider the probability that a packet is dropped. There are two ways to look at this situation. One is to take a "bird's eye" view of the switching element. The probability that *a* packet is dropped is then simply the probability of three arrivals.

The second way to look at this situation is from the viewpoint of a "tagged" arriving packet. That is, suppose we take a seat on an incoming packet. If we know this packet is arriving on an input in the current slot with probability one, what is the probability a packet is dropped? It is equal to the probability that there are two additional arrivals on the other inputs. By way of contrast the probability that the arriving tagged packet is dropped is one third of the probability that there are two additional arrivals as any of three arriving packets is equally likely to be dropped.

Policy B Again consider a 3 × 2 switching element with three inputs and two outputs. Now suppose that an arriving packet is equally likely to go to either output port. Thus 0, 1, 2, or 3 packets may each prefer to go to a given output/port in a given slot. However, under the protocol to be examined, at most one packet may exit an output in one slot because of the limited capacity of the output link and the next switching element it leads to. If more than one packet wants to go to a particular output port, one packet is chosen randomly and fairly to go to the output and the remaining packets are dropped.

Now the average throughput of a switching element is a weighted combination of the probability that a given number of arrivals occur in a slot, the probability that a certain pattern of packets prefer to go to each of the output ports in a slot for a given number of switching element arrivals and the number of arriving packets successfully going to the output ports in a slot. The first term is the same as for Policy A. Let us find the other terms.

Consider the probability that a certain number of packets were to go to a given output port in a slot for a given number of switching element arrivals. To find this consider Fig. 2.19. Here there are two arriving packets to a three input two output (3 × 2) switching element. Call the two arriving packets A and B. The four boxes are for the patterns of packet output port preference. Each box is divided into two boxes, one for each of the two output ports.

It can be seen that 50% of the time the packets go to the same output port and 25% of the time both packets prefer a specific output port. With two packets preferring the same output port only a single packet actually leaves the output port under the assumed protocol so that the mean throughput has the term

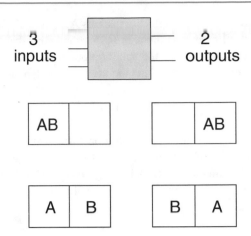

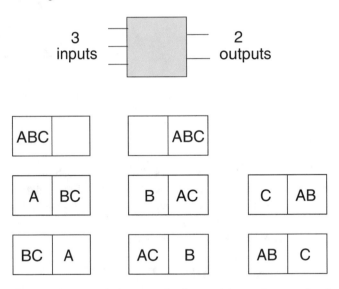

Fig. 2.19 Packet arrival patterns for two arriving packets to a 3 × 2 switching element

Fig. 2.20 Packet arrival patterns for three arriving packets to a 3 × 2 switching element

$$1 \cdot \left(\frac{1}{2}\right) \cdot 3p^2(1-p) \qquad (2.159)$$

The other 50% of the time when two packets arrive to the switching element each packet goes to a different output port (see Fig. 2.15) Thus two packets will exit the output ports. So one has another term for mean throughput of

$$2 \cdot \left(\frac{1}{2}\right) \cdot 3p^2(1-p) \qquad (2.160)$$

Another way to find the factor of 1/2 intuitively is to realize that if two packets arrive to the switching element, one will go to some output port. There is then a 0.5 probability that the second packet goes to the OTHER output port causing one packet to appear on each output port.

Now suppose that three packets arrive to the switching element. The possible patterns of arriving packets' output port preferences are shown in Fig. 2.20. There are eight possibilities. Call the arriving packets A, B, and C. One can see in the figure that there is 0.25 probability that all the arriving packets prefer the same output port. Naturally this leads to one departure. Thus the average throughput has the term

$$1 \cdot \left(\frac{1}{4}\right) \cdot p^3 \tag{2.161}$$

From Fig. 2.20 there is a 0.75 probability that two arriving packets prefer one output port and the third arriving packet prefers the other output port, leading to two packet departures from the switching element. Thus another average throughput term is

$$2 \cdot \left(\frac{3}{4}\right) \cdot p^3 \tag{2.162}$$

Finally, if only one packet arrives to the switching element, it definitely gets to the output port it prefers so that the final mean throughput term is

$$1 \cdot 3p(1-p)^2 \tag{2.163}$$

Putting the five terms together, one has the following expressions for the mean throughput of Policy B

$$\overline{Throughput} = \underbrace{1 \cdot 3p(1-p)^2}_{1-arrival} \tag{2.164}$$

$$= \underbrace{1 \cdot \left(\frac{1}{2}\right) \cdot 3p^2(1-p) + 2 \cdot \left(\frac{1}{2}\right) \cdot 3p^2(1-p)}_{2-arrivals} \tag{2.165}$$

$$= \underbrace{1 \cdot \left(\frac{1}{4}\right) \cdot p^3 + 2 \cdot \left(\frac{3}{4}\right) \cdot p^3}_{3-arrivals} \tag{2.166}$$

Now let us consider loss probability, the probability that a packet is lost, first from a bird's eye view. Let us define $P_{loss|n}$ as the probability of loss if there are n arrivals. Also $P(n)$ is the probability of n arrivals. Naturally loss only can occur if there are two or three arrivals. Then

$$P_{loss} = P_{loss|2}P(2) + P_{loss|3}P(3) \tag{2.167}$$

The probability of n arrivals, $P(2)$ and $P(3)$, is the same as usual. In Fig. 2.19, of the eight arriving packets, two are lost so that $P_{loss|2} = 0.25$. In Fig. 2.20, of the 24 arriving packets, 10 are lost so that $P_{loss|3} = 5/12$. Thus

$$P_{loss} = \left(\frac{1}{4}\right) \cdot 3p^2(1-p) + \left(\frac{5}{12}\right) \cdot p^3 \tag{2.168}$$

For Policy B let us again look at a "tagged" arriving packet. This is the view from a packet that we have a seat on that we KNOW is arriving. Suppose first that the question is what is the loss probability (of *some* packet) given our tagged packet arrives. Then from Fig. 2.19, half of the four boxes lead to loss so $P_{loss|2} = 0.5$. For three arriving packets, from Fig. 2.20, all of the eight boxes lead to loss so $P_{loss|3} = 1.0$. However since we "know" that our tagged packet is arriving, the probability of two arrivals is simply the probability of one more packet arriving. Similarly, for the same reason the probability of three arriving packets is the probability of two additional packets arriving.

Then

$$P_{loss} = \frac{1}{2} \cdot 2p(1-p) + 1.0 \cdot p^2 \tag{2.169}$$

Now say that we are interested in the probability that OUR tagged packet is lost. For two packets arriving to the switching element, again half of the time loss occurs but since two packets are involved and the chance that the tagged packet is the one dropped is 0.5 and so $P_{loss|2} = 0.5 \times 0.5 = 0.25$. If there are three arriving packets and all three packets prefer the same output (which occurs 2/8 of the time), the probability that the tagged packet is the one lost is 2/3 (i.e., two packets out of three are lost). If there are three arriving packets and two packets prefer the same output and the remaining packet prefers the other output (which occurs 6/8 of the time), the probability that the tagged packet is the one lost is 1/3 (one of three packets are lost). The probability of loss is modified in the equation above for

Thus

$$P_{loss} = \frac{1}{4} \cdot 2p(1-p) + 1.0 \cdot \left(\frac{2}{8} \cdot \frac{2}{3} + \frac{6}{8} \cdot \frac{1}{3}\right) \cdot p^2 \tag{2.170}$$

2.6.3 Networks

As indicated above, an engineer is interested in not just single switching elements but in networks of switching elements and networks for packet distribution. In fact a complete study of networks of switching elements requires a knowledge of queueing theory (next chapter) so only some basic situations are covered below.

Tree Networks

Tree topologies can be modeled as a graph with no loops (cycles). Usually one node is identified as a root. Trees are important as they can be embedded into arbitrary interconnection networks (graphs) to provide connectivity between nodes. For a set of N nodes, a spanning tree connects all nodes.

Practically, cable TV systems and hierarchical circuit switching telephone network are tree networks.

Tree networks are also related to multiplexing hierarchies. Consider a digital telephone network. A digital phone can produce 64 kbps of uncompressed voice (i.e., 8 bits/sample × 8000 samples/s). In the North American phone system 24 voice calls are interleaved to produce a 1.544 Mbps "T1" stream. Four T1 streams can be interleaved into a 6.312 Mbps T2 stream and on and on. Demultiplexing proceeds in the opposite manner. That is, a T2 stream is broken down into constituent T1 streams and each T1 stream is broken down into 24 phone channels.

In the following, two tree type problems are considered. The first problem involves capacity calculation. The second problem deals with average path distance.

Capacity Calculation

Consider a binary tree network as in Fig. 2.21. A binary tree has two children nodes per node. In this capacity allocation problem assume that the actual users are in the bottom level. The other nodes are switching elements (i.e., multiplexers and demultiplexers M/D).

Assume all traffic passes through the root. Suppose that the total capacity between each pair of bottom level nodes including both directions is 1 unit of capacity. How much capacity is needed in link X (the link just below the root on the left side)?

Intuitively one can see that the closer a link is to the root, the more capacity is needed. More specifically traffic passing through link X consists of two components. One component is traffic from the lower left user nodes to the lower right nodes. If there are N lower level nodes this traffic, with the usual uniform loading assumption, is $(N/2 \times N/2)$ or $(N/2)^2$. That is, the first component is the number of bottom level nodes on the left side of the tree times the number of bottom level nodes on the right side of the tree. This is because each pair of bottom level nodes is equally likely to generate one unit of traffic. The second component of traffic at link X is traffic from one lower left node to one of the other lower left nodes $(N/2) \times ((N/2) - 1)$. The term of "-1" occurs because a node does not send traffic to itself.

Thus

$$Load_x = \left(\frac{N}{2}\right)^2 + \left(\frac{N}{2}\right)\left(\frac{N}{2} - 1\right) \qquad (2.171)$$

Note that for a l level tree there are 2^l nodes in the lower level or

$$N = 2^l \qquad (2.172)$$

Substituting this equation into the previous one yields

$$Load_x = \left(\frac{2^l}{2}\right)^2 + \left(\frac{2^l}{2}\right)\left(\frac{2^l}{2} - 1\right) \qquad (2.173)$$

$$Load_x = (2^{l-1})^2 + (2^{l-1})(2^{l-1} - 1) \qquad (2.174)$$

$$Load_x = 2^{2l-2} + (2^{2l-2} - 2^{l-1}) \qquad (2.175)$$

$$Load_x = 2^{2l-1} - 2^{l-1} \qquad (2.176)$$

One can see here that the traffic load at link X grows as the square of N or approximately two to an exponent of $2l - 1$.

Assume now that shortest path routing is used in the tree of Fig. 2.21. That is, traffic is not necessarily routed through the root but through the shortest path involving M/D (Multiplexers/Demultiplexers).

Fig. 2.21 Tree network

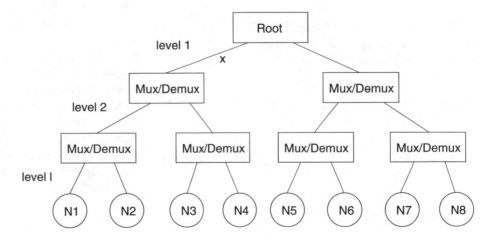

In this case the traffic load in link X consists only of the traffic from the lower left nodes to the lower right nodes (that must transit the root). Thus

$$Load_x = (N/2)^2 = \left(2^l/2\right)^2 = (2^{l-1})^2 = 2^{2l-2}$$

(2.177)

If there are T bps of traffic between each pair of nodes, the above load equation is valid if multiplied by T bps. Naturally, expressions for the load at any level may also be calculated in a similar manner. The fact that trees needed more capacity near the root under a uniform traffic assumption has led to the proposal for the use of "fat" trees as in [151].

Distance Calculation

Consider a binary tree now where each node is a user, except for the root. Assume all transmissions must pass through the root. The question in the distance calculation problem to be considered is to find the average distance (number of hops) between any pair of nodes for a very large tree. A "hop" is an approximate measure of distance. In transiting a link between two neighboring nodes, one makes one hop.

To solve this problem [227] one can see that the average distance between nodes is comprised of two equal components. The first component is the average distance from the nodes to the root. The second component is the average distance from the root to the nodes. If we solve for the first component we need to only double it for the complete answer

$$\overline{Distance} = 2 \times \overline{1\ way\ distance}$$

(2.178)

A little thought will show that for a large binary tree about 50% of the nodes are in the lower most level, about 25% of the nodes in the next lower level, about 12.5% are in the next lower level, etc.

Thus

$$\overline{1\ way\ distance} = 0.5 \times l + 0.25 \times (l-1) + 0.125 \times (l-2)\ldots$$

(2.179)

Here we have a weighed sum of the fraction of nodes at each level and the levels' distances to the root. This weighed sum can be written as

$$\overline{1\ way\ distance} = \sum_{i=0}^{l} (0.5)^{i+1}(l-i)$$

(2.180)

$$\overline{1\ way\ distance} = \sum_{i=0}^{l} (0.5)^{i+1}l - \sum_{i=0}^{l} i(0.5)^{i+1}$$

(2.181)

Each summation can be solved separately. We have (using a summation formula from the Appendix A) and letting $l \to \infty$

$$\sum_{i=0}^{l} (0.5)^{i+1}l = 0.5l \sum_{i=0}^{l} (0.5)^i$$

(2.182)

$$\sum_{i=0}^{l} (0.5)^{i+1}l = 0.5l \frac{1 - 0.5^{l+1}}{1 - 0.5}$$

(2.183)

$$\lim_{l \to \infty} \sum_{i=0}^{l} (0.5)^{i+1}l = l$$

(2.184)

For the second summation

$$\sum_{i=0}^{l} i(0.5)^{i+1} = 0.5 \sum_{i=0}^{l} i(0.5)^i$$

(2.185)

Letting $l \to \infty$ and using another Appendix A summation formula

$$\lim_{l \to \infty} 0.5 \sum_{i=0}^{l} i(0.5)^i = 0.5 \frac{0.5}{(1 - 0.5)^2} = 1$$

(2.186)

Thus

$$\overline{1\ way\ distance} = l - 1$$

(2.187)

That the average distance from any node to the root is $l - 1$ makes intuitive sense as 75% of the nodes are in the two lower most layers (i.e., 50% + 25%). Finally

$$\overline{Distance} = 2 \times \overline{1\ way\ distance} = 2l - 2$$

(2.188)

Knockout Switch

A great deal of effort since the 1980s has gone into designing high speed packet switches (i.e., switching computers). The knockout switch, designed by Yeh et al. [250], is one such high speed packet switch design. A system level design is shown in Fig. 2.22. The overall system has N inputs and N outputs. All N inputs go over buses to each of N bus interfaces (one for each output). A bus interface filters out packets that are destined for its output.

A key feature of the knockout switch is that each bus interface (Fig. 2.23) relays at most L packets arriving in a slot from the inputs to its output. Excess packets (beyond L) are dropped or erased. The L packets are sequentially fed to L shared buffers that lead to the bus interface's output. The term "knockout" is used because the interface contains a network

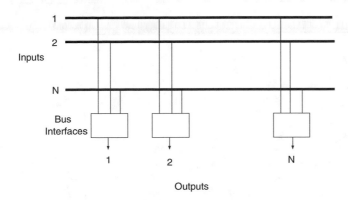

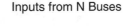

Fig. 2.22 Knockout switch system

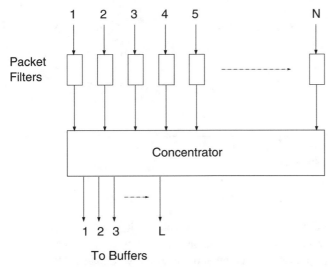

Fig. 2.23 Knockout switch bus interface for a single output

(concentrator) implementing a sports knockout tournament strategy to randomly choose, using a VLSI circuit, which packets to send to the output if more than L packets arrive [191, 204, 250].

Assume that packet arrivals on each input line are aligned with arrivals on other input lines (i.e., arrivals are synchronous). It turns out that if the traffic arrivals are uniform, it is statistically rare for more than L packets to arrive in a slot. To see this, let p be the independent Bernoulli arrival probability for each input. We assume uniform arrivals. That is, each input has the same arrival rate (p) and each packet is equally likely to go to any output. Then the probability that n packets arrive at a bus interface destined for that bus interface's output line is

$$P(n) = \binom{N}{n} \left(\frac{p}{N}\right)^n \left(1 - \frac{p}{N}\right)^{N-n} \quad n = 0.1, 2, \ldots N$$

$$(2.189)$$

Here we use (p/N) rather than the p of the earlier binomial distributions in this chapter (Sect. 2.3) as a packet arrives to an input with probability p but goes to a particular bus interface with uniform probability, $1/N$. Thus the relevant packet arrival probability at a bus interface for a particular output is p/N.

Let us find the (bird's eye view) probability of packet loss, P_{loss}, that is the probability a packet is dropped because the bus interface accepts some other L packets. The average number of packets lost by a bus interface is clearly the average number of packets arriving to the bus interface times the loss probability. Now Np packets arrive on average to the whole switch (N inputs each with arrival probability p). However $1/N$ of the Np packets goes to a single bus interface so that the average number of packets arriving to the interface unit is p. This is less than 1.0 but it is an average, more than one packet may arrive to the interface unit on occasion.

Thus the loss probability P_{loss} is

$$P_{loss} = \frac{1}{p} \times \overline{L} \qquad (2.190)$$

Here $\overline{L}$ is the average number of lost packets at an interface unit.

Then

$$\boxed{P_{loss} = \frac{1}{p} \left(\sum_{n=L+1}^{N} (n-L) \binom{N}{n} \left(\frac{p}{N}\right)^n \left(1 - \frac{p}{N}\right)^{N-n} \right)}$$

$$(2.191)$$

In this expression, as an example, if a bus interface accepts no more than two packets ($L = 2$) and five packets arrive to it ($n = 5$), the three lost packets ($n - L = 3$) weight the binomial probability of five arriving packets.

Figure 2.24 illustrates the loss probability versus different size switches for a $p = 0.9$ (90%) loading. Note that if a bus interface accepts at most eight packets, then the loss probability is less than one in a million for any size switch ($n > 8$).

This idea, that few packets are lost, breaks down if traffic is not uniform. If traffic is highly multiplexed, the uniformity assumption may be reasonable. If there is one or more "hot spot" output (outputs sinking an inordinately large number of packets), the loss probability will increase.

Crossbar Switch

One can use "interconnection networks" to do the switching in a high speed packet switch. An interconnection network [243] is a network of relatively simple switching elements used to connect a set of inputs to a set outputs. The connectiv-

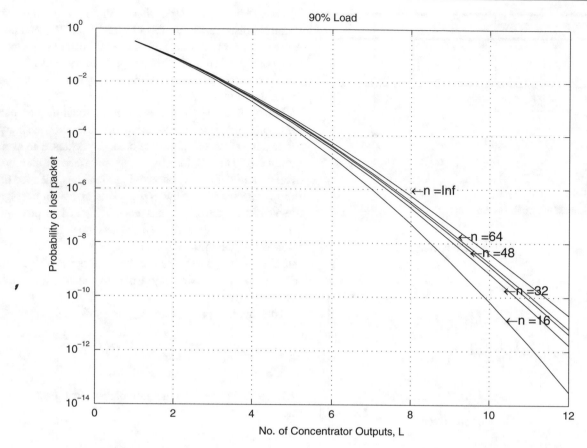

Fig. 2.24 Knockout loss probability versus switch size

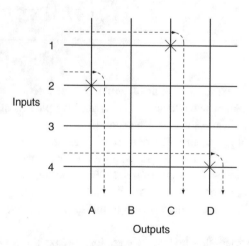

Fig. 2.25 Crossbar switch

the wires normally do not touch (Fig. 2.25). At each place where two wires cross in the grid there is a "crosspoint" switch that, on command, electrically connects the wires. If one is implementing circuit switching, closing one/some of the switches connects desired inputs to desired outputs as in the figure. Note that for point to point communication only one crosspoint per row and column may be closed at one time.

In using a cross-bar interconnection network for packet switching, crossbar switches are closed long enough to forward a packet from input to output. Packet transmissions through the cross-bar may be synchronized. That is, switches close in synchronism for one slot time to transmit fixed length packets. Buffers may be placed at each input, each output, or both. One places buffers at inputs to store arriving packets that cannot immediately get through the switch because of previously stored packets (there is queueing). Buffers may be placed at outputs if more packets may occasionally arrive during a slot for a given output than the output line (from the buffer) can transmit in one slot. Performance evaluations and design tradeoffs for buffered crossbar interconnection networks appear in [119, 131, 173, 191, 194, 253]. Major design tradeoff issues are implementation complexity/feasibility and throughput.

ity may be circuit switched based or packet switched based. We have seen the evaluation of isolated switching elements in Sects. 2.4.2 and 2.6.2. Networks of switching elements are optimized for such considerations as blocking probability, throughput, delay, and implementation complexity.

A fundamental interconnection network is the cross-bar. This can be visualized as N horizontal wires from inputs and N vertical wires going to outputs, arranged in a grid where

In this subsection we consider a crossbar without buffers, as in Fig. 2.26. As described in [178], let fixed length packets fit exactly in one time slot. In each slot a packet may arrive to one of a inputs with independent probability p (and not arrive with independent probability $1 - p$). There are b outputs. Arrivals and switch closings are synchronized (occur together each slot). Even if more than one packet arrives for a specific output, only one (randomly chosen) packet is actually forwarded to the output. Others are dropped. This is a consequence of the fact that only one switch on a vertical wire leading to an output may close during each slot.

The problem here is to find the throughput of the crossbar under a uniform load. The throughput is less than 1.0 (100%) as packets are dropped.

The answer to this problem involves a binomial distribution. In a manner similar to the previous knockout switch problem, the probability an input sends a packet to a specific output is p/b under this uniform load. That is, a packet would arrive on a given input with probability p and go to a specific output with probability $1/b$.

The probability that n packets from the inputs attempt to go to a specific output in a slot is

$$P(n) = \binom{a}{n} \left(\frac{p}{b}\right)^n \left(1 - \frac{p}{b}\right)^{a-n} \quad n = 0, 1, 2 \ldots a \quad (2.192)$$

The average throughput for one output line is the probability that one or more packets try to go to that output

$$\overline{Throughput}_{output} = P(n \geq 1) = 1 - P(0) \quad (2.193)$$

$$\overline{Throughput}_{output} = 1 - \left(1 - \frac{p}{b}\right)^a \quad (2.194)$$

The total throughput for the whole switch is b times the throughput of one output line or

$$\overline{Throughput}_{output} = b \left(1 - \left(1 - \frac{p}{b}\right)^a\right) \quad (2.195)$$

For a very large switch $(a, b \to \infty)$ under a heavy load $(p = 1)$ since

$$\lim_{x \to \infty} (1 - 1/x)^x = \frac{1}{e} \quad (2.196)$$

One has

$$\lim_{a,b \to \infty} b \left(1 - \left(1 - \frac{p}{b}\right)^a\right) = b \left(1 - \frac{1}{e}\right) \quad (2.197)$$

$$\lim_{a,b \to \infty} \overline{Throughput}_{output} = 0.632\, b \quad (2.198)$$

So under a uniform heavy load, a very large switch has a 63.2% average throughput. That is, 63.2% of the packets get through the cross-bar. If the load is not uniform, there will be deviation from this value.

Multiple Bus System

Using a bus to interconnect stations is called a shared media connection. The simple discrete time Ethernet model of Sect. 2.5.2 involves N stations sharing a bus. What if there are multiple buses, and each station is attached to several of them?

Suppose one has N stations, M buses, and R connections to different buses per station ($R \leq M$). Figures 2.27 and 2.28 show two possible architectures.

From the figure it is clear that only certain combinations of N, M, and R lead to symmetric networks. An important parameter of this problem is the number of connections per bus or NR/M. This is the number of system wide connections divided by the number of buses. For a symmetric system it must be an integer.

Let p be the probability a station attempts to transmit a packet on a given bus in a slot. Here we allow one station to attempt to transmit multiple (different) packets on

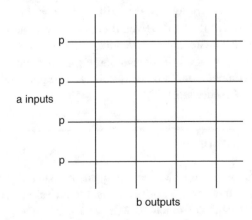

a inputs

b outputs

Fig. 2.26 Crossbar packet switch

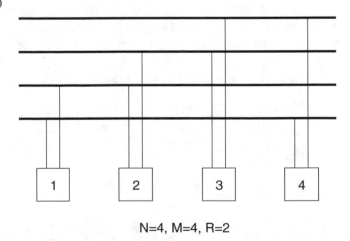

N=4, M=4, R=2

Fig. 2.27 A multibus architecture

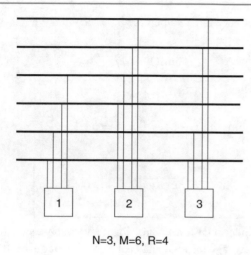

N=3, M=6, R=4

Fig. 2.28 Another multibus architecture

more than one bus in a slot. As in the earlier discrete time Ethernet example, only if one station attempts to transmit on a specific bus is there a successful transmission (more than one station attempting to transmit on a specific bus causes useless collisions).

For the throughput of one bus one has a binomial distribution

$$\overline{Throughput}_{bus} = \binom{NR/M}{1} p(1-p)^{\frac{NR}{M}-1} \quad (2.199)$$

For the whole system of M buses, the throughput is M times as big

$$\overline{Throughput}_{system} = M\left(\frac{NR}{M}\right)p(1-p)^{\frac{NR}{M}-1} \quad (2.200)$$

$$= NRp(1-p)^{\frac{NR}{M}-1} \quad (2.201)$$

If p is too small, many slots on buses will be idle. If p is too large, there will be many collisions. Both situations reduce throughput. The optimal choice of p can be found by taking the derivative of mean system throughput with respect to p.

Since, in calculus, for two function $f(p)$ and $g(p)$

$$\frac{df(p)g(p)}{dp} = f(p)\frac{dg(p)}{dp} + g(p)\frac{df(p)}{dp} \quad (2.202)$$

One has

$$\frac{d}{dp}\overline{Throughput}_{system}$$

$$= NR(1-p)^{\frac{NR}{M}-1} - NRp(1-p)^{\frac{NR}{M}-2}\left(\frac{NR}{M}-1\right) = 0 \quad (2.203)$$

With algebra one can show

$$\boxed{P_{optimal} = M/NR} \quad (2.204)$$

This makes intuitive sense. If p is M/NR and there are NR/M connections per bus, the total average offered load on a bus is $(M/NR) \times (NR/M) = 1.0$. Thus on average one station transmits on the bus at a time. This is an average. Sometimes, no stations or more than one station attempts to transmit on a bus. But with offered load averaging to one station access per slot per bus, throughput is maximized as this is the condition under which successful communication takes place.

This problem formulation is quite general. As in Sect. 2.5.2, if there is one bus, $M = 1$ and $R = 1$ so that mean system throughput is

$$\overline{Throughput}_{system} = Np(1-p)^{N-1} \quad (2.205)$$

If there are M buses and each station connects to every bus, $M = R$ and

$$\overline{Throughput}_{system} = MNp(1-p)^{N-1} \quad (2.206)$$

This M fold increase in throughput is consistent with the previous equation.

2.7   Conclusion

Since its inception several hundred years ago, probability has been a powerful tool to evaluate uncertainty in various systems. This chapter has introduced basic concepts such as the Bernoulli process, related distributions, and applied models in the context of important networking concepts and applications.

Once one has developed skills in applying probability to technological problems, the skills can be applied to new technologies as they arise. Even if one never evaluates a high speed packet switching architecture, one may work with ad hoc radio networks, TCP/IP versions, digital communication, parallel computation, or other systems and be glad one has a bag of probability tools.

2.8   Problems

1. Suppose a simple Markovian queue has state dependent arrival and service rates. That is, the arrival and service mean rates are functions $\lambda(n)$ and $\mu(n)$ (respectively) of the number of customers in the queue just prior

to an arrival or departure. Rewrite the development of Eqs. (2.15)–(2.20) in these terms.

2. Suppose a telephone switch accepts 400 calls/s during a certain time period, according to a Poisson process. Using the Poisson distribution find the probability of 0, 1, or 2 calls during a time interval of 1/400 s and during a 1/1000 s interval. Do the numbers make intuitive sense?

3. If 400 calls/s arrive to a telephone switch, according to a Poisson process what is the average number of calls arriving in 1/400 s? In 1/4 s? In 1 s?

4. Software in a telephone switch can process 400 calls/s. If the call arrival rate is tripled, how many more calls arrive in 0.25 s?

5. Show that Eq. (2.34) is correct for its application.

6. The geometric distribution of Eq. (2.36) decreases as i increases. Intuitively why?

7. Every time a telephone truck leaves the depot, there is a 1/500 probability that it breaks down. How many days from today, on average, can the first breakdown be expected? Hint: what statistical process and distribution is implied? Why might this not be a good reliability model?

8. A packet train is a series of consecutive time slots with packets. The probability of a packet arrival in a slot is p. Given that one packet arrives and starts the train, what is the average length of the train? Hint: Think of the events that end the packet train.

9. What value of p maximizes the variance in the number of arrivals described by a Bernoulli process? Why?

10. Prove that the mean (average) number of arrivals for a Bernoulli process with packet arrival probability p for N total slots is Np.

11. Packets arrive to an input line of a packet switch according to a Bernoulli process. The packet arrival probability is $p = 0.2$. What is the probability of five arrivals in ten slots in any pattern of arrivals?

12. In the previous problem, the probability of 3 arrivals in 10 slots, when $p = 0.5$ is the same as the probability of 7 arrivals, in 10 slots with $p = 0.5$. Why?

13. Consider three telephone circuits from the USA to Europe belonging to a company. Measurements reveal that during the business day each circuit is busy 20 min out of an hour, on average. Most calls are short (2 or 3 min or less).

    (a) Let p be the independent probability that a single circuit is busy. Find a numerical value of p and show how you arrived at it.

    (b) Write an expression for the probability a given call goes through (finds a free circuit).

    (c) Write an expression for the probability that, at a given instant, *exactly* one circuit is free.

    (d) Write an expression for the average number of busy channels. Hint: Write it as a weighted sum.

14. An online company has its own computerized consumer order system. There are three regional centers. Each regional center (New York, Dallas, and San Francisco) has two independent computers for reliability. Let p be the independent probability that a single computer is down. For the entire network to be considered to be functional, at least one computer must be up in each regional center. Find an expression for the probability that the entire network is functional.

15. For a Bernoulli arrival process, find the smallest value of p so that the probability of exactly four packets arriving in ten slots is at least 0.6.

16. Show that the variance of a binomial distribution is:

$$\sigma^2 = Np(1-p) \qquad (2.207)$$

17. Packets arrive according to a Bernoulli process to an input line of a packet switch. Find the probability that the fifth arrival occurs in exactly ten slots, if $p = 0.35$.

18. In the previous problem, what is the average number of slots holding five arrivals if $p = 0.35$.

19. Plot the throughput of the discrete time simple Ethernet model (Eq. (2.111)) versus offered load, p. Let $N = 10$. Also plot the probability of collision versus p (Eq. (2.113)). Comment on the shape of the curves.

20. For the simple Ethernet model of Sect. 2.5.2 show that the optimal (throughput maximizing) choice of p, the station transmission probability is $1/N$. See Eq. (2.111).

21. Referring to the Ethernet design equation, as the probability that only a single station acquires a channel increases, the average number of slots per contention interval decreases (see Eq. (2.121)). Intuitively, why is this so?

22. Find the Ethernet utilization (efficiency) for a 100 Mbps network of size 50 m and a 512 bit minimum frame size.

23. What is more realistic for designing a 10 Gbps Ethernet: reducing network physical size by a factor of 10 or increasing the minimum frame size by a factor of 8–10? Refer to Eq. (2.127) and the following discussion.

24. Explain why the window of vulnerability for pure Aloha is equal to 2T (see discussion above Eq. (2.134)).

25. Show that throughput is maximized at 18.4% for pure Aloha at $G = 0.5$ and at 36.8% for slotted Aloha at $G = 1.0$.

26. Which system slotted or pure Aloha has a larger mean delay as G becomes larger? Intuitively why?

27. What are the three quantities multiplied to form the terms in Eqs. (2.164) through (2.166)?

28. Explain Figs. 2.19 and 2.20.

29. Find the load in link x in the capacity calculation section for a binary tree (Sect. 2.6.3) if there are 2, 5, and 10 levels.

30. Consider a tree as in the capacity calculation subsection of Sect. 2.6.3 and Fig. 2.21. Calculate the capacity needed at an arbitrary link in the tree using the same uniform loading assumption as in the capacity calculation section.

31. For the Knockout switch how do you justify each bus interface accepting and transmitting at most L packets?

32. Find the expression for throughput in the multiple bus problem of Sect. 2.6.3 if a station transmits in a slot with probability p to a specific randomly chosen bus (with probability $1/M$). That is, a station attempts to transmit in only one bus in a slot.

33. Consider a diamond shaped network of four links and four nodes. The left corner node is A, the top corner node is B, the bottom corner node is C, and the right corner node is D. Sketch a diagram. Links AB, BD, AC, and CD carry at most one circuit each. Here p is the independent probability that a link is available (idle).

    (a) Find the probability that the upper path from A to D (ABD) is available (both links available). Call this probability q.

    (b) Find the probability at least one of the paths from A to D (upper and lower) is available.

34. Consider three nodes, A, B, and C. You can draw them horizontally. There are three links from A to B and three links from B to C. Each link carries at most one circuit. At any time p is the measured probability that each link is in use (and not accepting further circuits).

    (a) Find an expression for the average number of busy links (actually a number between 0 and 6). There is both a simple and a more elaborate answer.

    (b) Find an expression for the probability that at least one idle path (two consecutive idle links) exists from nodes A to C.

35. Consider now a model like the last problem but with four nodes, A, B, C, and D. There are three links each between A and B, B and C, and C and D. Again p is the independent probability that each of the (nine) links is busy (each link holds at most one circuit). We are interested in circuits (paths) from A to D. A circuit for a call from A to D uses any available link from A to B, B to C, and C to D.

    (a) Find the probability that a call from A to D is "blocked" (i.e., there is no available path from A to D).

    (b) Find the probability that at least two (i.e., two or three) paths from A to D are available.

36. Consider a small business with 10 phones connected to two outgoing lines through a PBX (private branch exchange switch). A phone seeks an outside line with independent probability p (and is idle with probability $1 - p$). Only two phones, at most, can utilize the two outside lines at a time (one phone per outside line).

    (a) Find an expression for the probability that n phones wish to seek an outside line.

    (b) Find the probability of "blocking" occurs (one or more calls cannot get through when they want to). This should be a function of p.

37. Consider a switching element with two inputs. Time is slotted and the independent probability of an arrival in a slot at an input is p. The probability of no arrival at an input in a slot is $1 - p$.

    Find an expression for the probability of three or more packets arriving in two consecutive slots over both inputs.

38. Consider a two-input, two-output switching element. Time is slotted. Arrivals occur to each input, each slot, with independent Bernoulli probability p.

    If only one packet arrives to the inputs, it is randomly routed to one of the outputs (with probability 0.5 of going to a specific output). If two packets arrive to the inputs (one on each input), each goes to a separate output. For the purposes of this problem it does not matter which packet goes to which output in the latter case of two arriving packets.

    (a) Find the probability of there being a packet on a specific (given) output in a slot. Show how you arrived at your answer.

    (b) Find an expression for the expected (average) number of packets at the outputs in a slot.

39. Consider a three-input, two-output switching element. One output has a connection which feeds its packets back to one of the inputs so that there are two external inputs and one external output. The probability of a packet arrival at an external input in a slot is p (and the probability of no arrival is $1 - p$).

    The switching element policy is that one of the up to three input packets is selected randomly to go to the external output. If two or more packets wish to go to the output, one is selected for the external output, one is selected for the feedback output and any remaining packet is erased (dropped).

    (a) Find an expression for q, the probability that a packet is fed back in a slot. This will be an implicit equation where q is a function for q.

    (b) Solve part (a) for q (i.e., find an explicit equation).

40. Consider a multicasting network of one input, two output switching elements. The network has a single input to the first switching element (stage). This switching elements two outputs each go to another (of two) switching elements (stage 2). The stage 2 switching element outputs each go to (one of four) switching

elements (stage 4). Thus there are seven elements, one network input, and eight network outputs.

For each switching element, if there is an input packet, a copy appears on each of the two element outputs with independent probability c.

Packet arrivals to the network input occur in a slot with independent probability p (with no arrival with probability $1 - p$).

(a) Find an expression for the probability that a copy appears at a network system output.

(b) Find an expression for the average number of copies produced at the network outputs in one slot.

41. Sketch a four-input, two-output switching element. Time is slotted and synchronized at both inputs and outputs. One slot holds at most one packet. Packet arrival processes to each input are Bernoulli (with arrival probability p for a packet in each slot and non-arrival probability $1 - p$ in each slot).

(a) Determine the probabilities of 0, 1, 2, 3, and 4 arrivals in a slot.

(b) Determine the probability of at least one arrival in a slot, simply.

42. Sketch a packet switching element with four inputs and two outputs. Time is slotted and synchronized at both inputs and outputs. One slot holds at most one packet. Packet arrival processes to each input are Bernoulli (with independent probability p for a packet in each slot and independent non-arrival probability $1 - p$ in each slot).

If one or two packets arrive in a slot, they are transmitted on the output(s). If 3 or 4 packets arrive, two packets are randomly chosen to be transmitted on the outputs (one on each at most) and the remaining packet(s) are lost (erased). Packets are assigned to outputs randomly.

(a) Determine from a "bird's eye" view, the probability that a packet is dropped.

(b) Determine the probability a given (tagged) arriving packet is dropped.

(c) Determine the mean throughput of the switching element. This is the mean number of packets transmitted on the output links.

43. Sketch a switching element with N inputs and three outputs. Time is slotted and synchronized at both inputs and outputs. One slot holds at most one packet. Packet arrival processes to each input are Bernoulli (with arrival probability p for a packet in each slot and non-arrival probability $1 - p$ in each slot).

If 1, 2, or 3 packets arrive to the inputs in a slot, the packets are transmitted on the outputs. If more than three packets arrive at the element input, in a slot, three are randomly chosen to be transmitted on the outputs and the remainder are lost. Packets are assigned to outputs randomly.

(a) Determine the mean switching element throughput (as a function of p and N).

(b) Determine the mean number of dropped packets during a slot as a function of p and N.

44. Sketch a network of four switching elements. All switching elements, A,B,C,D, have three inputs and one output. The inputs of element D are A, B and C's outputs. Thus, the overall system has nine inputs and one output. Time is slotted and synchronized at both inputs and outputs. One slot holds at most one packet. Packet arrival processes to each input are Bernoulli (with independent probability p for a packet arrival in each slot and independent non-arrival probability, $1 - p$ in each slot).

For each switching element if one packet arrives at its inputs in a slot it is transmitted on the single element output. If more than one packet arrives to element inputs in a slot, one packet is randomly chosen to be transmitted through the output and the remaining packets are lost.

(a) Determine the probability, q, that a packet is at the output of either elements A,B, or C in a slot.

(b) Find the mean throughput of the system.

45. Find and sketch two different attachment patterns from the ones illustrated for the multiple bus system in Sect. 2.6.3.

46. For the multiple bus system of Sect. 2.6.3 if p is 10% larger (smaller) than the optimal p, find the percentage change in throughput (let there be three stations, six buses, and four bus attachments per station).

47. **Computer Project**: Plot the Poisson distribution (Sect. 2.2) for $n = 0, 1, 2, 3$. Plot the probability $P_n(t)$ versus t when $\lambda(t) = 1$.

48. **Computer Project**: Plot the Knockout switch loss probability (Eq. (2.191)) on a log scale versus the number of concentrator outputs (1 through 12) for $N = 16, 32, 64$ and N very large with a 90% load ($p = 0.9$).

49. **Computer Project**: Plot the crossbar output throughput of Eq. (2.195) as a function of p for $a = b$ from 2 through 30 in steps of 2.

50. **Computer Project**: Plot the system throughput of a multiple bus system (Eq. (2.201)) versus p where $N = 12$, $R = 3$ and $M = 2, 3, 4, 6, 9, 12, 16$.

Abstract

Queueing theory models things waiting in lines. Such things include packets, telephone calls or computer jobs. Continuous time and discrete time single queues are reviewed. This includes M/M/1, Geom/Geom/1 and M/G/1 results. Networks of Markovian queues along with the mean value analysis (MVA) computational algorithm are discussed. Negative customer networks are examined. Recursive solutions for certain non-product form networks are covered. Stochastic Petri networks (SPN) with product form solutions are also considered. General solution techniques for these models are outlined.

3.1 Introduction

The concept of using mathematical models to evaluate the carrying capacity of communication devices began in the early years of the telephone industry. Around 20 years into the twentieth century the Danish mathematician A.K. Erlang applied the theory of Markov models developed by the Russian mathematician A.A. Markov 10 years earlier to predicting the capacity of telephone systems. Erlang's brainchild went on to be called queueing theory.

A "queue" is the British word for a waiting line. Queueing theory is the study of the statistics of things waiting in lines. Calls may wait at a telephone switch, jobs may wait for access to a processor, packets of data to be networked may wait in an output line buffer, planes may wait in a holding pattern, and a customer may wait at a supermarket checkout counter. For 100 years queueing theory has been used to study these and similar situations.

This chapter introduces basic queueing theory using Markov models. In fact advanced research in queueing theory involves non-Markov models (such as self-similar traffic) but Markov models remain the foundation of queueing theory and will allow the reader of this introductory text entry to the world of queues.

In Sect. 3.2 of this chapter two basic queueing models are examined. One is the continuous time M/M/1 model of Erlang. The second model is the more recent discrete time GEOM/GEOM/1 model. Section 3.3 discusses some important specific single queue models. Common performance measures are described in Sect. 3.4. Networks of queues and an associated computational algorithm are presented in Sects. 3.5 and 3.6, respectively.

The special cases of queues with negative customers, recursive solutions for the equilibrium state probabilities of non-product form networks, and stochastic Petri nets are discussed in Sects. 3.7, 3.8, and 3.9 respectively. Finally, Sect. 3.10 covers numerical and simulation solution techniques for models.

3.2 Single Queue Models

3.2.1 M/M/1 Queue

The M/M/1 is the simplest of all Markov based queueing models. In the standard queueing model notation the first position in the descriptor indicates the arrival process. The "M" here represents a Poisson arrival process (see Sect. 2.2) with the M representing the memoryless (Markovian) nature of the Poisson process. The second position in the M/M/1 descriptor indicates that the time spent in the server by a customer follows a negative exponential distribution

$$\mu e^{-\mu t} \tag{3.1}$$

Here μ is the mean service rate. This is the only continuous time distribution that is memoryless in nature, allowing for the "M" in the descriptor's second position. That is, if a customer has been in the server for x seconds, the distribution of time until it completes service is also negative exponen-

T. G. Robertazzi, L. Shi, *Networking and Computation*, https://doi.org/10.1007/978-3-030-36704-6_3

tially distributed as above. An intuitive explanation of the memoryless property of the negative exponential distribution is that the negative exponential distribution is a continuous time analog of a random coin flip, where the history of past flips has no influence on the future. In fact the negative exponential distribution is a good model for the duration of voice telephone calls. Finally, the last position in the M/M/1 descriptor indicates the number of servers fed by the waiting line. A server is where the actual processing takes place, as at the checkout counter at a supermarket. The usual assumption is a server holds at most one customer at a time. Here "queue" may refer to the waiting line only or the whole system of waiting line and server(s). This is usually clear from the context.

Queues and queueing networks have a schematic language that gives a higher level application view of the underlying Markov chain. Figure 3.1 shows a single server queue. The open box is the waiting line and the circle represents the server. Figure 3.2 illustrates a 4 server queue, as in some banks, with one common waiting line and four tellers.

The schematic queue of Fig. 3.1, with Markovian statistics (M/M/1) gives rise to the state transition diagram of Fig. 3.3. A state transition diagram is a stochastic finite state machine.

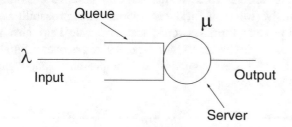

Fig. 3.1 Single server queue schematic

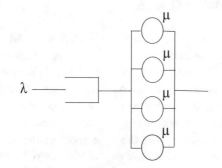

Fig. 3.2 Four server queue schematic

Fig. 3.3 M/M/1 queue state transition diagram

The circles are states $(0, 1, 2 \ldots$ customers in the entire queueing system). The transitions correspond to arrivals of customers (which change the state from n to $n+1$ with mean rate λ) and departures of customers from the server (which change the state from n to $n-1$ with mean rate μ).

Because the transitions only increment or decrement the number of customers in the queueing model by one, this type of model is called a birth death population process in statistics. More elaborate models allow "batch" arrivals or departures, that is multiple arrivals or departures at the same time instant. Naturally the corresponding transitions may jump several states.

This discussion raises an interesting point. In such a continuous time model only one transition is made (in zero time) at a time. That is one never has an arrival or departure at the same time instant. Of course in a continuous time model, time is fine enough that this is possible (if one has an arrival at $T = 2.72316$ s, a departure may be at $T = 2.72316001$ s).

The state transition diagram of Fig. 3.3 is like an electric circuit in the sense that something "flows" through the graph. That something is probability flux. The flux flowing over a transition at time t is simply equal to the product of the transition rate of the transition and the state probability at time t at the state at which the transition originates. Note that Markov chains are different from electric circuits in that for electric circuits flow direction is not preset (it depends on a voltage difference across an element) and a battery or voltage source guarantees non-zero flows in a circuit. In a Markov chain the normalization equation which holds that the sum of state probabilities is one guarantees non-zero flows.

Probability flux, intuitively, has units of [events/sec] and represents the mean number of times per second a transition is transited. A Markov chain may be viewed as a "lattice" on which the system state performs a random walk. Random walks are widely studied random processes [176]. Under the conditions of a continuous time Markov chain random walk, the system state stays a negative exponential distributed random time in a state with a mean time equal to the inverse of the sum of the outgoing transition rates. At the end of this time the system state enters a neighboring state on a random basis. In Fig. 3.3 the system state will leave an interior state n for state $n + 1$ with probability $\lambda/(\lambda + \mu)$ and leave state n for state $n - 1$ with probability $\mu/(\lambda + \mu)$. The pattern should be clear. The probability that the system state leaves a state on a particular transition is equal to that transition's transition rate divided by the sum of outgoing transition rates from the state.

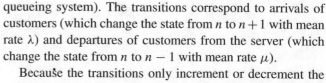

One can set up a series of differential equations to solve for the state probabilities for the M/M/1 model [140, 194]. An alternative approach that will be pursued here is to use the concept of global balance. Global balance, which is analogous to Kirchhoff's current law for electric circuits, holds that the difference between the total flow of probability flux into any state and the total flow of probability flux out of the state at any point in time is the change of probability at the state at that time.

Referring to Fig. 3.4 the total flow into state n is

$$\lambda P_{n-1}(t) + \mu P_{n+1}(t) \qquad (3.2)$$

Here $P_n(t)$ is the nth state probability at time t. The first term represents the probability flux from state $n-1$ to n and the second term represents the probability flux from state $n+1$ to n. Also the total flow out of state n to states $n-1$ and $n+1$ is

$$-(\lambda + \mu)P_n(t) \qquad (3.3)$$

Note here that a negative sign here indicates outward flow and a positive sign indicates inward flow. As discussed, the difference between the above two quantities represents the change in probability at state n as a function of time. Thus

$$\frac{dP_n(t)}{dt} = -(\lambda + \mu)P_n(t) + \lambda P_{n-1}(t) + \mu P_{n+1}(t)$$
$$n = 1, 2 \dots \qquad (3.4)$$

Using a similar argument, the boundary state ($n = 0$) can be modeled by

$$\frac{dP_0(t)}{dt} = -\lambda P_0(t) + \mu P_1(t) \qquad (3.5)$$

Naturally normalization holds that

$$P_0(t) + P_1(t) + P_2(t) + \cdots + P_n(t) + \cdots = 1 \qquad (3.6)$$

These equations represent a family of differential equation to model the time varying behavior of the M/M/1 queue. In fact such transient solutions can be quite complex, even

for the apparently simple M/M/1 queue. For instance, if one starts with ι customers in the queue at time $t = 0$, the probability that there are n customers in the queue at time t is a function of modified Bessel functions of the first kind [194].

The differential equation model of the M/M/1 queue allows transient modeling. However an important type of system behavior with a far simpler solution is steady state (or equilibrium) operation. In other words, if one waits long enough for transient effects to settle out while system parameters (like arrival and service rate) are constant, then system metrics approach a constant and the system exhibits a steady state type of behavior. The M/M/1 queueing system is still a stochastic system but performance metrics like mean throughput, mean delay, and utilization approach a constant. In this mode of operation the state probabilities are constant so

$$\frac{dP_n(t)}{dt} = 0 \qquad n = 0, 1, 2 \dots \qquad (3.7)$$

This leads the previous family of differential equations to become a family of linear equations

$$-(\lambda + \mu)p_n + \lambda p_{n-1} + \mu p_{n+1} = 0 \qquad n = 1, 2 \dots$$
$$(3.8)$$
$$-\lambda p_0 + \mu p_1 = 0 \qquad (3.9)$$

This last equation is a boundary equation for the state $n = 0$. In these linear equations the time argument for state probability has been deleted as the equilibrium probabilities are constant. These equations can also be simply found by writing a global balance equation for each state.

It turns out that to solve any Markov chain with N states, one writes a global balance equation for each state. Any one of the N equations is redundant (contributes no unique information) and can be replaced by the normalization equation

$$p_0 + p_1 + p_2 + \cdots + p_{N-1} + p_N = 1 \qquad (3.10)$$

These N equations and N unknowns can be solved for the equilibrium state probabilities using a standard linear equation solver algorithm. This is of eminent interest because useful performance metrics such as mean throughput, delay, and utilization are simple functions of the state probabilities.

The problem one runs into in trying to use this approach for performance evaluation is that even modestly sized models can have a vast number of states. Moreover direct linear equation solution has a computational complexity that is proportional to the cube of the number of states. This further compounds the problem. For instance, a closed (sealed) queueing network of nine queues and sixty customers has over seven billion states! One can see that modeling the Internet in detail, with this approach at least, with its thousands of nodes and millions upon millions of packets, is out of the question.

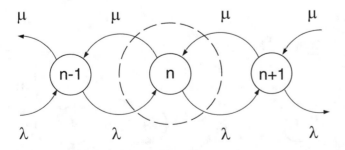

Fig. 3.4 Global balance boundary about state n

However a single M/M/1 queue with its one-dimensional state transition diagram is quite easy to solve, even with an infinite number of states. The way this can be accomplished is to use a different type of balancing called local balance. For the one-dimensional Markov chain one sets up boundaries between each adjacent pair of states (Fig. 3.5). In equilibrium, probability (flux) should not build up or be depleted at each state. Thus one can equate the flow of probability flux across each boundary from left to right to the flow of probability flux across the boundary from right to left to obtain the following family of equations:

$$\lambda p_0 = \mu p_1 \tag{3.11}$$

$$\lambda p_1 = \mu p_2 \tag{3.12}$$

$$\lambda p_2 = \mu p_3 \tag{3.13}$$

$$\cdot$$
$$\cdot$$

$$\lambda p_{n-1} = \mu p_n \tag{3.14}$$

A bit of algebra results in

$$p_1 = \frac{\lambda}{\mu} p_0 \tag{3.15}$$

$$p_2 = \frac{\lambda}{\mu} p_1 \tag{3.16}$$

$$p_3 = \frac{\lambda}{\mu} p_2 \tag{3.17}$$

$$\cdot$$
$$\cdot$$

$$p_n = \frac{\lambda}{\mu} p_{n-1} \tag{3.18}$$

$$\cdot$$

Suppose now that we chain the equations together. That is, substitute the first into the second, this into the third, and so on to obtain

$$p_1 = \left(\frac{\lambda}{\mu}\right) p_0 \tag{3.19}$$

$$p_2 = \left(\frac{\lambda}{\mu}\right)^2 p_0 \tag{3.20}$$

$$p_3 = \left(\frac{\lambda}{\mu}\right)^3 p_0 \tag{3.21}$$

$$\cdot$$

$$p_n = \left(\frac{\lambda}{\mu}\right)^n p_0 \tag{3.22}$$

$$\cdot$$

Now all the equilibrium probabilities are functions of λ and μ (which are known) and p_0 (which is unknown but soon will not be).

To solve for p_0, one can write the normalization equation

$$p_0 + p_1 + p_2 + \cdots + p_n + \cdots = 1 \tag{3.23}$$

Substituting the previous equations for the state probabilities

$$p_0 + \frac{\lambda}{\mu} p_0 + \left(\frac{\lambda}{\mu}\right)^2 p_0 + \cdots + \left(\frac{\lambda}{\mu}\right)^n p_0 + \cdots = 1 \tag{3.24}$$

Factoring out p_0 yields

$$p_0 \left(1 + \frac{\lambda}{\mu} + \left(\frac{\lambda}{\mu}\right)^2 + \cdots + \left(\frac{\lambda}{\mu}\right)^n + \cdots \right) = 1 \tag{3.25}$$

Or:

$$p_0 \left(\sum_{n=0}^{\infty} \left(\frac{\lambda}{\mu}\right)^n \right) = 1 \tag{3.26}$$

$$p_0 = 1 / \left(\sum_{n=0}^{\infty} \left(\frac{\lambda}{\mu}\right)^n \right) \tag{3.27}$$

This can be simplified further. Let $\rho = \lambda/\mu$. From Appendix A we know that

Fig. 3.5 Local balance boundaries between adjacent states

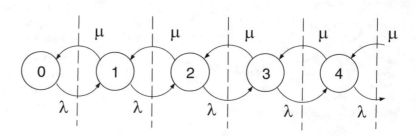

$$\sum_{n=0}^{\infty} \rho^n = \frac{1}{1-\rho} \qquad (3.28)$$

Here $0 \le \rho \le 1$. Then

$$p_0 = 1 / \left(\sum_{n=0}^{\infty} \left(\frac{\lambda}{\mu} \right)^n \right) = 1 / \left(\sum_{n=0}^{\infty} \rho^n \right) = 1 - \rho \qquad (3.29)$$

Also

$$p_n = \left(\frac{\lambda}{\mu} \right)^n p_0 \qquad (3.30)$$

$$\boxed{p_n = \rho^n (1 - \rho) \qquad n = 0, 1, 2, \dots . \qquad (3.31)}$$

This is an extremely simple and elegant expression for the equilibrium state probability of the M/M/1 queue. The particularly simple expression, even though there are an infinite number of states, is due in part to the fact that there is a simple expression for the associated infinite summation.

We can deduce an important fact about the M/M/1 equilibrium state probability distribution from the above expression. The term $(1 - \rho)$ is a constant and the term ρ^n decreases as N increases (since $\rho < 1$). Thus the probability that the M/M/1 queue is in a state n (holds n customers) decreases as n increases. Thus, for example, if λ equals 1 customer/second and μ equals 3 customers per second, the probability that the queue holds 100 customers is $(1/3)^{99}$ or 5.8×10^{-48} times smaller than the probability that there is one customer. With the geometric like decrease in state probability as N increases, an M/M/1 queue only has a relatively small number of customers most of the time.

Another important fact about the M/M/1 queue is that for an infinite sized buffer model as we have here, the arrival rate must be less than the service rate ($\lambda < \mu$ or $\rho < 1$) Otherwise customers would arrive faster than the queue could dispose of them and the queue size would increase without limit in an unstable mode of operation. The condition for infinite buffer M/M/1 queue stability is simply $\rho < 1$.

If the arrival process to an M/M/1 is a Poisson process, what can be said of the departure process? In fact a theorem due to Burke (Burke) shows it is also Poisson. Despite the appealing symmetry, this is not at all obvious. Recall that a process of rate λ is Poisson if and only if the inter-arrival (inter-event) times are independent negatively distributed exponential random variables. If a queue has customers, the inter-departure time is indeed negative exponentially distributed though with service rate μ. That is, the time between departures is simply the service completion time. But sometimes, the queue is empty!

In that case the time between a departure that empties the queue and the next departure is the sum of two random variables, the negative exponential arrival time (with rate λ), and the negative exponential service time of that first arriving customer (with rate μ). The distribution of the sum of two such random variables is not negative exponential!

But it turns out that in the totality of the output process, the statistics are Poisson. While a proof of this is beyond the scope of this book, the concept of "reversibility" can be used in this effort. A reversible process is one where the statistics are the same whether time flows forward or backward. To the best of the author's knowledge, no one has used this to create a time machine—but certainly such an effort would involve a great many states!

A very general result that applies to many queueing systems, including the M/M/1 queue is Little's law. It is usually written as:

$$\boxed{L = \lambda W} \qquad (3.32)$$

Here L is the average queue size (length), λ is the mean arrival rate, and W is the mean delay a customer experiences in moving through a queue. As an example, if forty cars an hour arrive to a car wash and it takes the average car 6 min (0.10 h) to pass through the car wash, there is an average of about 4 cars (40×0.10) at the car wash at any one time.

Little's law will even apply to a queueing system where the inter-arrival and service times follow specified arbitrary (general) distributions. Such a queue with one server is called a G/G/1 queue.

3.2.2 Geom/Geom/1 Queue

The Geom/Geom/1 queue is a discrete time queue [194, 242]. That is, time is slotted and in each slot there is 0 or 1 arrivals to the queue and 0 or 1 departures. Arrivals are Bernoulli so p is the independent probability of an arrival in a slot to the queue and $1 - p$ is the probability of no arrivals. We'll let s be the independent probability that a packet in the server departs in a slot. Thus, even though a packet may be in the server, it may not depart in a given slot.

The time to an arrival and the time for a departure to occur, measured in slots, are naturally given by a geometric distribution (see Sect. 2.3). The geometric distribution is the discrete time analog of the continuous time negative exponential distribution. In fact the geometric distribution is the only memoryless discrete time distribution, just as the negative exponential distribution is the only memoryless continuous time distribution. If one looks at the underlying Bernoulli process, one can see that, just as in a series of coin flips, the time to the next head (arrival) is a memoryless quantity.

Fig. 3.6 Geom/Geom/1 queue
state transition diagram

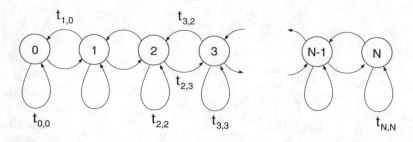

The Markov chain of the Geom/Geom/1 queue is shown in Fig. 3.6. Once again, we have equilibrium state probabilities $P_0, P_1, P_2 \ldots P_n \ldots$. Here capital P is used for the equilibrium state probabilities to distinguish them from the arrival probability p. Now though instead of transition rates we have transition probabilities (the probability that a transition from state i to state j is made from one slot to the next is $t_{i,j}$).

The probability of going from state n to state $n+1$, thereby increasing the number of packets in the queue by one, is the probability of an arrival and no service completion or $p(1-s)$. The probability of going from state n to state $n-1$, decreasing the number of packets in the queue by one, is the probability of a departure and no arrival or $s(1-p)$. Finally, for the interior states, the probability of staying in the same state between two slots is the sum of the probabilities of two events. One term is the probability of no arrival and no service completion or $(1-p)(1-s)$ and the other term is the probability of an arrival and a service completion in the same slot or ps.

The boundary states require some thought. Consider first state 0 and transition probability $t_{0,0}$. Then if one allows a packet to enter and leave an empty queue in the same slot (virtual cut through [135]) $t_{0,0} = (1-p) + ps$. This is because either no arrival or an arrival and immediate departure will leave the queue in the same state 0. Also $t_{0,1}$ is $p(1-s)$ as stated above.

Now one could also implement the more traditional store and forward arrival policy where an arriving packet to an empty queue must wait one slot before departing. Thus $t_{0,0}$ is $(1-p)$ and $t_{0,1}$ is p. That is, a packet arriving to an empty queue waits at least one slot prior to departure as the system always goes from state 0 to state 1 upon an arrival.

For the right boundary, let's suppose now that we have a finite buffer queue (N packets at most). In a single slot either departures precede arrivals or arrivals precede departures. The first case is preferable as a departing packet from a full (state N) queue will leave a space for a subsequent arrival in the same slot. In the latter case an arriving packet to a full queue may be turned away (lost or cleared) even though a space may become available later in the slot.

If departures precede arrivals, $t_{N,N}$ is

$$t_{N,N} = p(s + (1-s)) + (1-p)(1-s) \quad (3.33)$$

$$t_{N,N} = ps + p(1-s) + (1-p)(1-s) \quad (3.34)$$

$$t_{N,N} = ps + (1-s) \quad (3.35)$$

And

$$t_{N,N-1} = (1-p)s \quad (3.36)$$

For $t_{N,N}$ the queue stays in the same state if either there is an arrival and a service completion or no service completion. The queue enters state $N-1$ from state N if there is no arrival and a service completion.

It is important to realize that from state to state some transition must transpire, even if it results in no state change. The sum of all outgoing transition probabilities for a state must also sum to one. Thus

$$t_{N,N-1} + t_{N,N} = (1-p)s + ps + (1-s) = 1 \quad (3.37)$$

This is a good way to check if one has the correct transition probabilities for any discrete time Markov chain. The bookkeeping can become burdensome though for larger chains.

Finally, suppose arrivals occur before departures in a slot. Then $t_{N,N-1} = s$ and $t_{N,N} = 1 - s$. That is, if there is no service departure (with probability $1 - s$), the queue stays in state N.

Summarizing for a queue with virtual cut through and departures before arrivals one has

$$t_{0,0} = ps + (1-p) \quad (3.38)$$

$$t_{0,1} = p(1-s) \quad (3.39)$$

$$t_{n,n+1} = p(1-s) \qquad 1 \le n \le N-1 \quad (3.40)$$

$$t_{n,n-1} = s(1-p) \qquad 1 \le n \le N \quad (3.41)$$

$$t_{n,n} = (1-p)(1-s) + ps \qquad 1 \le n \le N-1 \quad (3.42)$$

$$t_{N,N} = ps + (1-s) \quad (3.43)$$

To solve for the state probabilities, one can draw vertical boundaries, between adjacent states, and equate the flow of probability flux from left to right to that from right to left. For a discrete time queue the probability flux flowing through a transition is the product of the transition probability of that transition and the state probability the transition originates from. This is not too different from the continuous time case.

One has

$$p(1-s)P_0 = s(1-p)P_1 \qquad (3.44)$$

$$p(1-s)P_1 = s(1-p)P_2 \qquad (3.45)$$

$$p(1-s)P_2 = s(1-p)P_3 \qquad (3.46)$$

$$\cdot$$
$$\cdot$$
$$\cdot$$

$$p(1-s)P_{n-1} = s(1-p)P_n \qquad (3.47)$$

$$\cdot$$
$$\cdot$$
$$\cdot$$

$$p(1-s)P_{N-1} = s(1-p)P_N \qquad (3.48)$$

With algebra we can solve for the equilibrium state probabilities recursively (where, again, we use capital "P" for the equilibrium state probability to distinguish it from small "p", the arrival probability).

$$P_1 = \frac{p(1-s)}{s(1-p)}P_0 \qquad (3.49)$$

$$P_2 = \frac{p(1-s)}{s(1-p)}P_1 \qquad (3.50)$$

$$\cdot$$
$$\cdot$$

$$P_n = \frac{p(1-s)}{s(1-p)}P_{n-1} \qquad (3.51)$$

$$\cdot$$
$$\cdot$$

$$P_N = \frac{p(1-s)}{s(1-p)}P_{N-1} \qquad (3.52)$$

Chaining the equations together one has

$$P_1 = \left(\frac{p(1-s)}{s(1-p)}\right)P_0 \qquad (3.53)$$

$$P_2 = \left(\frac{p(1-s)}{s(1-p)}\right)^2 P_0 \qquad (3.54)$$

$$P_n = \left(\frac{p(1-s)}{s(1-p)}\right)^n P_0 \qquad (3.55)$$

$$P_N = \left(\frac{p(1-s)}{s(1-p)}\right)^N P_0 \qquad (3.56)$$

Putting this together

$$P_n = \left(\frac{p(1-s)}{s(1-p)}\right)^n P_0 \qquad n = 1, 2 \ldots N \qquad (3.57)$$

To solve for P_0 for an infinite buffer queue one uses the normalization equation as with continuous time queues.

$$P_0 + P_1 + P_2 + \cdots + P_n + \cdots = 1 \qquad (3.58)$$

Substituting

$$P_0 + \left(\frac{p(1-s)}{s(1-p)}\right)P_0 + \left(\frac{p(1-s)}{s(1-p)}\right)^2 P_0 + \cdots = 1 \qquad (3.59)$$

Or

$$P_0\left(\sum_{i=0}^{\infty}\left(\frac{p(1-s)}{s(1-p)}\right)^i\right) = 1 \qquad (3.60)$$

$$P_0 = 1/\sum_{i=0}^{\infty}\left(\frac{p(1-s)}{s(1-p)}\right)^i \qquad (3.61)$$

Using a summation formula from Appendix A, one has for an infinite buffer Geom/Geom/1 queue with virtual cut through and departures occurring in a slot before arrivals in a slot

$$\boxed{P_0 = 1 - \frac{p(1-s)}{s(1-p)}} \qquad (3.62)$$

For a finite buffer Geom/Geom/1/N queue with virtual cut through switching and departures in a slot occurring before arrivals one does the same as above except the summation is finite. Using a different summation from Appendix A

$$P_0 = 1/\left(\sum_{i=0}^{N}\left(\frac{p(1-s)}{s(1-p)}\right)^i\right) \qquad (3.63)$$

$$P_0 = \frac{1 - \frac{p(1-s)}{s(1-p)}}{1 - \left(\frac{p(1-s)}{s(1-p)}\right)^{N+1}} \qquad (3.64)$$

For an infinite buffer system, one can substitute Eq. (3.62) into Eq. (3.57) to find the equilibrium state probabilities as a single function of p and s. In fact if

$$\rho = \frac{p(1-s)}{s(1-p)}, \qquad (3.65)$$

one then has

$$P_n = \rho^n (1 - \rho) \qquad n = 0, 1, 2 \ldots \qquad (3.66)$$

This is the same form as the earlier expression for the equilibrium state probabilities of the M/M/1 queue (Eq. (3.31))! Also, in much the same way as for an M/M/1 queue, the arrival probability p should be less than the service completion probability s for this infinite buffer discrete time queue's stability.

3.3 Some Important Single Queue Models

A number of useful continuous time, single queue models and solutions are presented in this section.

3.3.1 The Finite Buffer M/M/1 Queueing System

Here we have an M/M/1 queue with the assumption that it holds at most N customers (one customer in the server, $N - 1$ in the waiting line). If a customer arrives to a full queue, he/she is turned away. Such rejected customers are sometimes called lost or cleared from the system. Note that this type of system is called an M/M/1/N queue where the fourth symbol is the buffer size.

The Markov chain of this model appears in Fig. 3.7. We can draw vertical boundaries and equate the probability flux flowing from left to right to the flux flowing from right to left. Then with some simple algebra and solution chaining as in Eqs. (3.11) through (3.18) one has

$$p_1 = \left(\frac{\lambda}{\mu} \right) p_0 \qquad (3.67)$$

$$p_2 = \left(\frac{\lambda}{\mu} \right) p_1 = \left(\frac{\lambda}{\mu} \right)^2 p_0 \qquad (3.68)$$

$$p_3 = \left(\frac{\lambda}{\mu} \right) p_2 = \left(\frac{\lambda}{\mu} \right)^3 p_0 \qquad (3.69)$$

$$\cdot$$
$$\cdot$$

$$p_n = \left(\frac{\lambda}{\mu} \right) p_{n-1} = \left(\frac{\lambda}{\mu} \right)^n p_0 \qquad (3.70)$$

$$\cdot$$

$$p_N = \left(\frac{\lambda}{\mu} \right) p_{N-1} = \left(\frac{\lambda}{\mu} \right)^N p_0 \qquad (3.71)$$

Or

$$p_n = \left(\frac{\lambda}{\mu} \right)^n p_0 \qquad 1 \le n \le N \qquad (3.72)$$

Using the normalization equation

$$p_0 + p_1 + p_2 + p_3 + \cdots + p_N = 1 \qquad (3.73)$$

$$p_0 + \left(\frac{\lambda}{\mu} \right) p_0 + \left(\frac{\lambda}{\mu} \right)^2 p_0 + \left(\frac{\lambda}{\mu} \right)^3 p_0 + \cdots + \left(\frac{\lambda}{\mu} \right)^N p_0 = 1 \qquad (3.74)$$

$$p_0 = \frac{1}{1 + \left(\frac{\lambda}{\mu} \right) + \left(\frac{\lambda}{\mu} \right)^2 + \left(\frac{\lambda}{\mu} \right)^3 + \cdots + \left(\frac{\lambda}{\mu} \right)^N} \qquad (3.75)$$

$$p_0 = \frac{1}{\sum_{n=0}^{N} \left(\frac{\lambda}{\mu} \right)^n} \qquad (3.76)$$

Using a summation from Appendix A

$$p_0 = \frac{1 - \frac{\lambda}{\mu}}{1 - \left(\frac{\lambda}{\mu} \right)^{N+1}} \qquad (3.77)$$

To obtain numerical values for a particular set of parameters one uses the above formula to calculate p_0 and then substitutes p_0 into Eq. (3.72) for $p_1, p_2, p_3 \ldots$

This model is basically the same as that for the M/M/1 queue except for the number of states. Here there is a finite number of states, rather than the infinite number of states of the M/M/1 queue. The change in the number of states requires a renormalization which is allowed for in the denominator of Eq. (3.77). The numerator is the same as for the M/M/1 system. In fact, if the buffer size goes to infinity, this expression for p_0 reduces to $1 - \frac{\lambda}{\mu}$, the M/M/1 queue result. For a finite buffer M/M/1 queue the arrival rate, λ, can be greater than the service rate, μ, since if the queue fills up, customers are simply turned away.

Fig. 3.7 Finite buffer M/M/1/N Markov chain

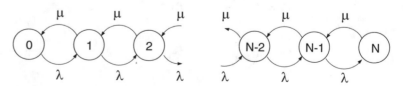

3.3.2 The M/M/m/m Loss Queueing System

Suppose now, as Erlang did, that one has a bank of m parallel servers with negative exponential service times and no waiting line (Fig. 3.8). That is, a (Poisson) arriving customer is placed in an empty server. If all servers are busy though, an arriving customer is cleared (lost) from the system. This model may represent m telephone circuits fed by a common pool of users. If a call does not immediately get through, it is "blocked." Naturally this model does not take into account retries of blocked calls.

The state transition diagram appears in Fig. 3.9. Note that if there are n customers in the system, the aggregate transition rate is $n\mu$ (n busy servers complete a servicing of a customer n times faster than 1 busy server). The solution technique for the equilibrium state probabilities is the same as usual. One draws vertical boundaries between states and equates the flow of probabilities flux across the boundaries

$$\lambda p_0 = \mu p_1 \tag{3.78}$$

$$\lambda p_1 = 2\mu p_2 \tag{3.79}$$

$$\lambda p_2 = 3\mu p_3 \tag{3.80}$$

$$\lambda p_n = (n+1)\mu p_{n+1} \tag{3.81}$$

$$\lambda p_{m-1} = m\mu p_m \tag{3.82}$$

With some algebra and solution chaining one has, where $n \leq m$

$$p_1 = \left(\frac{\lambda}{\mu}\right) p_0 \tag{3.83}$$

$$p_2 = \left(\frac{\lambda}{2\mu}\right) p_1 = \frac{1}{2}\left(\frac{\lambda}{\mu}\right)^2 p_0 \tag{3.84}$$

$$p_3 = \left(\frac{\lambda}{3\mu}\right) p_2 = \frac{1}{6}\left(\frac{\lambda}{\mu}\right)^3 p_0 \tag{3.85}$$

$$p_n = \left(\frac{\lambda}{n\mu}\right) p_{n-1} = \frac{1}{n!}\left(\frac{\lambda}{\mu}\right)^n p_0 \tag{3.86}$$

$$p_m = \left(\frac{\lambda}{m\mu}\right) p_{N-1} = \frac{1}{m!}\left(\frac{\lambda}{\mu}\right)^m p_0 \tag{3.87}$$

Using the normalization equation

$$p_0 + p_1 + p_2 + p_3 + \cdots + p_N = 1 \tag{3.88}$$

$$p_0 + \left(\frac{\lambda}{\mu}\right) p_0 + \frac{1}{2}\left(\frac{\lambda}{\mu}\right)^2 p_0 + \frac{1}{6}\left(\frac{\lambda}{\mu}\right)^3 p_0 + \cdots$$
$$+ \frac{1}{m!}\left(\frac{\lambda}{\mu}\right)^m p_0 = 1 \tag{3.89}$$

$$p_0 = \frac{1}{1 + \left(\frac{\lambda}{\mu}\right) + \frac{1}{2}\left(\frac{\lambda}{\mu}\right)^2 + \frac{1}{6}\left(\frac{\lambda}{\mu}\right)^3 + \cdots + \frac{1}{m!}\left(\frac{\lambda}{\mu}\right)^m} \tag{3.90}$$

$$p_0 = \frac{1}{1 + \sum_{n=1}^{m} \frac{1}{n!}\left(\frac{\lambda}{\mu}\right)^n} \tag{3.91}$$

In telephone system design an important performance measure is the probability that all servers are busy. This is the fraction of time that arriving customers are turned away. Naturally this performance measure is simply p_m. So

$$p_m = \frac{\frac{1}{m!}\left(\frac{\lambda}{\mu}\right)^m}{1 + \sum_{n=1}^{m} \frac{1}{n!}\left(\frac{\lambda}{\mu}\right)^n} \tag{3.92}$$

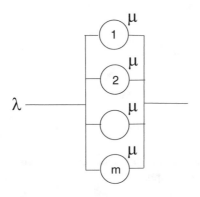

Fig. 3.8 M/M/m/m loss system schematic

Fig. 3.9 M/M/m/m loss system Markov chain

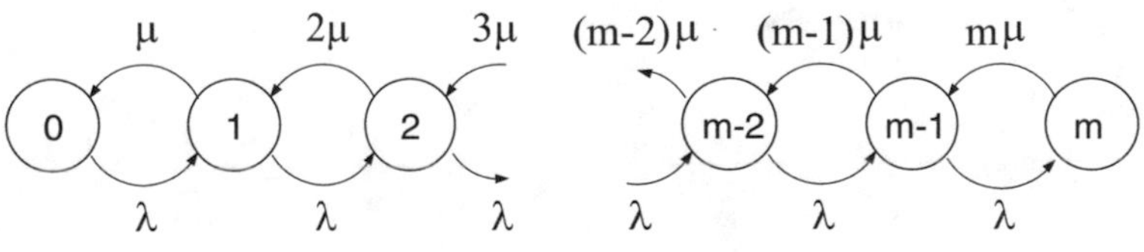

This formula is known as the Erlang B formula or Erlang's formula of the first kind in Europe.

A related queueing model arises if every arriving customer gets its own server (i.e., there is an unlimited number of servers). For this M/M/1/∞ system it is still true that

$$p_n = \frac{1}{n!}\left(\frac{\lambda}{\mu}\right)^n p_0 \qquad (3.93)$$

Also from the equation above for p_0, letting $m \to \infty$

$$p_0 = \frac{1}{1 + \sum_{n=1}^{\infty}\frac{1}{n!}\left(\frac{\lambda}{\mu}\right)^n} = e^{-\frac{\lambda}{\mu}} \qquad (3.94)$$

Here we have used a summation from Appendix A.

3.3.3　M/M/m Queueing System

Now suppose we have the model of the previous section with m servers and with a waiting line (Fig. 3.10). This is a good model of a telephone system where one queues to get an outside line. The Markov chain appears in Fig. 3.11. Note that the chain transitions are identical to those of the M/M/1/M model below state m. Above state m all departure transitions have value $m\mu$. This is because if there are m or more customers in the system at most m servers are busy leading to an aggregate departure rate of $m\mu$. The buffer here is unlimited (infinite) in size.

For the state probabilities at or below state m ($n \leq m$), one has from the M/M/m/m system

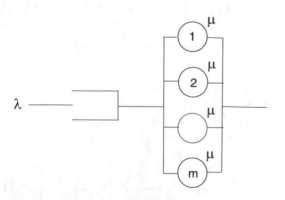

Fig. 3.10　M/M/m queueing system schematic

$$p_n = \frac{1}{n!}\left(\frac{\lambda}{\mu}\right)^n p_0 \qquad n \leq m \qquad (3.95)$$

For states above state m

$$\lambda p_m = m\mu p_{m+1} \qquad (3.96)$$

$$\lambda p_{m+1} = m\mu p_{m+2} \qquad (3.97)$$

$$\lambda p_{m+2} = m\mu p_{m+3} \qquad (3.98)$$

$$\cdot$$
$$\cdot$$

Or

$$p_{m+1} = \frac{1}{m}\left(\frac{\lambda}{\mu}\right) p_m \qquad (3.99)$$

$$p_{m+2} = \frac{1}{m}\left(\frac{\lambda}{\mu}\right) p_{m+1} = \frac{1}{m^2}\left(\frac{\lambda}{\mu}\right)^2 p_m \qquad (3.100)$$

$$p_{m+3} = \frac{1}{m}\left(\frac{\lambda}{\mu}\right) p_{m+2} = \frac{1}{m^3}\left(\frac{\lambda}{\mu}\right)^3 p_m \qquad (3.101)$$

$$\cdot$$
$$\cdot \quad \cdot$$

Or

$$p_n = \left(\frac{\lambda}{m\mu}\right)^{n-m} p_m \qquad n > m \qquad (3.102)$$

So

$$p_n = \left(\frac{1}{m!}\left(\frac{\lambda}{\mu}\right)^m\right)\left(\left(\frac{\lambda}{m\mu}\right)^{n-m}\right) p_0 \qquad n > m \qquad (3.103)$$

$$p_n = \frac{1}{m!m^{n-m}}\left(\frac{\lambda}{\mu}\right)^n p_0 \qquad n > m \qquad (3.104)$$

One can than substitute the two equations for p_n ($n \leq m$, $n > m$) into the normalization equation

$$p_0 + p_1 + p_2 + \cdots + p_n + \cdots = 1 \qquad (3.105)$$

Solving for p_0 one obtains after some algebra and using a summation from Appendix A

Fig. 3.11　M/M/m queueing system Markov chain

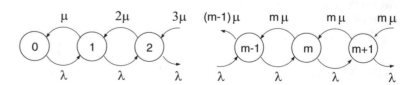

$$\rho_0 = \left[1 + \sum_{n=1}^{m-1} \frac{1}{n!} \left(\frac{\lambda}{\mu} \right)^n + \frac{1}{m!} \left(\frac{\lambda}{\mu} \right)^m \left(\frac{1}{1-\rho} \right) \right]^{-1}$$
$$(3.106)$$

Here $\rho = \lambda/m\mu$. In telephone system design another important performance measure, for this M/M/m model, is the probability that a call does not get a server immediately but must wait. This is equal to the sum of the probabilities that there are $m, m+1, m+2 \ldots$ customers.

$$Prob[queueing] = \sum_{n=m}^{\infty} p_n \qquad (3.107)$$

Since only states $n > m$ are involved the resulting Erlang C (or formula of the second kind in Europe) formula can be found as

$$Prob[queueing] = \left(\sum_{n=m}^{\infty} \frac{1}{m! m^{n-m}} \left(\frac{\lambda}{\mu} \right)^n \right) p_0 \quad (3.108)$$

$$Prob[queueing] = \frac{1}{m!} \left(\frac{\lambda}{\mu} \right)^m \left(\frac{1}{1-\rho} \right) p_0 \qquad (3.109)$$

Or

$$Prob[queueing] = \frac{\frac{1}{m!} \left(\frac{\lambda}{\mu} \right)^m \left(\frac{1}{1-\rho} \right)}{\left[1 + \sum_{n=1}^{m-1} \frac{1}{n!} \left(\frac{\lambda}{\mu} \right)^n + \frac{1}{m!} \left(\frac{\lambda}{\mu} \right)^m \left(\frac{1}{1-\rho} \right) \right]}$$
$$(3.110)$$

Again, $\rho = \lambda/m\mu$.

3.3.4 A Queueing Based Memory Model

The previous queueing models are well suited to model customers (e.g., jobs, packets) waiting in a line for some service. A clever model due to Kaufman [133, 202] can also take a form of memory requirements into account. In this model the aggregate arrival stream is Poisson with mean arrival rate λ. An arriving customer belongs to the ith of k customer classes with independent probability q_i. A customer of the ith class has a distinct temporal (service time) requirement τ_i and memory space requirement b_i. There are C units of memory. An arriving customer with a memory requirement of b_i units is accommodated if there are at least b_i (not necessarily contiguous) units of free memory; otherwise, it is blocked and cleared from the system. The residency time distribution of a class can have a rational Laplace transform with mean $1/\mu_i$. Thus the time distribution belongs to a class that includes, but is more general than, a simple negative exponential distribution.

The genius of Kaufman's model lies in the simplicity of the solution for the equilibrium state probabilities. Following Kaufman's notation, let

$$\underline{n}_i = (n_1, n_2, \ldots n_{i-1}, n_i, n_{i+1}, \ldots n_k) \qquad (3.111)$$

This is the (population) vector of the number of customers of each of the K classes in the queueing system. Then let

$$\underline{n}_i^+ = (n_1, n_2, \ldots n_{i-1}, n_i + 1, n_{i+1}, \ldots n_k) \qquad (3.112)$$

$$\underline{n}_i^- = (n_1, n_2, \ldots n_{i-1}, n_i - 1, n_{i+1}, \ldots n_k) \qquad (3.113)$$

Here $\underline{n}_i^+$ has one additional customer in the ith queue compared to $\underline{n}_i$ and $\underline{n}_i^-$ has one less customer in the ith queue compared to $\underline{n}_i$.

Let Ω be the set of allowable states which depends on the resource sharing policy being used. Also

$$\delta_i^+(\underline{n}) = \begin{cases} 1 \text{ if } \underline{n}_i^+ \in \Omega \\ 0 \text{ otherwise} \end{cases} \qquad (3.114)$$

$$\delta_i^-(\underline{n}) = \begin{cases} 1 \text{ if } \underline{n}_i^- \in \Omega \\ 0 \text{ otherwise} \end{cases} \qquad (3.115)$$

Here $\underline{n}_i^+ \in \Omega$ means that the state $\underline{n}_i^+$ is a member of Ω. Naturally

$$\underline{n} \cdot \underline{b} = \sum_{i=1}^{k} n_i b_i \qquad (3.116)$$

Here $\underline{n} \cdot \underline{b}$ is the amount of memory space that is occupied in state $\underline{n}$.

The Markov chain state transition diagram is more complex than in the previous cases. To solve for the equilibrium state probabilities one can start by writing a global balance equation for each state $\underline{n}$ belonging to set Ω.

$$\left[\sum_{i=1}^{k} \lambda_i \delta_i^+(\underline{n}) + \sum_{i=1}^{k} n_i \mu_i \delta_i^-(\underline{n}) \right] p(\underline{n})$$

$$= \sum_{i=1}^{k} \lambda_i \delta_i^-(\underline{n}) p(\underline{n}_i^-) + \sum_{i=1}^{k} (n_i + 1) \mu_i \delta_i^+(\underline{n}) p(\underline{n}_i^+) \quad (3.117)$$

Here $\lambda_i = \lambda q_i$. The left side of this equation is associated with the probability flux leaving state $\underline{n}$. The right side of this equation is associated with the probability flux entering state $\underline{n}$. The δ terms account for state space boundaries. At an interior state all δ's have a value of 1.0. More importantly in terms of a solution for the equilibrium state probabilities one can write a set of local balance equations for each state $\underline{n}$ belonging to Ω

$$\lambda_i \delta_i^-(\underline{n}) p(\underline{n_i^-}) = n_i \mu_i \delta_i^-(\underline{n}) p(\underline{n}) \qquad i = 1, 2 \ldots k$$
$$\text{(3.118)}$$

The left side of this equation represents the probability flux due to an arriving class i customer causing the system to enter state $\underline{n}$. The right side represents a class i customer departing the system and causing the system to leave state $\underline{n}$. Again, the δ's account for state space boundaries. These local balance equations can be solved recursively

$$p(\underline{n}) = \frac{\lambda_i}{n_i \mu_i} p(n_1, \ldots n_{i-1}, n_i - 1, n_{i+1}, \ldots n_k) \quad \text{(3.119)}$$

Proceeding with the recursion to zero out the ith term

$$p(\underline{n}) = \frac{a_i^{n_i}}{n_i!} p(n_1, \ldots n_{i-1}, 0, n_{i+1}, \ldots n_k) \qquad \text{(3.120)}$$

Here $a_i = \lambda_i / \mu_i$. If one does this for each class, one arrives at the equilibrium state probability solution

$$p(\underline{n}) = \left(\prod_{i=1}^{k} \frac{a_i^{n_i}}{n_i!} \right) p(0, 0, 0 \ldots 0)$$

$$p(\underline{n}) = \left(\prod_{i=1}^{k} \frac{a_i^{n_i}}{n_i!} \right) G^{-1}(\Omega) \qquad all \quad \underline{n} \in \Omega \quad \text{(3.121)}$$

In the above equation $G^{-1}(\Omega)$ is the inverse normalization constant from the normalization equation

$$G(\Omega) = \sum_{\underline{n} \in \Omega} \left(\prod_{i=1}^{k} \frac{a_i^{n_i}}{n_i!} \right) \qquad \text{(3.122)}$$

Normalization constants are correction factors used so that the sum of a finite number of state probabilities does indeed sum to one. The solution above can be inserted into either the local or global balance equation and the balancing verified to prove we have the correct solution. One can compute the probability that a class i arriving customer is blocked from

$$p_{b_i} = \sum_{\underline{n} \in B_i^+} p(\underline{n}) \qquad \text{(3.123)}$$

Here

$$B_i^+ = \{ \underline{n} \in \Omega : \quad \underline{n_i^+} \quad not \quad in \quad \Omega \} \qquad \text{(3.124)}$$

Here B_i^+ is the set of states where the system is in a blocking state.

3.3.5 M/G/1 Queueing System

The previous continuous queueing system had Poisson arrivals and negative exponentially distributed service times. With these assumptions one has a memoryless system and the solution for the equilibrium state probabilities is straightforward. With a memoryless system the state of the queueing system at any point in time is simply the number of customers in the queueing system at the time. This makes calculation tractable. One need not account for the time since the last arrival or the time the customer in the server has been in service so far.

Assuming non-exponential style arrival or departure times does indeed make analysis more complex. However there are some special cases where relatively simple results are available. One such case is the M/G/1 queue. Here we have a single queue where there are Poisson arrivals and a "general" (arbitrary) service time distribution. Even if we assume any possible (a general) service time distribution, as the M/G/1 queue does, it is possible to develop a fairly simple formula for the average (expected) number of customers in the queueing system. This is the Pollaczek and Khinchin mean value formula published by these researchers in 1930 [185] and 1932 [136], respectively. Though they arrived at this result by complex means, David Kendall [134] in 1951 published a simple deviation which we will outline [107, 111, 140, 194].

Kendall's Approach and Result

Kendall's approach is to use a Markov chain "imbedded" at the departure instants. This is based on the idea that for some queueing systems, the queue behavior in equilibrium at an arbitrary point in time, t, is the same as the behavior at the departure points [176]. A proof of this can first show that the equilibrium state probabilities "seen" at the departure instants are identical to the equilibrium state probabilities seen at the arrival instants. However since arrivals are Poisson (and occur at random times that are independent of the queueing system state) one may go further and show the equilibrium state probabilities at the arrival instants are identical to those at any point in time.

Kendall's approach is to write a recursion for the number of customers at the departure instants. If the queue is non-empty at the ith departure instant ($n_i > 0$)

$$n_{i+1} = n_i - 1 + a_{i+1} \qquad n_i > 0 \qquad \text{(3.125)}$$

Here n_i is the number of customers in the queue immediately after the ith departure instant. The "−1" accounts for the departure at the $(i + 1)$st instant. Also a_{i+1} is the number of customers that arrive into the system between the ith and $(i + 1)$st departure instants. Naturally if the queue is empty after the ith departure instant, one has

$$n_{i+1} = a_{i+1} \qquad n_i = 0 \qquad (3.126)$$

We would like to combine these two equations into a single equation. This can be done with a unit step function

$$u(n_i) = \begin{cases} 1 & n_i > 0 \\ 0 & n_i = 0 \end{cases} \qquad (3.127)$$

So

$$\boxed{n_{i+1} = n_i - u(n_i) + a_{i+1}} \qquad (3.128)$$

Kendall's approach is to square both sides of the recursion and then take expectations of both sides

$$E[(n_{i+1})^2] = E[(n_i - u(n_i) + a_{i+1})^2] \qquad (3.129)$$

Expanding the right side one has

$$E\left[(n_{i+1})^2\right] = E\left[n_i^2\right] + E\left[(u(n_i))^2\right] + E\left[a_{i+1}^2\right]$$

$$- 2E[n_i u(n_i)] + 2E[n_i a_{i+1}] - 2E[u(n_i)a_{i+1}] \qquad (3.130)$$

If one solves for each of these terms [107, 140, 194], one arrives with some algebra at

$$E[n] = \frac{2\rho - \rho^2 + \lambda^2 \sigma_s^2}{2(1 - \rho)} \qquad (3.131)$$

$$E[n] = \rho + \frac{\rho^2 + \lambda^2 \sigma_s^2}{2(1 - \rho)} \qquad (3.132)$$

These are two forms of the Pollaczek-Khinchin mean value formula. All one needs is the arrival rate λ, the utilization ρ (mean arrival rate divided by mean service rate of the general distribution $b(t)$ or $\rho = 1 - p_0$), and the variance of the service distribution used. Thus for any distribution of service time, only its first two moments are needed to evaluate the mean number of customers in the queue. This is surprising as an arbitrary service time distribution needs higher moments to completely specify it.

Naturally the mean delay a customer experiences in passing through the queue can be calculated from the P-K formula and Little's law.

The M/G/1 State Transition Diagram

An interesting question is finding the topology of the Markov chain imbedded at departure instants. We will do this by first creating a matrix of state transition probabilities. Let

$$\underline{P} = [P_{rs}] = P[n_{i+1} = s | n_i = r] \qquad (3.133)$$

To move from r customers in the queueing system immediately after the ith departure instant ($n_i = r$) to s customers immediately after the $(i + 1)$st departure instant there should be $s - r + 1$ arrivals between the two departure instants. We need the "+1" terms as the queue loses one customer at the $(i + 1)$st departure instant. Thus

$$[P_{rs}] = k_{s-r+1} = k_{\#arrivals} \qquad (3.134)$$

Continuing with Kendall's notation one has the following (infinite size) matrix

$P = [P_{rs}]$	0	1	2	3	4	.
0	k_0	k_1	k_2	k_3	k_4	.
1	k_0	k_1	k_2	k_3	k_4	.
2	0	k_0	k_1	k_2	k_3	.
3	0	0	k_0	k_1	k_2	.
4	0	0	0	k_0	k_1	.
.		.	.	.	.	.

This infinite size matrix is a stochastic matrix. That is, the sum of entries in any row is one. This is because the probability of going from a specific state (row number) to *some* state (column number) is 1.0.

To see how the entries were placed in the table, consider entry $[p_{2,4}]$ (row 2, column 4). To go from 2 customers immediately after a departure instant to having four customers immediately after the next departure instant requires three arrivals (or k_3) as one customer departs after the $(i + 1)$st departure instant.

A partial Markov chain (state transition diagram) is shown in Fig. 3.12. Here transitions leaving state n are illustrated.

How does one compute the k_j's? One has

$$k_j = \int_0^\infty Prob[j \; arrivals \, | \, time \; s]b(s)ds \qquad (3.135)$$

Here we integrate over the probability that there are j arrivals in a time s between consecutive departures where this probability is weighted by the distribution of s. That is, $b(s)$ is the general service time distribution. Since arrivals are Poisson

Fig. 3.12 M/G/1 (imbedded
Markov chain) state transition
diagram

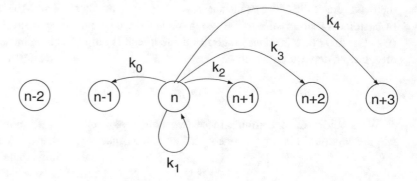

$$k_j = \int_0^\infty \frac{(\lambda s)^j e^{-\lambda s}}{j!} b(s) ds \qquad (3.136)$$

There are two ways to solve for the equilibrium state probabilities of this M/G/1 imbedded Markov chain. One is by drawing vertical boundaries between adjacent states, equating flow in both directions across the boundaries and creating a (somewhat complex) recursive series of equations starting from state 0 ([194]). In theory one can solve for all state probabilities as a function of p_0 and then normalize the probabilities.

The problem with this numerical approach is that it is approximate in that only a finite number of transitions can be used leaving each state, on a computer, not the infinite number of states the mathematical model calls for. An alternative approach to solve for the chain's state probabilities is to use moment generating functions. Moment generating functions are a frequency spectrum like description of probability distributions.

Advanced books on queueing theory [107, 140, 194] show that if $K(z)$ is the moment generating function of the k_j

$$K(z) = \sum_{j=0}^\infty k_j z^j, \qquad (3.137)$$

then

$$\Pi(z) = \frac{(1-\rho)(1-z)K(z)}{K(z) - z} \qquad (3.138)$$

Here $\Pi(z)$ is the moment generating function of the chain equilibrium state probabilities. In general $\Pi(z)$ can be inverted (analytically or sometimes only numerically) to find the equilibrium state probabilities.

3.4 Common Performance Measures

A great deal of effort has been expended in the previous pages of this chapter in determining the equilibrium state probabilities. The main reason we are interested in these probabilities is that many common performance measures

are simple functions of these state probabilities. While this is often not the only or the most efficient means of computing these performance measures, it is the most direct way. In this subsection continuous time queue performance measures will be described. Analogous expressions hold for discrete time queues. For instance, for a queue with a single server the fraction of time that the server is busy, or the utilization of the server is

$$U = 1 - p_0 \qquad (3.139)$$

If a single queue holds at most N customers, the probability an arriving customer is "blocked" from entering the queue because it is full, or the blocking probability is

$$P_B = p_N \qquad (3.140)$$

The mean (average) number of customers in an infinite size buffer, $\bar{n}$, is

$$\bar{n} = \sum_{n=1}^\infty n p_n \qquad (3.141)$$

This is a weighted average of each possible number of customers and the probability in equilibrium that there are that number of customers in the queue. Note that it would make no difference if the index of the summation started at $n = 0$ since the zeroeth term would have a value of zero.

The mean "throughput" or flow of customers through a single server, infinite buffer size queue is

$$\bar{T} = \sum_{n=1}^\infty \mu(n) p_n \qquad (3.142)$$

In this equation $\mu(n)$ is the state dependent service rate of the server when there are n customers. Again, the throughput expression is a weighted sum of the (state dependent) service rate when there are n customers multiplied by the equilibrium probability that there are n customers. Since the throughput of an empty queue is zero, the summation index starts at $n = 1$. If one has a finite buffer queue of size N, then the equations for $\bar{n}$ and $\bar{T}$ are the same except that ∞ is replaced by N.

Finally, consider an infinite size buffer queue so that the mean arrival rate equals the mean throughput. Then from Little's law, the mean delay, $\bar{\tau}$, a customer experiences in moving through the queue (and server) is

$$\bar{\tau} = \frac{\bar{n}}{\lambda} = \frac{\bar{n}}{\bar{T}} = \frac{\sum_{n=1}^{\infty} n p_n}{\sum_{n=1}^{\infty} \mu(n) p_n} \qquad (3.143)$$

Note that the units of this ratio are [customers] divided by [customers/sec] or [seconds], which makes sense for mean delay.

3.5 Markovian Queueing Networks

The queueing modeled in the previous sections involves a single queue. What of networks of queues? Networks of queues can be either open (with external arrivals and departures from the network) or closed (sealed) with a fixed number of customers circulating in the network. In fact for both types of continuous time queueing networks the elegant solutions of the previous sections for equilibrium state probabilities can be generalized into what is referred to as the product form solution.

For a continuous time network of M Markovian queues we seek an expression for $p(\underline{n})$ the equilibrium probability that the network is in state $\underline{n}$. Here $\underline{n}$ is a vector

$$\underline{n} = (n_1, n_2, \ldots, n_{i-1}, n_i, n_{i+1}, \ldots, n_M) \qquad (3.144)$$

In this vector n_i is the number of customers in the ith queue in state $\underline{n}$. We assume a Markovian system (i.e., Poisson arrivals for open networks and independent negative exponential random service times for both open and closed networks). For Markovian systems the system state at any time instant is completely summarized by the number of customers in each network queue at that time instant. Again, the system is memoryless so that one does not have to include

the times customers have been in service, or the times since the last arrivals, into the system state.

Because even moderately sized networks of queues have very large numbers of states, the use of global balance equations to solve for the equilibrium state probabilities would be impractical as one would have to solve too large sets of linear equations. However the existence of tractable product form solutions for Markovian queueing network equilibrium probabilities is in fact closely related to the concept of local balance. Recall that global balance equates the total flow of probability flux into a state on all incoming transitions to the total flow out of the state on all outgoing transitions. Local balance, on the other hand, equates the incoming and outgoing flows in subsets of a state's transitions.

Local balance exists only for certain classes of queueing networks, including open and closed Markovian networks where the queues have unlimited buffers and routing between queues is random. These are the two network models addressed in this section. It turns out that if local balance exists, so does a product form solution and the reverse is true as well.

In terms of the history of product form results, Jackson [126] in 1957 was the first to solve the open network problem. Ten years later, in 1967, Gordon and Newell [105] provided a solution of the closed network problem. A classic paper generalizing these results is by Baskett, Chandy, Muntz, and Palacios [22] in 1975.

The following two subsections discuss open and closed Markovian queueing networks in turn. We use a similar method to [145] and [208].

3.5.1 Open Networks

Consider the open queueing network in Fig. 3.13. A Markovian system of M queues, an external source of Poisson arriving customers and an external destination for customers is pictured. Buffers are unlimited in size.

Fig. 3.13 Open queueing network schematic representation

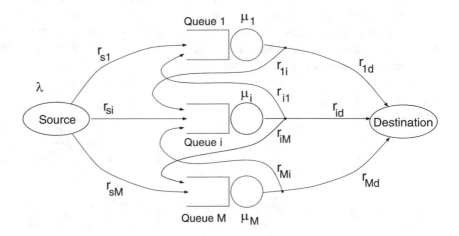

Customers are generated at the external source with mean rate λ. The probability that a customer generated at the external source randomly is chosen to enter queue i is r_{si}. Also, the (negative exponential distribution based) service rate of the ith queue is μ_i. The probability that a customer departing queue i enters queue j is r_{ij}. Customers may reenter the same queue they depart from (with probability r_{ii} for the ith queue). The probability that a customer leaving the ith queue proceeds to the external destination is r_{id}. In addition to the state vector (see (3.144) above), let

$$\underline{1}_i = (0, 0, \ldots, 0, 1, 0 \ldots, 0) \qquad (3.145)$$

Here there is a "1" in the ith position.

Using this notation one can write a global balance equation for state $\underline{n}$ in an open network as

$$\left(\lambda + \sum_{i=1}^{M} \mu_i\right) p(\underline{n}) = \sum_{i=1}^{M} \lambda r_{si} \, p(\underline{n} - \underline{1}_i) \qquad (3.146)$$

$$+ \sum_{i=1}^{M} \mu_i r_{id} \, p(\underline{n} + \underline{1}_i) + \sum_{i=1}^{M} \sum_{j=1}^{M} \mu_j r_{ji} \, p(\underline{n} + \underline{1}_j - \underline{1}_i)$$

This global balance equation equates the net flow of probability flux out of the state (left side) to the net flow into the state (right side). More specifically, the left side of the equation is associated with customers entering or leaving queues, causing the network to leave state $\underline{n}$. The right-hand side of the equation has three terms associated with the three ways that the network may directly enter state $\underline{n}$. One is via an external arrival to the ith queue when the network is in state $\underline{n} - \underline{1}_i$ (almost state $\underline{n}$ except that there are $n_i - 1$ customers in the ith queue). The second is via a departure from queue i to the destination when the network is in state $\underline{n} + \underline{1}_i$ (almost state $\underline{n}$ except that the ith queue has $n_i + 1$ customers). The last terms correspond to a transfer between the jth and ith queues starting from state $\underline{n} + \underline{1}_j - \underline{1}_i$ and through a single transfer ending in state $\underline{n}$.

To obtain a local balance equation leading to a solution for the equilibrium probabilities, let us first consider what are called the traffic equations of the network—a simple way to calculate queue mean throughput in open networks. Let θ_i be the mean throughput of queue i. Then

$$\theta_i = r_{si}\lambda + \sum_{j=1}^{M} r_{ji}\theta_j \qquad i = 1, 2 \ldots M \qquad (3.147)$$

This equation states that the ith queue mean throughput equals the sum of the mean external arrival rate to the ith queue and the mean rates of customer transfers from other queues to the ith queue. The M traffic equations are linear equations that can be solved by standard means for the M queues' throughputs. As an example, consider Fig. 3.14.

The traffic equations of this network are

$$\theta_1 = \lambda + \frac{3}{4}\theta_2 \qquad (3.148)$$

$$\theta_2 = \theta_1 \qquad (3.149)$$

So solving

$$\theta_1 = \lambda + \frac{3}{4}\theta_1 \qquad (3.150)$$

$$\theta_1 = \theta_2 = 4\lambda \qquad (3.151)$$

Note that since there is a feedback, that is a customer may visit a queue several times, queue throughput can be larger than the arrival rate.

Now, to generate a local balance equation, the traffic equation may be solved for λr_{si} which is then substituted into the global balance equation (3.146). After some algebraic manipulation (see [145, 194]) one has the following local balance equation

$$\theta_i \, p(\underline{n} - \underline{1}_i) = \mu_i \, p(\underline{n}) \qquad (3.152)$$

This equation states that for queue i the flow out of state $\underline{n}$ due to a departure (right side) equals the flow due to an arrival to queue i (left side) when the network is in state $\underline{n} - \underline{1}_i$ (state $\underline{n}$ with one less customer in queue i). Here the queues' total arrival rates are equal to the throughputs, the θ_i's. This balancing is not at all an obvious fact and is part of the amazing nature of local balance. It should be noted that local balance occurs in certain queueing networks (and generally not in electric circuits) because state transitions (unlike impedances) are labeled in a patterned manner that makes local balance possible. The equation can be rewritten as

Fig. 3.14 Open queueing network example

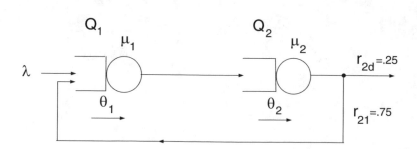

$$p(\underline{n}) = \frac{\theta_i}{\mu_i} p(n-1_i) \qquad (3.153)$$

Expanding this

$$p(\underline{n}) = \frac{\theta_i}{\mu_i} p(n_1, n_2, \ldots, n_{i-1}, n_i - 1, n_{i+1}, \ldots, n_M) \qquad (3.154)$$

Repetitively using the local balance equation results in

$$p(\underline{n}) = \left(\frac{\theta_i}{\mu_i}\right)^{n_i} p(n_1, n_2, \ldots, n_{i-1}, 0, n_{i+1}, \ldots, n_M) \qquad (3.155)$$

Here the ith term has been zeroed out. This can be done for each term resulting in our first version of the product form solution.

$$p(\underline{n}) = \left(\prod_{i=1}^{M} \left(\frac{\theta_i}{\mu_i}\right)^{n_i}\right) p(0, 0, 0, 0, 0) \qquad (3.156)$$

$$p(\underline{n}) = \left(\prod_{i=1}^{M} \left(\frac{\theta_i}{\mu_i}\right)^{n_i}\right) p(\underline{0}) \qquad (3.157)$$

To solve for $p(\underline{0})$, one can use the normalization equation

$$\sum_{\underline{n}} p(\underline{n}) = \sum_{\underline{n}} \left(\prod_{i=1}^{M} \left(\frac{\theta_i}{\mu_i}\right)^{n_i}\right) p(\underline{0}) \qquad (3.158)$$

The index of the product term does not depend on $\underline{n}$ so that the summation and product can be interchanged.

$$\sum_{\underline{n}} p(\underline{n}) = \prod_{i=1}^{M} \left(\sum_{n_i=0}^{\infty} \left(\frac{\theta_i}{\mu_i}\right)^{n_i}\right) p(\underline{0}) \qquad (3.159)$$

Using a summation formula from Appendix A

$$\sum_{\underline{n}} p(\underline{n}) = \prod_{i=1}^{M} \left(1 - \frac{\theta_i}{\mu_i}\right)^{-1} p(\underline{0}) \qquad (3.160)$$

Since $\sum_{\underline{n}} p(\underline{n}) = 1$

$$p(\underline{0}) = \prod_{i=1}^{M} \left(1 - \frac{\theta_i}{\mu_i}\right) \qquad (3.161)$$

Writing this out in a final form

$$p(\underline{n}) = \prod_{i=1}^{M} p_i(n_i) \qquad (3.162)$$

$$p_i(n_i) = \left(1 - \frac{\theta_i}{\mu_i}\right)\left(\frac{\theta_i}{\mu_i}\right)^{n_i} \qquad (3.163)$$

One can see that the name "product form" comes from the fact the expression for the equilibrium probability of state $\underline{n}$ for an open Markovian network of queues with unlimited buffers and random routing is a product of terms. Each term is analogous to the expression for the state probabilities of an M/M/1 queue (Eq. (3.31)) except that λ is replaced by θ_i. The product form solution for the previous example is thus

$$p(n_1, n_2) = \left(1 - \frac{4\lambda}{\mu_1}\right)\left(\frac{4\lambda}{\mu_1}\right)^{n_1}\left(1 - \frac{4\lambda}{\mu_2}\right)\left(\frac{4\lambda}{\mu_2}\right)^{n_2} \qquad (3.164)$$

Note that if the buffers were finite and the routing was not random (e.g., a join the shortest queue policy for a bank of parallel queues) then there would be no local balance and no product form solution.

One can compute numerical probabilities using the product form equations (3.162) and (3.163), though one must first solve the traffic equations for the θ_i's. It is interesting to note that the form of Eq. (3.162) above is similar to a joint distribution of independent random variables. Does this mean number of customers in the queues in an open Markovian network are independent of one another? This is only true at one instant at time. As the queueing theorist R. Disney [75] points out, if one compared the system state at two close instants, one would see significant correlations between the number of customers in different queues between the two instants.

3.5.2 Closed Networks

Consider now a closed Markovian queueing network. It is a "sealed" system with a fixed number of customers, N, circulating through it. The service rate of the ith queue is μ_i. Routing between queues is again random (r_{ij} is the probability that a customer departing the ith queue enters the jth queue). Buffers are big enough to always be able to accommodate all customers. A global balance equation for state $\underline{n}$ can be written as

$$\sum_{i=1}^{M} \mu_i p(\underline{n}) = \sum_{i=1}^{M} \sum_{j=1}^{M} \mu_j r_{ji} p(\underline{n} + 1_j - 1_i) \qquad (3.165)$$

The formula equates the total probability flux leaving state $\underline{n}$ (left side) to that entering the state (right side). More specifically the left side is associated with departures from queues causing the network to leave state $\underline{n}$. The right side is associated with transfers from the jth to the ith queue when the network is in state $\underline{n} + 1_j - 1_i$ (essentially state $\underline{n}$ with one extra customer in queue j and one less customer in queue i) causing it to transit to network state $\underline{n}$.

Once again traffic equations can be written

$$\theta_i = \sum_{j=1}^{M} r_{ji}\theta_j \qquad i = 1, 2 \ldots M \qquad (3.166)$$

Each of the above M equations equates the mean throughput of queue i, θ_i, to the sum of the mean customer flows from each of the queues entering queue i. Note that there is no external arrival term as there is for the traffic equations of open networks. Also, these traffic solutions do not have a unique solution. If $(\theta_1, \theta_2 \ldots, \theta_M)$ is a solution. so is $(c\theta_1, c\theta_2 \ldots, c\theta_M)$. What these equations yield are relative, not absolute, mean queue throughputs. To obtain a local balance equation, one combines the traffic equation with the global balance equation [194]. With some algebra [194], the following local balance equation is obtained

$$\mu_i p(\underline{n}) = \theta_i p(\underline{n} - \underline{1}_i) \qquad (3.167)$$

The formula equates the flow of probability flux out of state $\underline{n}$ due to a departure from the ith queue to the flow into state $\underline{n}$ from state $\underline{n} - \underline{1}_i$ due to an arrival to queue i.

This local balance equation can be written as

$$p(\underline{n}) = \frac{\theta_i}{\mu_i} p(\underline{n} - \underline{1}_i) \qquad (3.168)$$

Or

$$p(\underline{n}) = \frac{\theta_i}{\mu_i} p(n_1, n_2, \ldots, n_{i-1}, n_i - 1, n_{i+1}, \ldots, n_M) \qquad (3.169)$$

Utilizing the local balance equation repetitively, the ith term can be zeroed out

$$p(\underline{n}) = \left(\frac{\theta_i}{\mu_i}\right)^{n_i} p(n_1, n_2, \ldots, n_{i-1}, 0, n_{i+1}, \ldots, n_M) \qquad (3.170)$$

Then each term can be zeroed out

$$p(\underline{n}) = \left(\prod_{i=1}^{M}\left(\frac{\theta_i}{\mu_i}\right)^{n_i}\right) p(\underline{0}) \qquad (3.171)$$

The inverse of $p(\underline{0})$ is known as the normalization constant or $G(N)$. A physicist would call it a partition function. From the normalization equation

$$\sum_{\underline{n}} p(\underline{n}) = p(\underline{0})G(N) = 1 \qquad (3.172)$$

$$G(N) = \sum_{\underline{n}}\left(\prod_{i=1}^{M}\left(\frac{\theta_i}{\mu_i}\right)^{n_i}\right) \qquad (3.173)$$

The expression for $G(N)$ can be deduced from what we know of the form of the expression for $p(\underline{n})$ (Eq. (3.171)) and the normalization equation. The product form solution for closed Markovian queueing networks with random routing and ample buffers is then

$$p(\underline{n}) = \frac{1}{G(N)}\prod_{i=1}^{M}\left(\frac{\theta_i}{\mu_i}\right)^{n_i} \qquad (3.174)$$

An open queueing network has an infinite number of states as it can have a potentially unlimited number of customers. A closed queueing network has a finite number of states as it has a finite number of customers. Again, the normalization constant can be thought of a correction factor so that if one sums over the finite (but possibly large) number of states of a closed Markovian queueing network the sum of the state probabilities is one. Note $1/G(N)$ does not factor into $p_1(0)p_2(0)p_3(0)\ldots p_M(0)$ as it does for open networks so closed networks are more difficult to solve. An algorithm for solving closed queueing networks is discussed in the next section.

A final thought that may occur to the reader is if one has several incoming and outgoing transitions incident to a state, is there a simple way to know which transitions are paired for local balance? In fact the rule is that the flow of probability flux entering a state due to an arrival to a specific queue balances with (equals) the flow of probability flux leaving the state due to a departure from the same queue. This can be seen in (3.152) and (3.167).

3.6 Mean Value Analysis for Closed Networks

Even moderately sized queueing networks have a great many states. Consider a closed network of M queues and N customers. Finding the number of states is equivalent to finding the number of ways N identical customers can be in M queues. This number is:

$$\text{Number of States} = \binom{M + N - 1}{N} \qquad (3.175)$$

As an example, suppose there are eight queues, and sixty customers. Then

$$\binom{67}{60} = 869,648,208 \text{ states} \qquad (3.176)$$

Using the product form solution for closed networks, Eq. (3.174), would require the multiplication of nine constants to compute each of the state probabilities or a total of more than seven billion multiplications! For larger closed networks or a network where there are customer classes the problem is even worse.

Is there a better way to compute closed Markovian (negative exponential service times) network performance measures? The answer is yes. The trick is not to calculate the state probabilities explicitly.

One approach, the convolution algorithm (see J. Buzen [55]) and M. Reiser and H. Kobayashi [188], uses clever recursions to calculate the normalization constant G(N) (see Eq. (3.173)) and then performance measures [54, 194]. It is computationally efficient. Another computationally efficient approach is mean value analysis, due to M. Reiser and S. Lavenberg [189]. The mean value analysis algorithm exactly computes each network queue's mean throughput, mean delay, and the mean number of customers, all through the use of some clever queueing principle based recursions. Mean value analysis, unlike the convolution algorithm, does not compute normalization constants as part of its solution method. Two versions of the mean value algorithm are discussed below. Both involve state independent servers. The first is for a cyclic network. In the second case the algorithm is generalized to a closed network with random routing.

3.6.1 MVA for Cyclic Networks

Again we have M Markovian queues and N customers. The network is cyclic. That is, the first queue's output is connected to the input of the second queue, the second queue's output is connected to the input of the third queue …and the last queue's output is connected to the input of the first queue. The service rate of the ith queue is μ_i. The service time is negative exponentially distributed.

Now the average delay a customer undergoes at the ith queue, $\overline{\tau_i}$, has two parts. One part is the delay at the server and the second part is the sum of the service times of each customer in the queue ahead of the customer in question. Since the average service time of any customer is $1/\mu_i$ where μ_i is the service rate, one has

$$\overline{\tau_i} = \frac{1}{\mu_i} + \frac{1}{\mu_i}$$

$$\times \text{ (avg. number of customers in queue at arrival)}$$
$$(3.177)$$

The fact that this expression does not take into account the time that the customer in the server has been in service is due to the memoryless nature of the negative exponential

service time. That is, because the system is memoryless the remaining time in service of the customer in the server of the ith queue at the time of a customer arrival always follows a negative exponential distribution with mean $1/\mu_i$.

Reiser and Lavenberg's insight was to realize (and prove) that in a closed Markovian network the number of customers in a queue a customer "sees" on its arrival to the queue has the same distribution as that for the network in equilibrium with one customer less. Thus

$$\overline{\tau_i}(N) = \frac{1}{\mu_i} + \frac{1}{\mu_i} \times \overline{n_i}(N-1) \qquad (3.178)$$

In this equation $\overline{\tau_i}(N)$ is the average delay for the ith queue when the network has N customers. Also, $\overline{n_i}(N)$ is the average number of customers in the ith queue when there are N customers in the network.

For the rest of the mean value analysis algorithm, one can use Little's law (see (3.32)). Firstly, applying Little's law to the entire cyclic network,

$$\overline{T}(N) \sum_{i=1}^{M} \overline{\tau_i}(N) = N \qquad (3.179)$$

Or

$$\overline{T}(N) = \frac{N}{\sum_{i=1}^{M} \overline{\tau_i}(N)} \qquad (3.180)$$

Here $\overline{T}(N)$ is the average throughput in the network. Since the queues are arranged in cyclic fashion $\overline{T}(N)$ is the same for every queue and is equal to each queue's arrival rate.

Secondly, for the ith queue

$$\overline{n_i}(N) = \overline{T}(N)\overline{\tau_i}(N) \qquad (3.181)$$

Starting with $\overline{n_i}(0) = 0$ for $i = 1, 2 \ldots M$ one can use Eqs. (3.178), (3.180), and (3.181) to recursively compute the mean delay, the mean throughput, and the mean number of customers in each queue. One has for the mean value analysis algorithm for cyclic networks

MVA Algorithm for Cyclic Networks
For $i = 1, 2, 3 \ldots M$

$$\overline{n_i}(0) = 0 \qquad (3.182)$$

For $N = 1, 2, 3 \ldots$
   For $i = 1, 2, 3 \ldots M$

$$\overline{\tau_i}(N) = \frac{1}{\mu_i} + \frac{1}{\mu_i} \times \overline{n_i}(N-1) \qquad (3.183)$$

$$\overline{T}(N) = \frac{N}{\sum_{j=1}^{M} \overline{\tau_j}(N)} \qquad (3.184)$$

$$\overline{n_i}(N) = \overline{T}(N)\overline{\tau_i}(N) \tag{3.185}$$

Example: M Identical Cyclic Queues

This canonical example appears in [208]. There are M cyclic queues all with service rate μ. Since all of the queues have the same service rate, the mean number of customers, the mean delay and the mean throughput for each queue is the same.

Now with one customer and M queues

$$\overline{n_i}(0) = 0 \tag{3.186}$$

$$\overline{\tau_i}(1) = \frac{1}{\mu} \tag{3.187}$$

$$\overline{T}(1) = \frac{1}{M\frac{1}{\mu}} = \frac{\mu}{M} \tag{3.188}$$

$$\overline{n_i}(1) = \frac{\mu}{M} \times \frac{1}{\mu} = \frac{1}{M} \tag{3.189}$$

With two customers and M queues

$$\overline{\tau_i}(2) = \frac{1}{\mu} + \left(\frac{1}{\mu} \times \frac{1}{M}\right) = \frac{1}{\mu}\left(\frac{M+1}{M}\right) \tag{3.190}$$

$$\overline{T}(2) = \frac{2}{M \times \frac{1}{\mu} \times \left(\frac{M+1}{M}\right)} = \frac{2\mu}{M+1} \tag{3.191}$$

$$\overline{n_i}(2) = \frac{2\mu}{M+1} \times \frac{1}{\mu} \times \left(\frac{M+1}{M}\right) = \frac{2}{M} \tag{3.192}$$

With three customers and M queues

$$\overline{\tau_i}(3) = \frac{1}{\mu} + \left(\frac{1}{\mu} \times \frac{2}{M}\right) = \frac{1}{\mu}\left(\frac{M+2}{M}\right) \tag{3.193}$$

$$\overline{T}(3) = \frac{3}{M \times \frac{1}{\mu} \times \left(\frac{M+2}{M}\right)} = \frac{3\mu}{M+2} \tag{3.194}$$

$$\overline{n_i}(3) = \frac{3\mu}{M+2} \times \frac{1}{\mu} \times \left(\frac{M+2}{M}\right) = \frac{3}{M} \tag{3.195}$$

A pattern can be observed in these results. The general solution for N customers over M queues is

$$\overline{\tau_i}(N) = \frac{1}{\mu}\left(\frac{M+N-1}{M}\right) \tag{3.196}$$

$$\overline{T}(N) = \frac{N\mu}{M+N-1} \tag{3.197}$$

$$\overline{n_i}(N) = \frac{N}{M} \tag{3.198}$$

3.6.2 MVA for Random Routing Networks

In this more general case, the network allows random routing between the queues. Again, the independent probability that a customer leaving the ith queue enters the jth queue is r_{ij}.

Now the θ_i's, the solutions of the traffic equations (3.166) have to be taken into account. We choose (the non-unique) θ_i's so that one queue's average throughput is 1.0. This is the reference queue. Then:

MVA Algorithm for Random Routing Networks
For $i = 1, 2, 3 \ldots M$

$$\overline{n_i}(0) = 0 \tag{3.199}$$

For $N = 1, 2, 3 \ldots$
    For $i = 1, 2, 3 \ldots M$

$$\overline{\tau_i}(N) = \frac{1}{\mu_i} + \frac{1}{\mu_i} \times \overline{n_i}(N-1) \tag{3.200}$$

$$\overline{T}(N) = \frac{N}{\sum_{j=1}^{M} \theta_j \overline{\tau_j}(N)} \tag{3.201}$$

$$\overline{n_i}(N) = \overline{T}(N)\theta_i\overline{\tau_i}(N) \tag{3.202}$$

Here $\overline{T}(N)$ is the actual unique throughput of the reference queue when there are N customers. The mean throughput of the ith, non-reference queue is $\theta_i\overline{T}(N)$. For a cyclic network $\theta_i = 1$ for all i and the above equations simplify to the earlier equations.

Example: Three Queues with Random Routing

Let us run the mean value analysis algorithm for a closed Markovian queueing network with random routing for the three queue networks illustrated in Fig. 3.15. The service rates of Q1, Q2, and Q3 are 4.0, 2.0, and 8.0, respectively. The reference queue is queue 1. The routing probabilities are $r_{12} = 0.25$ and $r_{13} = 0.75$. Therefore the relative throughputs are $\theta_1 = 1.0$, $\theta_2 = 0.25$, and $\theta_3 = 0.75$.

With one customer for this queueing network

$$\overline{n_i}(0) = 0 \qquad\qquad i = 1, 2, 3 \tag{3.203}$$

$$\overline{\tau_1}(1) = 0.25 \tag{3.204}$$

$$\overline{\tau_2}(1) = 0.50 \tag{3.205}$$

$$\overline{\tau_3}(1) = 0.125 \tag{3.206}$$

$$\overline{T}(1) = \frac{1}{1.0 \times 0.25 + 0.25 \times 0.50 + 0.75 \times 0.125} = 2.1333 \tag{3.207}$$

Fig. 3.15 A random routing network. Here $\mu_1 = 4.0$, $\mu_2 = 2.0$, $\mu_3 = 8.0$, $r_{12} = 0.25$, and $r_{13} = 0.75$

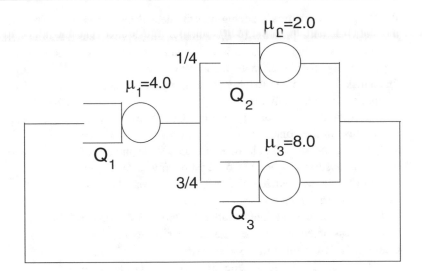

$$\overline{n_1}(1) = 2.1333 \times 1.0 \times 0.25 = 0.53333 \quad (3.208)$$

$$\overline{n_2}(1) = 2.1333 \times 0.25 \times 0.50 = 0.26666 \quad (3.209)$$

$$\overline{n_3}(1) = 2.1333 \times 0.75 \times 0.125 = 0.20000 \quad (3.210)$$

With two customers

$$\overline{\tau_1}(2) = 0.25 + 0.25 \times 0.5333 = 0.38333 \quad (3.211)$$

$$\overline{\tau_2}(2) = 0.50 + 0.50 \times 0.2666 = 0.63333 \quad (3.212)$$

$$\overline{\tau_3}(2) = 0.125 + 0.125 \times 0.2 = 0.15 \quad (3.213)$$

$$\overline{T}(2) = \frac{2}{1.0 \times 0.38333 + 0.25 \times 0.63333 + 0.75 \times 0.15}$$
$$= 3.05732 \quad (3.214)$$

$$\overline{n_1}(2) = 3.05732 \times 1.0 \times 0.38333 = 1.17197 \quad (3.215)$$

$$\overline{n_2}(2) = 3.05732 \times 0.25 \times 0.63333 = 0.484076 \quad (3.216)$$

$$\overline{n_3}(2) = 3.05732 \times 0.75 \times 0.15 = 0.343949 \quad (3.217)$$

With three customers

$$\overline{\tau_1}(3) = 0.25 + 0.25 \times 1.17197 = 0.54299 \quad (3.218)$$

$$\overline{\tau_2}(3) = 0.50 + 0.50 \times 0.484076 = 0.74204 \quad (3.219)$$

$$\overline{\tau_3}(3) = 0.125 + 0.125 \times 0.343949 = 0.16799 \quad (3.220)$$

$$\overline{T}(3) = \frac{3}{1.0 \times 0.54299 + 0.25 \times 0.74204 + 0.75 \times 0.16799}$$
$$= 3.51085 \quad (3.221)$$

$$\overline{n_1}(3) = 3.51085 \times 1.0 \times 0.54299 = 1.9064 \quad (3.222)$$

$$\overline{n_2}(3) = 3.51085 \times 0.25 \times 0.74204 = 0.65130 \quad (3.223)$$

$$\overline{n_3}(3) = 3.51085 \times 0.75 \times 0.16799 = 0.44234 \quad (3.224)$$

3.7 Negative Customer Queueing Networks

There are applications where it would be useful to have a queueing model where customers can "disappear" from a network. It may be desired for a customer to disappear if older messages are canceled by newer ones or if one is modeling a real time system.

Models with "negative customers" do model this type of activity. In such models, there are normal, "positive" customers and negative customers. A negative customer arriving to a queue will cancel a positive customer. That is, both will instantly disappear from the system. Much in the spirit of matter and anti-matter, positive and negative customers annihilate each other.

It should be noted that the application that E. Gelenbe, the original creator of negative customer models [95, 96] had in mind was neural network modeling. With a neuron modeled by a queue, positive customers represent excitation signals and negative customers represent inhibition signals.

How is the generation of negative customers modeled? Positive customers arrive to the ith queue according to a Poisson process with mean arrival rate Λ_i. Negative customers arrive to the ith queue according to a Poisson process with mean λ_i.

A second way that a negative customer may be generated is as a queue departure. That is, a positive customer leaving the ith queue enters the jth queue as a negative customer with independent probability r_{ij}^- and enters the jth queue as a positive customer with independent probability r_{ij}^+.

Also, a positive customer leaving the ith queue departs from the network with probability d_i. Finally, the service rate of customers in the ith queue is independent and negative exponentially distributed with mean service rate μ_i. Note that a negative customer arriving to an empty queue instantly disappears from the network.

Equilibrium results for negative customer networks only make sense in the context of open networks as for a closed networks the network would be empty in a finite amount of time as all positive customers would eventually be destroyed. It should also be realized the model described above is Markovian (memoryless).

It turns out that Gelenbe and co-authors developed a negative customer network model that has a product form solution. This model will now be presented. An alternate model due to Chao and Pinedo appears in [61] (see also [194]). A survey of work on negative customer models is [15].

3.7.1   Negative Customer Product Form Solution

Let's define the effective utilization of the ith queue as

$$q_i = \frac{\lambda_i^+}{\mu_i + \lambda_i^-} \tag{3.225}$$

The λ_i^+ and λ_i^- here are the solutions of the following traffic equations

$$\lambda_i^+ = \sum_j q_j \mu_j r_{ji}^+ + \Lambda_i \tag{3.226}$$

$$\lambda_i^- = \sum_j q_j \mu_j r_{ji}^- + \lambda_i \tag{3.227}$$

Note if the definition of the q_i is substituted into the traffic equations it can be seen that they are nonlinear, unlike the linear traffic equations of networks with only positive customers (see Sect. 3.5). Here λ_i^+ is the mean arrival rate of positive customers into the ith queue from both queue transfers (first term) and external arrivals (second term). Similarly λ_i^- is the mean arrival rate of negative customers to the ith queue due to queue transfers and external arrivals. From the definition of q_i it can be seen that the flow of arriving negative customers to a queue increases its effective service rate ($\mu_i + \lambda_i^-$).

Naturally from above

$$\sum_j r_{ij}^+ + \sum_j r_{ij}^- + d_i = 1 \qquad 1 \le i \le M \tag{3.228}$$

That is, a customer departing the ith queue either enters another jth queue as a positive or negative customer or departs from the system. There are M queues in the network.

To solve for an expression for the equilibrium state probabilities one needs to first set up a global balance equation for a state. Some notation in terms of the number of positive customers needed in each queue is

$$\underline{n} = (n_1, n_2, \ldots n_i \ldots n_M) \tag{3.229}$$

$$\underline{n}_i^+ = (n_1, n_2, \ldots n_{i-1}, n_i + 1, n_{i+1}, \ldots n_M) \tag{3.230}$$

$$\underline{n}_i^- = (n_1, n_2, \ldots n_{i-1}, n_i - 1, n_{i+1}, \ldots n_M) \tag{3.231}$$

$$\underline{n}_{ij}^{+-} = (n_1, n_2, \ldots n_i + 1, \ldots n_j - 1, \ldots n_M) \tag{3.232}$$

$$\underline{n}_{ij}^{++} = (n_1, n_2, \ldots n_i + 1, \ldots n_j + 1, \ldots n_M) \tag{3.233}$$

The first vector simply represents the state $\underline{n}$. The second vector indicates the network population after an external positive customer arrival to the ith queue while the network is in state $\underline{n}$. The third vector indicates the network population after a departure from queue i while the network is in state $\underline{n}$. The fourth vector corresponds to the state prior to a positive customer departure from queue i to queue j that brings the state to state $\underline{n}$. Finally, the last vector for $\underline{n}_{ij}^{++}$ corresponds to the state prior to a negative customer leaving the ith queue for the jth queue and bringing the network to state $\underline{n}$. Note that in all of this negative customers have lifetimes of zero seconds as once created they instantly cancel a positive customer and/or disappear from the network.

Also let $1[y]$ be an indicator function that has a value of 1.0 if y is greater than zero and has a value of zero otherwise.

Then the global balance equation for an interior state is

$$p(\underline{n}) \sum_i [\Lambda_i + (\lambda_i + \mu_i)1[n_i > 0]] \tag{3.234}$$

$$= \sum_i [p(\underline{n}_i^+)\mu_i d_i + p(\underline{n}_i^-)\Lambda_i 1[n_i > 0] + p(\underline{n}_i^+)\lambda_i$$

$$+ \sum_j (p(\underline{n}_{ij}^{+-})\mu_i r_{ij}^+ 1[n_j > 0] + p(\underline{n}_{ij}^{++})\mu_i r_{ij}^-$$

$$+ p(\underline{n}_i^+)\mu_i r_{ij}^- 1[n_j = 0])]$$

Gelenbe found that the expression for the equilibrium state probability, $p(\underline{n})$, that satisfies this global balance equation, is

$$p(\underline{n}) = \prod_{i=1}^M (1 - q_i)q_i^{n_i} \tag{3.235}$$

It can be seen that this expression is similar in form to the earlier one of Sect. 3.5.1 for open Markovian networks with only positive customers. That the equilibrium state probabilities has such a product form is not at all an obvious result.

Example: Tandem Network

Consider a tandem (series) of two queues, queue 1 followed by queue 2 (Fig. 3.16). Positive customers arrive to queue 1 with arrival rate Λ_1. No negative customers arrive to queue 1.

There is a Poisson stream of external negative customer arrivals to queue 2 with mean rate λ_2. Also, a customer departing queue 1 enters queue 2 as a positive customer with probability r_{12}^+ and as a negative customer with probability r_{12}^-. The service rates of queues 1 and 2 are μ_1 and μ_2, respectively.

The traffic equations are

$$\lambda_1^+ = \Lambda_1 \tag{3.236}$$

$$\lambda_1^- = 0 \tag{3.237}$$

$$\lambda_2^+ = q_1 \mu_1 r_{12}^+ \tag{3.238}$$

$$\lambda_2^- = q_1 \mu_1 r_{12}^- + \lambda_2 \tag{3.239}$$

Then from Eq. (3.225)

$$q_1 = \frac{\Lambda_1}{\mu_1} \tag{3.240}$$

$$q_2 = \frac{q_1 \mu_1 r_{12}^+}{\mu_2 + q_1 \mu_1 r_{12}^- + \lambda_2} \tag{3.241}$$

or

$$q_2 = \frac{\Lambda_1 r_{12}^+}{\mu_2 + \Lambda_1 r_{12}^- + \lambda_2} \tag{3.242}$$

where

$$p(\underline{n}) = \prod_{i=1}^{2} (1 - q_i) q_i^{n_i} \tag{3.243}$$

It can be observed that queue 1 is solely a positive customer queue. Thus queue 1 has the standard positive customer queue utilization.

Networks of queues with negative customers have been generalized over the years to such instances as networks where the customers may leave in batches, to networks with disasters (a single arriving negative customer causes all positive customers in a queue to be removed from the network) and to multiple classes of positive and negative customers. See [15] for a review.

3.8 Recursive Solutions for State Probabilities

Not every queueing model of interest is a product form network. Non-product form models are defined by exclusion: they are models that do not have a product form solution. Such realistic queueing features, as blocking, priority classes, and finite buffers, give rise to non-product form models. Any non-product form model may be solved, if it is not too large, by solving the model's global balance equations. However some non-product form queueing models can be efficiently solved recursively for the equilibrium state probabilities. This was first discussed by [112].

There are three ways that one can create recursions to generalize non-product form model equilibrium state probabilities. One is to draw boundaries that segment the state transition diagram into two parts. One can then sometimes write solvable equations that balance the flow of probability flux moving across the boundary in both directions.

The other two methods for generating recursions involve two ways of writing global balance equations [234]. Figure 3.17 illustrates a "Type A" structure in the state transition diagram. Here one can write a global balance equation for a state with (previously found) known state probability where there is only one incident transition to the state from a state with unknown state probability. In solving the global balance equation the unknown state probability is found. One may then continue by solving the global balance equation for that state with (now) known state probability.

For "Type B" structure (Fig. 3.18) one writes a global balance equation for a state with unknown state probability, with incident transitions from states with known probabilities and with departing transitions to states with known and/or unknown state probabilities. Once one solves the global balance equation for its single unknown state probability

Fig. 3.16 A negative queueing network example

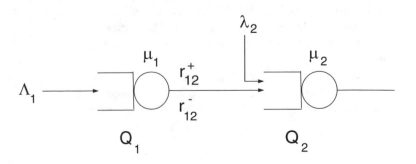

Fig. 3.17 Type A state transition
diagram structure

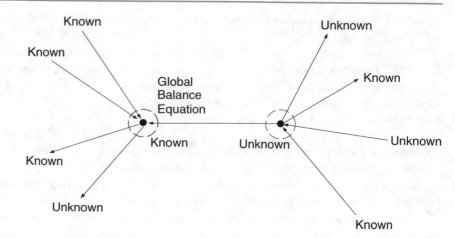

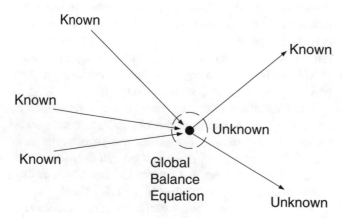

Fig. 3.18 Type B state transition diagram structure

one can move on to another state with unknown probability connected via an outgoing transition and continue the process. Note that states with unknown probability reached by departing transitions do not have their state probabilities enter into the global balance equation calculation.

It is possible to generate these type A and B recursions from subsets of states at a time, rather than single states. A more detailed discussion of such models appears in [194]. An example is now presented.

Example: Voice/Data Integrated Protocol

Let's consider a link that can either carry a single voice call or packet transmissions, but not both at the same time. The continuous time state transition diagram appears in Fig. 3.19. The horizontal axis indicates the number of packets in the transmission buffer. The buffer size is N. The vertical axis indicates either no voice call present (0) or a single voice call present (1).

A voice call arrives with Poisson rate λ_1 and completes service in a negative exponentially distributed amount of time with service rate μ_1. Packets arrive with Poisson rate

λ_2 and each completes service in a negative exponentially distributed amount of time with service rate μ_2.

In this protocol one of several packets at a time is only transmitted if there is no voice call [208]. Otherwise the packets are buffered until the voice call finishes. A voice call is only accepted if the packet buffer is empty.

Let $p(0, 0) = 1.0$. The state probabilities can be normalized when the recursions are finished. A global balance equation at state $(0, 1)$ is

$$(\lambda_2 + \mu_1)p(0, 1) = \lambda_1 p(0, 0) \tag{3.244}$$

So

$$p(0, 1) = \frac{\lambda_1}{\lambda_2 + \mu_1} p(0, 0) \tag{3.245}$$

Moving from left to right through the state transition diagram, bottom row equilibrium state probabilities can be calculated by drawing vertical boundaries through the state transition diagram and equating the flow of probability flux from left to right to that from right to left.

$$\mu_2 p(i, 0) = \lambda_2[p(i - 1, 0) + p(i - 1, 1)] \tag{3.246}$$

Or

$$p(i, 0) = \frac{\lambda_2}{\mu_2}[p(i - 1, 0) + p(i - 1, 1)] \tag{3.247}$$

Top row states follow a type B structure. Their equilibrium state probabilities can be solved from left to right by writing the global balance equation

$$(\lambda_2 + \mu_1)p(i, 1) = \lambda_2 p(i - 1, 1) \tag{3.248}$$

or

$$p(i, 1) = \frac{\lambda_2}{\lambda_2 + \mu_1} p(i - 1, 1) \tag{3.249}$$

Fig. 3.19 Integrated voice/data protocol recursive solution state transition diagram

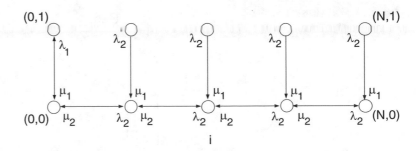

After computing p(0,0) and p(0,1), one can cycle through (3.247) and (3.249) where $i = 1, 2, 3 \dots N - 1$.

Finally, at the right boundary using global balance equations for states $(N,0)$ and $(N,1)$ and some simple algebra results in

$$p(N, 0) = \frac{\lambda_2}{\mu_2} [p(N - 1, 0) + p[(N - 1, 1)] \qquad (3.250)$$

$$p(N, 1) = \frac{\lambda_2}{\mu_1} p(N - 1, 1) \qquad (3.251)$$

Thus the above can be used to solve the state probabilities in terms of reference probability $p(0, 0)$. The probabilities can then be normalized by dividing each by the sum of the unnormalized probability values.

3.9 Stochastic Petri Nets

Petri networks are a graphical means of representing serialization, concurrency, resource sharing, and synchronization. Stochastic Petri networks use stochastic timing for events. Markovian Petri networks use memoryless distributions for timing. Just as Markovian queueing network schematic diagrams give rise to Markov chains, Markovian Petri network schematic diagrams give rise to Markov chains.

Petri networks first appeared in the doctoral dissertation of C.A. Petri [181] in Germany in 1962.

3.9.1 Petri Net Schematics

Stochastic Petri network schematics consist of a six tuple.

$$\underline{P} = (P, T, I, O, M, Q) \qquad (3.252)$$

As an example suppose we have two processors (P1 and P2) connected through a computer bus to a common memory (CM). Only one processor may use the bus to access the common memory at a time. A Petri network schematic for common memory access appears in Fig. 3.20. More details on such multiprocessors appear in [165, 166].

Referring to the diagram, P is a set of "places" which are drawn as circles. We use T to represent a set of "transitions" which are drawn as horizontal bars. The input function, I, maps each transition to one or more places. The output function, O, maps each place to one or more transitions. The input and output functions are represented by directed arcs. A "marking", M, assigns zero or more tokens (illustrated by dots) to each place. A specific marking is a state of the Petri net. Finally, Q is the set of transition rates associated with the transitions.

How does the Petri net schematic operate? When there is at least one token in each place incident to a transition, the transition is "enabled." The transition can then "fire" after some period of time. Firing involves removing one token from each place which is incident to the transition of interest and adding a token to each place that outgoing arcs lead to from the transition. Naturally the firing of a transition, by changing the marking, leads to a new network state.

In a Markovian Petri net (often called a "stochastic" Petri net) the time between when a transition is enabled for firing and when it actually fires is an independent negative exponentially distributed random variable. Such transitions are drawn as unfilled rectangles as in the figure. Immediate transitions fire in zero time once enabled and are drawn as filled rectangles. Generalized stochastic Petri networks contain both immediate and negative exponential timed transitions. All of the transitions in Fig. 3.20 have Markovian timing.

In Fig. 3.20 the marking indicates that P1 and P2 are idle (not trying to connect to the common memory) and the bus is free. The sequence of places Pi idle, Bus request, and CM access comprise a linear task sequence. Either or both of the linear task sequences can proceed through the firing of the q^{10} and q^{20} transitions to requesting the bus. Since a processor accessing the bus removes the bus free place token while it accesses the memory, only one processor can access the common memory (bus) at a time. Once a common memory access is finished the accessing processor becomes idle and the bus becomes free again.

This Petri net can be seen to model concurrency in having two parallel task sequences, and to model serializability in the serial nature of each task sequence. Resource sharing is

Fig. 3.20 A stochastic Petri net
for common memory access from
two processors over a bus

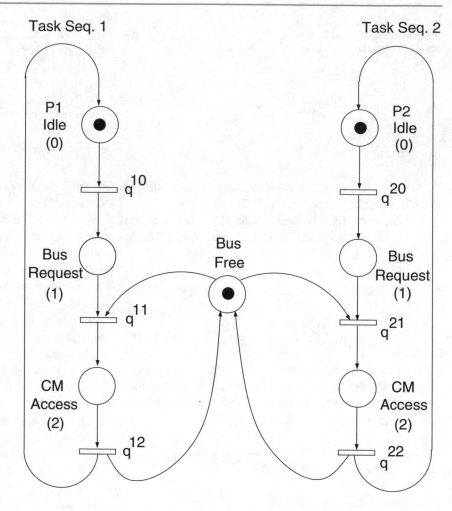

modeled through the bus free place. Finally, by requiring that
transitions q^{11} and q^{21} fire only when there is a bus request
and the bus is free, synchronization is modeled.

3.9.2 Petri Net Markov Chains

The regular structure of the Petri net schematic of Fig. 3.20
lends itself to creating a corresponding Markov chain that is
Cartesian coordinate based. The Markov chain is drawn in
Fig. 3.21. The horizontal axis corresponds to task sequence
1 and the vertical axis corresponds to task sequence 2. The
coordinates 0, 1, and 2 represent a token being in the P_i idle,
Bus request, and CM access places, respectively. Note that
the "wraparound" character of the Markov chain is naturally
embedded on the surface of a torus [150, 194, 234].

Stochastic Petri nets can be solved through simulation or
by Markov chain solution. The Markov chain in the figure
is a non-product form Markov chain. One can solve the
set of global balance equations for the equilibrium state
probabilities.

However if transitions ((2,0) to (2,1)) and ((0,2) to (1,2))
are removed from the chain, a different protocol results
which has a product form solution. The corresponding Petri
net is shown in Fig. 3.22. The inhibitor arcs that have been
added implement a complementary dependency. That is,
the condition for the transition that the inhibitor arcs are
attached to, to fire, includes there being no token in the
place that the inhibitor arc originates from. In the figure
the inhibitor arc in task sequence 1 is connected to the CM
access place in task sequence 2 ("T.S. 2") and vice versa.
Thus one can only move from a processor being idle to a bus
request if there is a non-zero probability that one can cycle
completely around the processors' task sequence without the
need for a state (marking) change in the other task sequence.
This precludes a form of blocking (a bus request not being
satisfied because the other processor is currently accessing
the common memory) that does not allow a product form
solution.

The global balance equation for the modified Markov
chain is

$$(q^{1,i} + q^{2,j})p(i, j) = q^{1,i-1}p(i-1, j) + q^{2,j-1}p(i, j-1)$$

$$(3.253)$$

Fig. 3.21 Markov chain of the Petri net example

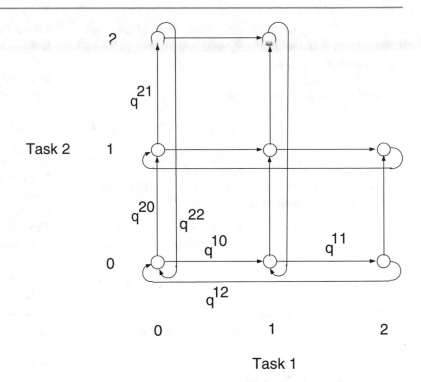

Fig. 3.22 Modified Petri net with product form solution

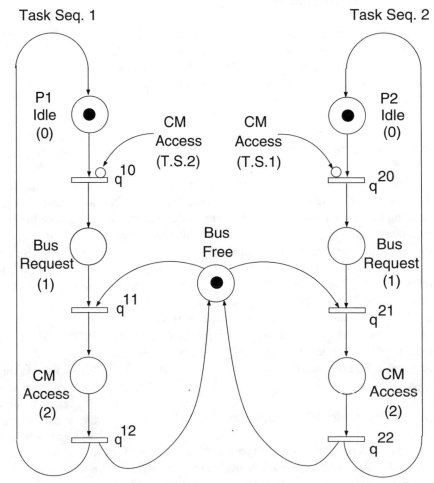

Here $q^{k,l}$ is the transition rate of the lth Petri schematic transition in the kth task sequence. Also, $p(i, j)$ is the equilibrium state probability of the (i, j)th state in the state transition diagram.

For the local balance equations one has

$$q^{1,i-1} p(i - 1, j) = q^{1,i} p(i, j) \qquad (3.254)$$

$$q^{2,j-1} p(i, j - 1) = q^{2,j} p(i, j) \qquad (3.255)$$

The product form solution for the Markov chain without the two deleted transitions is

$$p(i, j) = \frac{q^{1,0}}{q^{1,i}} \frac{q^{2,0}}{q^{2,j}} p(0, 0) \qquad (3.256)$$

This product form solution satisfies both the global and local balance equations.

Petri nets are a flexible tool for system modeling. More on stochastic Petri network modeling appears in [194].

3.10 Solution Techniques

A variety of solution techniques are available for solving queuing and stochastic Petri network models of networks (and computer systems). Each technique has advantages and disadvantages in terms of its modeling ability, ease of implementation and ease of use.

3.10.1 Analytical Solutions

Analytical solutions involve developing a mathematical model that can be solved to produce a closed form formula yielding the desired result. In fact the formula may be solved on a computer (as in the Erlang B or C formula) but usually the amount of computation required is trivial.

Because of the simplicity and the intuitive insight they offer, analytical closed form solutions are the most desirable of solutions. Unfortunately analytical solutions tend to be available only for simpler models. While it is sometimes possible to derive analytical solutions for more complex models, often the skills of a highly educated Ph.D. are required to find the solution.

3.10.2 Numerical Computation

Beyond implementing a simple formula on a computer (or even calculator), some approaches involve the numerical solution of mathematical equations modeling a system. Three examples are:

- *Linear Equation Solution:* In theory any Markovian system can be exactly solved for its equilibrium state probabilities by solving its linear global balance equations. As mentioned, N states give rise to N equations with one of them replaced by the normalization equation to obtain a set of N equations with a unique and correct solution. The difficulty, of course, is that even moderate size systems have so many states (and thus equations) as to make this not a computationally feasible approach. Moreover the fact that the computational complexity of general linear equation solution is proportional to the cube of the number of equations compounds this problem.

- *Transient Models:* Transient models involve the operation of a system over a limited time span (say from 0 to 2 s). Thus this is not a model in equilibrium. In theory, standard positive customer memoryless queueing networks can be solved in continuous time by coupled linear differential equations and in discrete time by coupled linear difference equations. Software packages for such systems of equations are available though their use is only practical for smaller systems.

- *Generating Functions:* As has been stated, moment generating functions provide a frequency domain like representation of equilibrium state probabilities. The moment generating function of a linear set of equilibrium state probabilities is

$$P(z) = \sum_{n=0}^{\infty} p_n z^n \qquad (3.257)$$

This expression can be seen to be similar, though not identical, to the z transform of digital signal processing. Producing a closed form moment generating function expression is an exercise in analysis. However such expressions, which are functions of z, can be numerically inverted on a computer to produce the equilibrium state probabilities.

It is of more interest than ever to produce complete probability distributions, rather than simply low order moments, because of the interest in low probability events such as the overflow of buffers [168].

3.10.3 Simulation

Simulation is a middle approach between mathematical models and experimentation to determine a system's performance. In a discrete event simulation a program mimics the actual system (e.g., calls are initiated and terminated, packets transmitted, buffers overflow, etc. . .) along with the timing of events. Part of the program collects statistics on the operation of the modeled system which are synthesized into performance results.

Simulation is cost effective in capturing realistic modeling features (e.g., blocking in networks, priority classes, non-Markovian statistics) that preclude analytical or even numerical solutions. Generally one does not have to be as sophisticated mathematically to produce a simulation as to produce analytical solutions. Both steady state and transient system operation can be simulated.

The size of a system to be simulated is limited by the available computer power. Very large systems such as the Internet cannot be simulated in extreme detail. Moreover since simulations can produce voluminous performance results, it is often easier to discern systematic trends and tradeoffs with analytic solutions.

Naturally a simulation of a queueing or stochastic Petri network involves many random quantities. These are generated by a "pseudo-random" number generator. These software programs generate random like sequences of numbers that follow the same sequence for a given "seed" number. Moreover after a very large number of pseudo-random numbers the sequence repeats. Though random like, the sequence is really deterministic. This aids in reproducing results. That is, for the same seed and no parameter changes a simulation will produce exactly the same results every time it is run. Changing the seed produces statistically similar, though not identical, results.

Pseudo-random numbers are usually uniformly distributed between 0 and 1. If one needs the probability of a packet arrival in a slot to be 0.2, there is an arrival if the pseudo-random number is between 0.0 and 0.2 and there is no arrival if the pseudo-random number is between 0.2 and 1.0.

How does one generate random numbers following non-uniform distributions? Say $f(x)$ is the continuous probability distribution we want. Then $F(x)$ is the cumulative distribution function [104].

$$F(x) = \int_{-\infty}^{x} f(z)dz \qquad (3.258)$$

Here z is a dummy integration variable. To generate a random variable with distribution $f(x)$, one takes a pseudo-random uniformly distributed variable y and lets

$$x = F^{-1}(y) \qquad (3.259)$$

The function $F^{-1}(y)$ is the functional inverse of the cumulative distribution function. For each uniformly distributed y, an x is generated according to the above formula that follows the distribution $f(x)$. An analogous procedure can create discrete random variables of any distribution.

Often a simulation is run many times, each time with a different seed, and the results averaged. Confidence intervals provide a means of expressing the amount of variability

in such results. A 98% confidence interval consists of an interval with two end points such that 98% of the time the quantity of interest is within the interval.

In considering confidence intervals (e.g., 90, 95, 98, 99%), one should realize that the smaller the percentage, the tighter the upper and lower limits are and the smaller the confidence interval is. Confidence intervals are often plotted as vertical lines (bars) superimposed on performance curves where the length of the line expresses the size of the confidence interval at each data point. The smaller the bars, the smaller is the variability in the curve.

Finally a sensitivity analysis determines the degree to which performance changes if a parameter value changes slightly.

3.11 Conclusion

Queueing theory and stochastic Petri network theory cover intriguing problems which have captured the imagination of many researchers and developers. The ubiquity of calls and packets waiting in buffers and of concurrency, serialization, synchronization and resource sharing means that such problems will be of interest for some time to come.

3.12 Problems

1. What are the statistical assumptions behind an M/M/1 queue? A Geom/Geom/1 queue?
2. What does it mean to say that an M/M/1 queue is memoryless?
3. Compare Markov chains to electric circuits.
4. Give an example of the use of Little's law.
5. Describe the difference between local and global balance.
6. What is the computational problem with solving global balance equations for large Markov chains?
7. If an M/M/1 queue input is Poisson, what can one say about the output process?
8. Can the arrival rate be greater than the service rate for a M/M/1 finite buffer queue? Explain.
9. Explain in which situation the use of the Erlang B formula is appropriate. Do the same for the Erlang C formula.
10. Give and explain an application for the queueing based memory model of Sect. 3.3.4.
11. What does it mean to use a Markov chain "imbedded" at departure instants for the M/G/1 queue analysis?
12. What are the statistical assumptions behind the Markov queueing results of Sect. 3.5?
13. What do the traffic equations of Sect. 3.5 model?

14. What queueing relationship is the mean value analysis algorithm based on? Does the MVA algorithm compute normalization constants?

15. What happens when a negative customer enters a queue with at least one positive customer? What happens when a negative customer enters an empty queue?

16. Can non-product form queueing networks be simply solved? Always or sometimes?

17. What types of actions do Petri nets model?

18. How is a Petri net marking related to its state?

19. What is the advantage of analytical studies of queueing and stochastic Petri nets?

20. Will simulation results always match experimental results?

21. Name three types of models discussed in this chapter having product form solutions.

22. Consider the Markov chain of Fig. 3.23 with three classes of packets and a buffer that holds one packet at one time.
    (a) Using boundaries, find p_1, p_2, and p_3 as functions of p_0, λ, and μ. Also find an expression for p_0.
    (b) Find the blocking probability (i.e., the probability an arriving customer is turned away).

23. Consider a finite buffer Geom/Geom/1/N discrete time queue. Assume that a customer arriving to an empty queue must wait at least one slot for service. Assume also (particularly if the buffer is full) that departures occur before arrivals in a slot.
    (a) Draw and label the state transition diagram.

(b) Solve for P_n in terms of p, s, and P_0 and also solve for P_0.

(c) Solve for P_0 if an infinite size buffer is used.

24. Consider a discrete time Geom/Geom/2/4 queue (two servers and holding at most four customers). A packet can enter an empty queue and depart during the same slot. Departures occur before arrivals in a slot. Draw and carefully label the state transition diagram.

25. Consider a single server finite buffer queue with the following state probabilities:

n	$p(n)$
0	0.15
1	0.20
2	0.35
3	0.30

Find the mean number of customers, the mean throughput and the mean delay through the system.

26. Consider a finite buffer M/M/1/N queue where the number of customers, N, is 3. Let $\lambda = 3.0$ and $\mu = 4.0$.
    (a) Draw and label the state transition diagram and calculate, numerically, p_0, p_1, p_2, and p_3.
    (b) Calculate the average (mean) delay through the queue (for customers that are not blocked, of course).
    (c) What is the average time that a customer waits in the waiting line before entering the server?

Fig. 3.23 A specific Markov chain

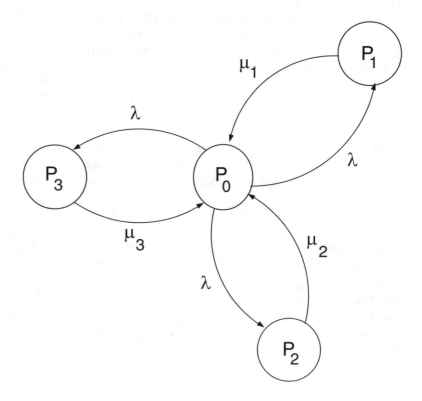

27. Consider an M/M/1 queue with $\lambda = 2.0$ and $\mu = 7.0$.
    (a) If the buffer is infinite in size, find the numerical value of utilization.
    (b) If now one has a finite buffer queue with the same parameters and a maximum capacity of $N = 4$, find the numerical utilization.
    (c) One of the values of the previous two parts is greater. Intuitively, why is this so?

28. Consider an M/M/2/4 queueing system with two servers and a maximum capacity of 4 customers. Find the blocking probability formula for arriving customers. It is a function of λ and μ. Note: The answer is different from the Erlang B formula though the technique to find it is similar.

29. Consider a Markovian queueing system of three parallel servers without a queue. The arrival rate to the system of queues is $\lambda = 10.0$ and the service rate, μ, of each server is 6.0.
    (a) Write an expression for the average number of empty servers as a function of the state probabilities.
    (b) Calculate numerical values for the equilibrium state probabilities. Show all steps in doing this. Substitute these probabilities into the answer of (a) to find a numerical value for the average number of empty servers.

30. A small company has three outside telephone lines. An average call lasts 12 min and 9 calls per hour are generated. Calls that do not immediately get a telephone line are queued (the network rings the caller when a line is available under a FIFO discipline).
    (a) What queueing model is described above?
    (b) What formula can be used to find the probability of queueing?

31. Consider a D/D/1 queue where one customer arrives per second and the server can process two customers a second. The arrival process and service process is deterministic (the time between events is a constant). Assuming that the queue is empty at $t = 0$ s, sketch the number of customers in the queue versus time for 0 to 3 s (label the graph accurately). From the graph and intuition, what is the average number of customers in the queue over an extended period of time?

32. For an M/M/∞ queueing system (where every arriving customer gets its own server of rate μ), what is the throughput? Use intuition, rather than calculation, to answer this question.

33. In a queue with discouraged arrivals as the waiting line gets longer, fewer new customers enter the queue. Assume that for a Markovian queue the service rate is μ and the state dependent arrival rate is

$$\lambda(n) = \frac{\lambda}{n+1} \qquad n = 0, 1, 2, 3 \ldots$$

   (a) Draw and carefully label the state transition diagram.
   (b) Find p_n as a function of p_0, λ, and μ.
   (c) Develop a closed form expression for p_0.

34. Consider the cyclic queueing system of Fig. 3.24.
    (a) Draw the state transition diagram if there are $N = 3$ customers. Let the state variable be the number of customers in the upper queueing system. Redraw the state transition diagram if λ is 1.0 and μ is 2.0.
    (b) Solve for p_1, p_2, and p_3 in terms of p_0. Solve for p_0 numerically.
    (c) Find the mean throughput of the upper queueing system.

35. Consider a queueing based memory model system as in Sect. 3.3.4. Let there be two classes of customers. Class 1 customers have a mean arrival rate of 10 requests per

Fig. 3.24 A cyclic queueing system

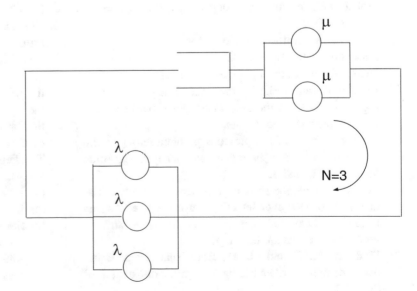

second, a mean service rate of 20 requests serviced per second and 1 Mbyte of memory is needed per request. Class 2 customers have a mean arrival rate of 5 requests per second, a mean service rate of 15 requests serviced per second and 3 Mbyte of memory is needed per request. The system has 1000 Mbytes of total memory.

(a) Write an expression for the probability that the system is in state $\underline{n} = (n_1, n_2)$ where $\underline{n} \in \Omega$. That is, $\underline{n}$ is an allowable state.

(b) How do the size of the memory requests enter into the model and solutions?

36. Consider an M/D/1 queueing system. This is an M/G/1 system where the service time is deterministic (the same constant service time for all customers). Modify the Pollaczek-Khinchin mean value formula for this case.

(a) How does it differ from the formula for the expected number of customers for an M/M/1 queue ($E[n] = \rho/(1 - \rho)$)?

(b) Tabulate the expected number of customers for the M/M/1 and M/D/1 systems for $\rho = 0.1, 0.2, 0.4, 0.6, 0.8, 0.9$, and 0.99. Which system has the larger expected number of customers? How large is the difference as $\rho \to 1.0$?

37. Consider an open queueing network as in Fig. 3.14 but with an extra feedback path from the output of Q2 to its input. The independent routing probabilities for customers departing Q2 are $r_{21} = 0.25$, $r_{22} = 0.5$, and $r_{2d} = 0.25$.

(a) Solve the traffic equations for the mean throughputs of Q1 and Q2 as a function of the mean arrival rate, λ.

(b) Write out the product form solution for the equilibrium probability $p(n_1, n_2)$.

38. Consider a closed network of two queues, Q1 and Q2, in a loop where there is also a feedback path from the output of Q2 to its input. The independent routing probabilities for customers departing Q2 are $r_{21} = 0.3$ and $r_{22} = 0.7$.

Solve the traffic equations for the relative mean throughputs of Q1 and Q2. Let Q1 be the reference queue with mean throughput equal to 1.0.

39. Develop equation (3.152) for closed Markovian queueing networks from the associated global balance equation and the traffic equations.

40. For the network of the previous problem let $\mu_1 = 2.0$ and $\mu_2 = 4.0$. Run the mean value analysis algorithm for $N = 1, 2$, and 3.

41. Consider a Gelenbe style negative customer network as in Fig. 3.16. However let Q2 receive positive (Λ) and negative (λ) external arrivals. Solve for q_1, q_2 and the product form solution for $p(\underline{n})$.

42. Find Eq. (3.250) and (3.251) from balance type equations for the voice/data integrated protocol example of Sect. 3.8.

43. Consider two finite buffer queues in tandem (series). Let the first queue hold at most N customers and the second queue hold at most one customer.

(a) Draw the state transition diagram. The arrival rate is δ and the state dependent service rate of the first queue is λ_i where i is the number of customers in the first queue. Also the service rate of the second queue is μ. Let the horizontal axis represent the number of customers in the first queue and let the vertical axis represent the number of customers (0 or 1) in the second queue.

(b) Write recursive equations, as in Sect. 3.8, for the network's equilibrium state probabilities. There are two possible sets of equations.

44. Draw and label a Petri net of the following situation. Sometimes a patron in a library cannot find a book on the shelves. The front desk assigns two pages to look for the book, one on each of the two floors of the library. Assuming that the book is found by one of the pages, that page finds the other page and they both return to the front desk with the book.

45. Consider a "Dining Philosophers" stochastic Petri net. In this classic distributed system problem [74] five philosophers are seated around a circular table. Between each philosopher on the table is placed a single chopstick. A philosopher needs two chopsticks to eat. If a philosopher picks up a chopstick from both sides of him/her, the philosophers on either side of him/her cannot eat.

(a) Draw and clearly label the Petri net of this situation. Each philosopher may be either thinking or dining, represented by places. With both chopsticks on the table on either side of a thinking ith philosopher, he/she picks them up at rate q^{i0}. The ith dining philosopher releases both chopsticks at rate q^{i1}. Each chopstick's availability is represented by its own place.

(b) Draw and label the state transition diagram.

46. Consider a Markovian Petri net of a *single* user submitting a job to a computer system. The user has three states: idle (0), job request (1), and job being processed (2). The user can only move from a job request to the job being processed if two independent resources, the memory and the CPU, are free (available). Once the job is processed, resources are released. The timing of all transitions is Markovian (negative exponential). The Petri net is safe (i.e., all places have at most one token).

(a) Draw and clearly label the Petri net.

(b) Draw the state transition diagram.

(c) Solve the state transition diagram for the three (equilibrium) state probabilities. Provide a closed form solution for the equilibrium probability that the processor is idle (not processing a job).

Abstract

In this chapter fundamental algorithms used in networking are described. This discussion starts with a consideration of routing, a network layer function. Two shortest paths routing algorithms and a bottleneck bandwidth routing algorithm are presented. This is followed by an exposition of some different types of routing strategies. Protocol verification and model checking are examined. Error codes, both error detecting and error correcting, are studied. The latter part of the chapter has sections on line codes, network coding, and quantum cryptography.

4.1   Introduction

Networking is a collection of computer controlled processes. It is not too surprising that algorithms (i.e., computer based procedures) for networking are crucial for networking to function successfully. This chapter begins with a discussion of both shortest path and bottleneck routing algorithms. This is followed by a discussion of different types of routing strategies. A discussion of protocol verification concepts is included. Error codes, means of adding mathematically computed redundant bits to message streams to assure accurate conveyance and preservation of data, are also covered. The latter part of the chapter presents expositions of line codes (used to pre-process data for good digital reception), network coding (a technique to boost the transmission capacity of networks by transmitting functions of data), and quantum cryptography (cryptography based on quantum physics principles).

4.2   Routing

4.2.1   Introduction

Some types of computer networks do not have a "routing problem" as there is a single path between nodes. Routing internal to a token ring or Ethernet are examples. However, in a wide area or metropolitan area network with multiple potential routes between each source and destination pair, there definitely is a routing problem.

For the purpose of routing, and for many other purposes, networks are usually represented as graphs. That is, the nodes in a graph will model packet switches, telephone switches, or computers. The edges of a graph represent links (either wired or wireless).

How many potential routes are there in a network graph between a source and destination? Let us do an illustrative example. Consider the rectangular graph of Fig. 4.1. We wish to find the number of direct routes (without loops) from node A to node Z.

Consider the indicated path. It can be seen that it consists only of movements up (U) in the graph or to the right (R). Thus the path can be represented by the "word."

$$U\,R\,R\,U\,U\,R\,U\,R \qquad (4.1)$$

Notice that there are four U's and four R's. A little thought will show that any direct path from node A to node Z consists of eight letters with four U's and four R's in some pattern. Thus the number of possible paths is

$$\binom{8}{4} = 70 \quad \text{paths} \qquad (4.2)$$

© Springer Nature Switzerland AG 2020

T. G. Robertazzi, L. Shi, *Networking and Computation*, https://doi.org/10.1007/978-3-030-36704-6_4

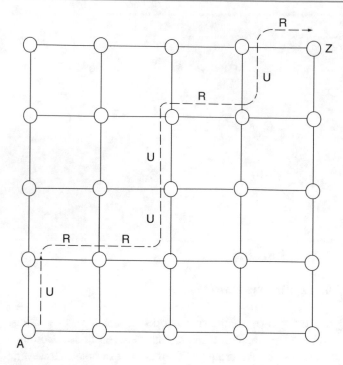

Fig. 4.1 A mesh network with a specific shortest path from node A to node Z

Table 4.1 Number of paths

N	No. of paths
10	48,620
20	3.5×10^{10}
30	3.0×10^{16}
40	2.7×10^{22}

That is quite a large number of paths for such a small graph. We can generalize this. If we have a rectangular graph of $N \times N$ nodes, then by the same reasoning the number of direct paths is

$$\binom{2(N-1)}{N-1} \text{ paths} \tag{4.3}$$

We have in tabular form.

One can see from Table 4.1 that the number of direct paths increases exponentially. It does not take a very big network to get a number of potential paths exceeding Avogadro's number!

If we measure the "cost" of a path by the distance along it, one can see that the "direct" paths of the previous examples are the shortest (distance wise) paths from node A to node Z. Such "shortest paths" are very desirable if one is routing packets or circuits.

However, distance is not the only way to measure cost. A link's "cost" may be in terms of quantities such as mean delay or monetary cost. Assume each link in a graph has a fixed cost. Then a "shortest path" between two nodes is a set

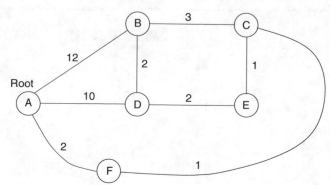

Fig. 4.2 A network graph with link costs indicated

Table 4.2 Dijkstra's algorithm

	N	B	C	D	E	F
1	{A}	12	∞	10	∞	2
2	{A,F}	12	3	10	∞	(2)
3	{A,C,F}	6	(3)	10	4	2
4	{A,C,E,F}	6	3	6	(4)	2
5	{A,C–F}	6	3	(6)	4	2
6	{A–F}	(6)	3	6	4	2

of consecutive links connecting the two nodes such that the sum of costs of each of its links is the smallest possible over all possible routes between the two nodes. In fact there are relatively efficient algorithms for finding shortest paths. Two such algorithms are discussed in the next two sections.

4.2.2 Dijkstra's Algorithm

Consider the network of Fig. 4.2 where the link numbers are distances. We wish to find all of the shortest paths from root node A to all of the other nodes. A table is created (Table 4.2) where one row will be added at a time. The set N in the table is the set of nodes for which we "know" a shortest path. The other column entries correspond to the current distance found for each node to node A.

The initial row has numerical entries for all nodes that are direct neighbors (one hop away) from node A. For Fig. 4.2 this is nodes B, D, and F.

Dijkstra's genius here lies in proving [226] that the smallest entry in each row not selected yet, corresponds to an optimal shortest path distance. The smallest distance in row 1 is 2 for node F. In each row one attempts to improve the paths to the root node A for direct neighbors of the selected optimal node (indicated by parenthesis) of that row.

This can be written as a recursion [208]

$$D(v_j) = \min[D(v_j), D(w_i) + l(w_i, v_j)] \tag{4.4}$$

Here w_i is the node selected in the ith row and v_j are its direct neighbors. The current distance from node v_j to the root is $D(v_j)$. The distance from node w_i to v_j, over a single link, since they are direct neighbors, is $l(w_i, v_j)$. The equation says that in the current iteration (row) the new distance from node v_j to the root is the minimum of the old distance from node v_j to the root or the distance resulting from a route from node v_j to node w_i and from w_i back to the root. In other words, in each row we seek to improve the routes of nodes v_j which are adjacent neighbors of the selected node w_j by going through w_j.

So in our Table 4.2 node F is selected in row 2 and brought into set N. Node B is not an adjacent neighbor of node F, so node B's entry is unchanged from the previous row. Since C is a direct neighbor of F with an entry of infinity, C's entry can be improved by going to the root from C to F to A with a total cost of 3. No other entry in this row can be improved.

For the next row (row 3) node C's entry has the smallest value so it is selected. Nodes B and E are direct neighbors of C and their distances can be improved, so we can enter their new distances in the table to the root through node C (6 and 4, respectively).

In the fourth row node E is selected and thus its direct neighbor D's entry improves from 10 to 6. In rows 5 and 6, nodes D and B are selected, respectively. However, no entries change in these two rows. It should be pointed out that in row 5 both D and B have entries of 6. It does not matter which one is selected first to be part of set N. Row 6 indicates the shortest path distances from each of the nodes to root node A.

Note that the Dijkstra's algorithm procedure generates N rows in the algorithm table for an N node network. Note also that the algorithm naturally finds the shortest paths not just between a pair of nodes but from a root node to all of the other nodes.

If one needs the shortest distances between every pair of nodes, the algorithm is run N times, each time with a different root node. If the actual paths are desired, these may be carried along in the table in each entry as pointers to the next node along the current path back to the root (as is done in the next section). Also, it is possible for the links to be bidirectional with different costs in each direction [204]. Finally, it should be noted that the routes found by the Dijkstra's (or Ford Fulkerson) algorithm form a spanning tree. A spanning tree is a graph without loops that touches every node in the original graph. A shortest paths spanning tree for the network used in the previous example and its routing solution appears in Fig. 4.3.

A use of the algorithm table is to generate a "routing table" that indicates which nodal output port leading to a direct neighbor to use to route packets/circuits to a distant destination. A routing table will be stored in each network node for routing purposes. The routing table for Fig. 4.2's node A appears in Table 4.3.

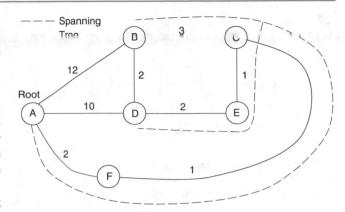

Fig. 4.3 A shortest paths spanning tree originating from the root (node A) superimposed on the network graph of Fig. 4.2

Table 4.3 "Routing" table

Destination	Nearest neighbor
B	F
C	F
D	F
E	F
F	F

In this particular example all of the routes from node A to other nodes go through a single node, node F.

In some situations it may be desired to find the k shortest paths between nodes A and Z that do not share any links (i.e., are link disjoint). To do this the shortest path algorithm is run to find the shortest path. The shortest path is recorded and its links are deleted from the graph. The shortest path algorithm is run again on the reduced graph to produce the second link disjoint shortest path which is recorded and its links are then removed from the graph. The process continues until all k link disjoint shortest paths are produced, if they exist.

4.2.3 Ford Fulkerson Algorithm

The Ford Fulkerson algorithm is a bit different from the Dijkstra's algorithm in the details. However, its goal is much the same as the Dijkstra's algorithm. The Ford Fulkerson algorithm will also find all of the shortest paths from a root node to each of the other nodes and create a spanning tree of routes in doing so.

In our version of the Ford Fulkerson algorithm table (Table 4.4) for Fig. 4.2 each entry has two parts. The first part is a pointer to the next node along the (current) path back to the root for that entry's node. The second part is the current distance along the path. The initialization row is filled with entries of $(\cdot, \infty)$ to indicate no path is selected yet. On

Table 4.4 Ford Fulkerson algorithm

	B	C	D	E	F
Initial	$(\cdot,\infty)$	$(\cdot,\infty)$	$(\cdot,\infty)$	$(\cdot,\infty)$	$(\cdot,\infty)$
1	(A,12)	(B,15)	(A,10)	(D,12)	(A,2)
2	(A,12)	(F,3)	(A.10)	(C,4)	(A,2)
3	(C,6)	(F,3)	(E,6)	(C,4)	(A,2)
4	(C,6)	(F,3)	(E,6)	(C,4)	(A,2)

a computer infinity is just a number much larger than any likely entry.

The entries are filled in from left to right, top to bottom. For a node's entry one attempts to improve the previous entry for that node by routing through the direct neighbor with the best total path back to the root.

The basic recursion for each algorithm table entry is [208]

$$D(v) = \min_{w_j}[D(w_j) + l(w_j, v)] \qquad (4.5)$$

Here v is a node whose entry one is trying to compute and $D(v)$ is the distance along the path from node v to the root. Also, w_j is the jth adjacent neighbor of node v. Finally, $l(w_j, v)$ is the single link distance between direct neighbors v and w_j. Thus the recursion says that the new entry of distance, $D(v)$, for node v is the minimum over all of v's adjacent neighbors (the w_j's) previous distances for node w_j to the root plus the cost of getting from node v to w_j.

The initial entries one puts in the table are for direct neighbors of the root. In Fig. 4.2 node B is a direct neighbor of the root so its entry in row 1 is (A,12). That is, go from node B to node A with a cost of 12. Each entry is based on the most recent information in the table (generally located to the left of the entry in the same row and to the right of the entry in the row above). So in row 1 node C can use node B's existing entry and one has a route from C to B to A with a cost of 15. The process continues. As an example, for row 3's node D entry, the previous entry in the row above is (A,10). However, at this point in the table there is an entry for node E in the row above of (C,4). Thus one can go from node D to E to C to F to A with a cost of 6 so the new entry for D is (E,6). If there is ever two equally good choices for the next node on the path back to the root, either one can be chosen.

One can see that the algorithm terminates when there are no changes in two consecutive rows. The number of rows in the completed algorithm table depends on the problem and will be smaller if columns are labeled left to right from nodes closest to the root to those furthest from the root.

Again, the shortest distances between every pair of nodes can be found by running the algorithm with a different root each of N times. Links may also be bidirectional.

This and the previous section describe bare bones shortest path routing algorithms. Implementing them in a distributed fashion on a dynamic network presents challenges. As rout-ing tables are updated, packets may loop (e.g., travel in circular paths through the network). If the cost function in a packet switched network is mean delay, and routing costs are based on this, oscillations are possible. That is, lightly loaded links attract traffic and become heavily loaded, while traffic avoids heavily loaded links which then become lightly loaded. See Bertsekas [36] and Schwartz [208] for discussions of such problems and solutions for them.

4.2.4 Optimizing Bottleneck Bandwidth

Shortest path algorithms are useful for finding paths that minimize distance, monetary cost, or path delay. It involves a single optimization criterion. A different single optimization criterion problem is to seek paths that maximize the mini-mum (i.e., bottleneck) bandwidth of the links along the path. Here "bandwidth" is used loosely to describe what is actually data rate and bottleneck data rate. For instance, in Fig. 4.4 if one has a serial path from A to Z, then the largest bandwidth path from A to Z that is possible is 4 Gbps. The 4 Gbps link limits the flow of data along the path and is the "bottleneck."

One can use a modified version of the Dijkstra's algorithm to find the bottleneck bandwidth on all the paths from a root node to all of the other nodes in a network graph [13, 64, 108, 236]. The bandwidth on the links that the algorithm operates on may in fact be residual bandwidth available on the links, while other link bandwidth is already assigned [236]. Note that if the algorithm is run and one then makes a path assignment using some/all of the residual bandwidth in some links, other bottleneck bandwidth maximizing paths on the network found in the same algorithm run may not be capable of being assigned if they share links with the original path. This because the residual bandwidth on shared links will be reduced as a result of its assignment to the first path.

For an example of using Dijkstra's algorithm to find maximal bottleneck bandwidth paths, consider the associated figure showing a network of six nodes (Fig. 4.5).

In this figure the link numbers are available bandwidth in Gbps. Node A is the root node. We seek the maximal bandwidth paths from the root node to all of the other nodes. We use Dijkstra's algorithm (see earlier subsection) with pointers. That is, each entry in the table starts with a pointer to the next node along the current path to the root from the node listed at the top of the column. The second entry is the current bottleneck bandwidth for that path (Table 4.5).

Dijkstra's algorithm for this problem works pretty much like the shortest path version of the earlier section with one major difference. We now highlight a node entry with aster-isks in each row with the largest value of current bottleneck bandwidth that has not yet been highlighted. For the earlier shortest path algorithm the smallest entry was highlighted.

Fig. 4.4 Bottleneck bandwidth of 1 Gbps for a serial path

can be improved by routing through node B so there are no other changes in this row.

In row 4 node E's entry is highlighted. Now node C's entry was originally (A,2) (go to A directly with a bandwidth of 2) and now can be improved to (E,5) (go from C to E to D to A) with a maximum bottleneck bandwidth of 5 Gbps.

One continues highlighting a node in each row which has the largest current bottleneck bandwidth value that has not been highlighted yet. However, at this point no other changes result in the table. In the last row one has all the paths with maximal bottleneck bandwidth from the root node A to all of the other nodes. With the pointers optimal bottleneck bandwidth paths may be recovered from the table entries. For instance, node E's final entry says go to node D and node D's final entry says go directly to node A. Thus the optimal path to maximize bottleneck bandwidth from node E to A is E to D to A. Note that the bottleneck bandwidth maximizing paths may not be unique.

Taking a general view, we have just optimized one criterion. In fact one may really want to find a routing that is optimal for dual criteria or even multiple criteria in some sense. There has been work on dual criteria routing such as bandwidth and hop count (i.e., number of links a path transits) [108] and also on bandwidth and delay [236]. Many composite routing problems are possible [64].

In this treatment of routing algorithms we have illuminated some basic concepts and just scratched the surface. At this point some routing techniques will be discussed (see [204] for an alternative treatment).

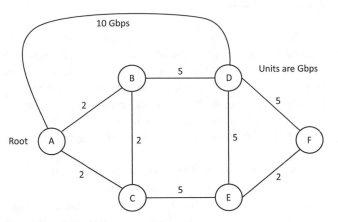

Fig. 4.5 Bottleneck bandwidth example

Table 4.5 Dijkstra's Bottleneck algorithm

	N	B	C	D	E	F
1	{A}	(A,2)	(A,2)	(A,10)	(·,0)	(·,0)
2	{A,D}	(D,5)	(A,2)	*(A,10)*	(D,5)	(D,5)
3	{A,B,D}	*(D,5)*	(A,2)	(A, 10)	(D,5)	(D,5)
4	{A,B,D,E}	(D,5)	(E,5)	(A,10)	*(D,5)*	(D,5)
5	{A,B,C,D,E}	(D,5)	*(E,5)*	(A,10)	(D,5)	(D,5)
6	{A–F}	(D,5)	(E,5)	(A,10)	(D,5)	*(D,5)*

Again, the set N in the second column keeps track of the nodes that have been highlighted.

So in the first row entries are entered in the table for adjacent neighbors of root node A (that is, for nodes B, C, and D) using the direct links. Nodes E and F are not adjacent neighbors of node A so we enter a null pointer and zero bottleneck bandwidth.

In row 2 *(A,10)* in column D is highlighted with asterisks. We improve any node's bottleneck bandwidth on a path from the node to the root through node D for which the node is an adjacent neighbor of node D. For instance, node B is an adjacent neighbor of D so in changing node B's entry, node B's current path now goes through D to the root. This improves its bottleneck bandwidth to 5 Gbps from 2 Gbps. Nodes E and F are also adjacent neighbors of node D. Non-trivial entries are now placed in the columns for nodes E and F. For instance, for node E we have (D,5), so the current maximal bottleneck bandwidth path goes from node E to D to A with a current bottleneck bandwidth of 5 Gbps.

In row 3, there are three nodes (B, E, and F) with a path to the root with a bandwidth of 5 Gbps that can be highlighted. It does not matter which we choose. We arbitrarily choose the first (node B). None of node B's adjacent neighbors entries

4.2.5 Table Driven Routing

In table driven routing, information on routes is stored in tables at each node. The tables are updated using shortest path algorithms based on events (e.g., a link going down) or periodically (e.g., every X seconds or every Y events). Usually a combination of event based updating and periodic updating is used. Event based updating is sometimes referred to as inside updating and periodic updating is sometimes referred to as outside updating.

In a packet switched network using a datagram mode of operation, a node reads an incoming packet's header for its destination address. The destination address is looked up in a "routing table" such as Table 4.3 (different from the algorithm tables of the previous sections) to find which output port of the node the packet should be sent out over. The packet is then placed in the buffer for that output port.

The routing table used here has been previously constructed from an algorithmic routing table. Note also that if the output port speeds are B bps per output port, the nodal processor needs to be able to place packets in N output buffers at rate NB bps to keep all output links continually busy.

In a circuit switching based network a circuit entering a node has an identification number. A table lookup based on the identification number allows the circuit to be continued (switched) out the appropriate output port. Virtual circuit based packets also have an identification number for each virtual circuit that is carried in the packet header. A node determines this identification number and uses a table lookup to see which output port a packet in the virtual circuit stream should be sent to.

4.2.6   Source Routing

Nodes do not maintain routing tables under source routing. Rather, a source node will insert the route (i.e., nodes to be visited) for a packet into its header. Each node visited by the packet refers to this list to determine the next node to send the packet to. The obvious question is how does the source know the path to insert into the packet header?

A centralized approach is for the network to have a path server [204]. A path server monitors the network and computes shortest paths. A node wishing to send a packet contacts the path server for a path. Like any centralized scheme, two main drawbacks of this approach are that the path server is a single point of failure (i.e., if it goes down the whole network is down) and if the path server receives too many requests it may be a performance bottleneck.

A distributed scheme is to use what are called path discovery packets. A source wishing to send a packet "floods" (see section below) the network with many path discovery packets that simply travel through the network without a specified path. Each node receiving a path discovery packet, before sending it to a neighboring node, puts its own node identification number at the end of a list of nodes visited by the packet in the packet's header.

The theory is that one or more of the flooded packets will eventually reach the intended destination node (which is also indicated in the path discovery packet's header). The destination node then has one or more routes to the source. A packet with either the first route received, or some choice of the "best" route if several path discovery packets are received, can then be sent back to source. The packet carrying the route can be source routed using the reversed list of nodes visited. Upon receiving the list, the original source node can launch a packet, or several packets, to the destination.

It should be pointed out that if a discovered path is used too long by a source, network conditions may have changed and it may not be a good, or even feasible, route. On the other hand, too frequent path discovery burdens the network with flooding overhead.

4.2.7   Flooding

Flooding is a technique to get a packet(s) to a destination(s) without any, or very little, routing knowledge. In the simplest version of flooding a node originating a flood sends copies of the packet(s) it wishes to send to all of its neighbors. A neighbor receiving such a packet(s) copies them out to all of its neighbors. That is, it will send copies of the (distinct) packets on all of its output ports except the one the packet(s) arrived at.

Flooding may be used in situations where it is desired to broadcast a message to all of the nodes in a network (see Sect. 4.2.10 below). It may be also used to get a message to a specific node, or a set of specific nodes, when there is no routing information available. It is also a good policy in a very unreliable network (where nodes and links go down frequently). However, the large number of packets generated is a large overhead, especially if just a small number of nodes need to be contacted.

To reduce the number of packets generated, there have been strategies for flooding developed that flood only in limited directions (i.e., towards a destination). This is easier to implement if nodal geographic coordinates are known. This may be possible, even for mobile networks, if location systems such as GPS (Global Positioning System) are used.

4.2.8   Hierarchical Routing

Hierarchical routing is a technique that allows routing table size reduction. It involves the way in which nodal addresses are assigned. Telephone numbers are an example of a hierarchical addressing scheme. In the USA, for instance, the first three digits of a ten digit phone number is the geographic area code. The next three digits indicate the switching exchange and the last four digits indicate the actual phone number within the indicated exchange.

The beauty of this system can be explained in terms of a simple example. Suppose that someone in San Francisco wishes to call a number in Manhattan, New York. The routing table used by long distance facilities in San Francisco need only store one entry for the millions of phones in the 212 Manhattan area code. Moreover only a single entry is needed for the ten thousand phones in the destination local exchange. It should be mentioned that country codes add an extra level to the telephone hierarchy. The actual switching hierarchy of long distance facilities and local exchanges is in fact a physical realization of the hierarchical addressing system.

Other than telephone networks, hierarchical routing can be done in other types of networks. One might have sets of wireless nodes grouped into "clusters" and clusters grouped into "super-clusters." The address 2.7.12 might indicate the twelfth node in the seventh cluster in the second super-cluster. There is a single entry in super-cluster routing tables for the second super-cluster nodes. Within clusters in the second super-cluster there is a single entry for the seventh cluster's nodes.

Hierarchical routing is very effective, particularly for large networks, at reducing routing table size. It also provides some intuitive structure to the address space, which is useful. On the downside, some hierarchical paths between nodes may be longer than direct connections, though this is less of a problem for large, dense, networks. There can also be a problem if a network under hierarchical routing grows with time (adds more nodes) and the space for the entries at each hierarchy level is limited—as in the current problem of proliferating telephone area codes in the USA.

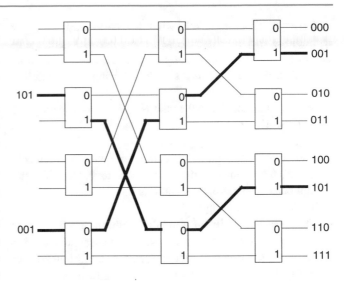

Fig. 4.6 An 8×8 delta network with two paths from specific inputs to outputs indicated

4.2.9 Self-routing

A number of networks with special structured topologies are known to have a useful property called "self-routing." As we shall see, self-routing means that packets can be routed from node to node in a network using only (part of) the destination address. Generally these self-routing networks or "switching fabrics" are implemented in VLSI to serve as the routing heart of a packet switch. This form of switching is also known as space division switching as there are spatially separate paths through the switching network.

As an example consider the eight input, eight output (8×8) delta network of Fig. 4.6. The inputs are on the left and the outputs are labeled in binary on the right. The nodes are called switching elements. Each switching element has two inputs on the left and two outputs, labeled as "0" (upper output) and "1" (lower output), on the right. Thus we have 2×2 switching elements. The fact that the same switching element circuitry can be replicated on a chip many times makes this a useful approach for VLSI implementation.

The wiring between switching elements is done in a patterned manner that allows self-routing. Specifically, a packet, say at the input, has the binary address for the intended output port for that packet placed in the packet's header. A switching element in the jth vertical column of switching elements receiving a packet will route it to the element's output with the same bit label (0 or 1) as the bit in the jth position (read left to right) of the destination binary address.

For instance, in Fig. 4.6 a packet is launched from the third input from the top for output port 101. The switching element in the first vertical column of switching elements that the packet enters will route it to its lower (1) output,

the switching element in the second column that the packet enters will route it to its upper (0) output, and the element in the last column that the packet enters will route it to its lower (1) output and hence to the destination output port. A similar path is shown for a packet at the seventh input going to output 001.

Why does this work? Study the wiring pattern of the 4×4 module in the second and third column, third and fourth row. This module is also a 4×4 self-routing delta network. For this module it can be seen that the wiring is such that packets arriving at the 4×4 module inputs with a 0 in the middle bit are routed to the upper output (third stage) switching element. Packets with a 1 in the middle bit are routed to the lower output switching element of the 4×4 delta module. In both cases, the third bit then directs the packet to the proper output. This all matches the output port address pattern.

Looking again at the overall delta network, it can be seen that the upper and lower groups of four switching elements in the second and third columns consist of two (upper and lower) 4×4 delta networks. The wiring between the first and second stage is such that if the first bit in the output port address is a 0 the packet is sent from the first stage to the upper 4×4 delta network, no matter which input it originates from. If the first bit is a 1 an arriving packet to any first stage input is sent to the lower 4×4 delta network. Again, this all matches the output port address pattern. Larger switches can be recursively constructed in this manner.

It can be seen in this example that there is no need for routing tables or a centralized routing control. Routing decisions are made locally at each switching element based on a single destination port address bit in the packet header.

Such interconnection network as the delta network was first studied in the context of telephone circuit switching and later applied to packet switching, particularly as ATM technology was developed [192]. Delta networks are a subclass of the more general Banyan networks. Delta networks include omega, flip, cube, shuffle exchange and baseline networks.

As Ahmadi and Denzel relate [8], the major features of these networks are

- They have $\log_b N$ stages (columns) of switching elements and N/b switching elements for each stage. Here N is the number of inputs/outputs and b is 2.
- They are self-routing.
- They can be built in modular fashion using smaller switches. As was said, in Fig. 4.4 in the second and third columns, the first and second (and third and fourth) row elements form 4×4 self-routing delta networks.
- They can be used in either synchronous or asynchronous mode.
- Their regular structure makes them attractive for VLSI implementation.

While the $\log_b N$ complexity of the number of switch points seems better than the N^2 complexity of a crossbar switch (see Chap. 2), the individual crosspoint circuit complexity for a crossbar is simpler so that in terms of chip area, Banyan and crossbars have been found in some studies to be comparable [92, 225].

With multiple packets being routed simultaneously through such self-routing interconnection networks, at times it will happen that the two packets entering a switching element's inputs have the same switching element output as their next destination. While one packet may be accommodated at a time, the other will have to wait (be buffered). This phenomena is called blocking and networks that exhibit it are called blocking networks.

To mitigate the amount of blocking in a switching network and boost switch throughput there are several approaches [8]

- Use higher internal link speeds compared to external link speeds.
- Place buffers at each switching element.
- Use a hand shaking protocol between each stage and a "back pressure" strategy to slow down blocked packet movement.
- Use parallel networks so that one has a multiplicity of paths between each input and output. Alternately, one can implement multiple links between each switching element.
- Pre-process load through networks such as distribution or sorting networks. Distribution networks distribute load in a uniform manner to a switching network. Sorting networks sort packets by output port address. As part of

a larger pre-processing system, sorting networks can be used to minimize blocking.

4.2.10 Multicasting

There are a number of ways in which packets can be multicasted (sent to multiple destinations). Packets can be flooded though this is somewhat indiscriminate and incurs a large overhead. Alternately, packets can be individually addressed to each destination. This is more efficient than flooding. Even more efficient than this at conserving network bandwidth is to put multiple addresses into individual packet headers. Assuming that nodes receiving such packets know the topology of the network, a multidestination packet arriving at a node is divided into a number of smaller multidestination packets, one for each nodal output port. The addresses in a divided packet are for nodes reachable by the output port the divided packet goes out over. The procedure repeats as packets arrive at nodes until copies are delivered to all individual destinations.

Finally, routers in networks can maintain a spanning tree(s) (see Fig. 4.3) for multicasting purposes. A special broadcast address allows packets to be forwarded at each node only to the spanning tree links.

Note that there are tradeoffs in these techniques between the bandwidth consumed and the additional network control overhead needed to implement the more efficient schemes.

4.2.11 Ad Hoc Network Routing

Ad hoc networks are wireless networks of (usually) mobile nodes that hop packets from node to node along the way to their destinations. This is called multiple hop transmission. Energy conservation is an important part of ad hoc network design. In fact because of the nonlinear relation between transmission energy and radio propagation distance, it is more energy efficient for a packet to make several smaller hops rather than one large hop.

Routing algorithms for ad hoc networks can be divided into topology versus position based algorithms [167].

Topology based algorithms can be further divided into proactive and reactive approaches. Proactive topology based routing algorithms use classical table based routing strategies. Information is continually maintained on paths that are available. A downside is that there is a large overhead in table update messages in maintaining the information for unused paths if there are frequent topology changes.

Reactive topology based algorithms only maintain routes that are currently in use. Naturally some sort of route discovery is necessary before a packet is transmitted. There may still be heavy update traffic with topology changes. Reactive protocols include DSR, TORA, and AODV [172, 179, 226, 220].

Position based routing makes use of the geographic locations of nodes in making routing decisions. A location service such as GPS may be used. The position of a destination is placed in the packet header. A packet is forwarded closer and closer to the destination. A well-known position based routing algorithm is GPSR (Greedy Perimeter Stateless Routing) [132].

4.3   Protocol Verification

Protocols are the rules of operation of computer networks. A protocol specification is a set of rules for communicating between processes on different machines. Simple protocols may be expressed as state machines (see any text on digital logic) in state machine diagram form. However, realistic protocols may have many states and transitions— too many to draw in a simple diagram. Thus an important question is whether a complex protocol has errors that may cause problems at some point in its operation such as deadlock. The problem of checking a protocol specification for logic errors is called protocol verification or protocol validation. It is advantageous to catch such logic errors early in the design process, rather than once a system is implemented.

To make this discussion more concrete, consider the state machine representation of two communicating processes in Fig. 4.5 [254]. A channel connects both processes in each of both directions.

Transmitted message identification numbers appear next to the transitions in the diagram. A negative ID number indicates a sent message and a positive ID number indicates a received message.

Suppose that both processes start in state 1. Process 1 can send message 1 leading it to state 2. Process 2 can then receive message 1, leading it to its state 2. In a similar manner, message 3 can be sent and received (bringing both processes to their respective state 3's) and then message 4 can be sent and received (bringing both processes back to their state 1's). Alternately, both processes can move from state 1 to state 2 and back to state 1 by sending and receiving packets 1 and 2 in sequence.

The state machine of Fig. 4.7 represents a "correct" protocol with no errors. However, a number of errors can arise in actual protocols. Among these are

- *Unspecified Reception:* A message in the channel may be received but not as initially specified in the design. Therefore the system behavior at this point cannot be predicted accurately.
- *Deadlock:* The system is stuck or frozen in some state.

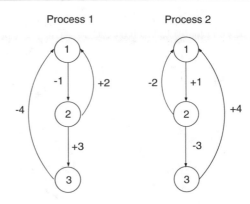

Fig. 4.7 State machine representation of a correct protocol. From Yuang [254] © IEEE

- *Livelock:* Messages are continually exchanged with no work accomplished.
- *State Ambiguity:* A state in a process can stably coexist (i.e., is reachable with empty channels) with several states in another process. This is not necessarily an error but one must be careful with these situations.
- *Overflow:* The number of messages in a channel or buffer grows in an unbounded fashion.
- *Non-executable Interaction:* A transmission or reception that is indicated in the specification but is never executed in reality. Also called dead code.

Let us consider an example of the specification of two communicating processes that have some errors (from [254]). The state machine diagram is shown in Fig. 4.8. A "reachability diagram" that illustrates all possible states is shown in Fig. 4.9. Each rectangular entry is a state. The entries in (row, column) positions (1,1) and (2,2) indicate the state of process 1 and process 2, respectively. The entry (1,2) indicates which message is in the channel from process 1 to process 2. Likewise, the entry (2,1) indicates which message is in the channel from process 2 to process 1. An "E" indicates an empty channel.

The diagram begins in global state GS 1. Both processes are in their respective state 1's and both channel directions are empty. Consider the left branch of the reachability diagram. Message 1 is put on the channel from process 1 to process 2 and process 1 moves into state 2 (GS 2). This message is received by process 2, leading to both processes being in their respective state 2's and both channel directions are empty (GS 3). Next, message 2 is put on the channel from process 2 to process 1 (GS 4) and process 1 receives it (GS 5). Now both processes are in their respective state 3's and both channels are empty (GS 5).

But there is now a problem. Referring to the state machine diagram (Fig. 4.8), process 1 expects to receive message 3 and process 2 expects to receive message 1, neither of which

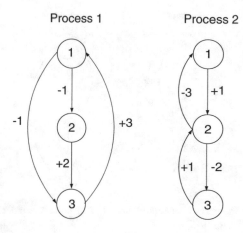

Fig. 4.8 State machine representation of a protocol with errors. From Yuang [254] © IEEE

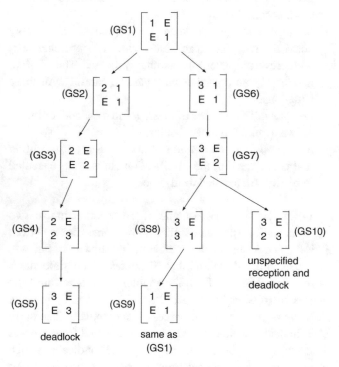

Fig. 4.9 Reachability tree of state machine protocol in Fig. 4.8. From Yuang [254] © IEEE

will be sent. This is a deadlock! The system is permanently stuck in this state.

Now consider the right branch of the reachability diagram. From GS 1, message 1 is put on the channel from process 1 to process 2 and process 1 moves into state 3 (GS 6). Process 2 receives the message (GS 7). With the system in GS 7, process 1 is in its state 3, process 2 is in its state 2, and both channel directions are empty. Two things can happen now. Process 2 can launch message 3 to process 1 (GS 8) followed by its reception by process 1, leading to the system returning to original state GS 1 (which is perfectly fine). Alternately,

from GS 7 process 2 can launch message 2 to process 1 leading to GS 10.

In global state GS 10 both processes are in their respective state 3's and message 2 is on the channel from process 2 to process 1. Again we have a deadlock. Process 1 is expecting message 3 and process 2 is expecting message 1, neither of which will be sent. Moreover as there is a message on the channel to process 1 that process 1 is not prepared to receive, we also have an unspecified reception. Note also that the transition from state 3 to state 2 in process 2 is never executed. It is a non-executable interaction (dead code).

Through this example one can appreciate that if the reachability diagram has hundreds or thousands of states, which can be true of even a system of moderate complexity, finding deadlocks, and other errors is a challenging algorithmic problem. Speaking generically, one can implement

- *Exhaustive Search:* Search the entire state space though this impractical for larger protocols.
- *Local Search:* Search local parts of the state space.
- *Probabilistic Search:* Search states with a high probability of occurring.
- *Divide and Conquer:* Break the problem into smaller parts.

As an example, the SPIN model checker software has been highly optimized over the years to be very efficient at finding protocol errors [116,117]. Specifically it is optimized to prove safety and liveness properties using techniques such as breadth first search and in the process it may uncover errors. Another software example is S-Taliro which is a tool optimized to find errors by solving an optimization problem in conjunction with a Matlab environment. It should be pointed out that an algorithm can either be designed to prove that there are no errors or to find errors. Most algorithms find errors as it is more tractable.

4.4 Error Codes

4.4.1 Introduction

The normal movement of electrons at any temperature causes thermal noise in electrical circuits. Lightning strikes cause impulses of noise (impulse noise) that interferes with radio transmission. Two wires that are physically close may be electromagnetically coupled, causing the signals in each to mix and thus cause cross-talk interference. Finally, there is optical noise in fiber optic cables due to the indivisible nature of photons.

When a binary stream of data is transmitted, such mechanisms in the channel may cause 1's to become 0's or 0's

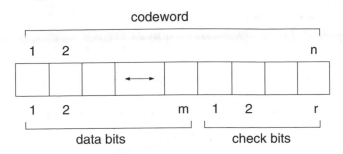

Fig. 4.10 A codeword with message and check bits

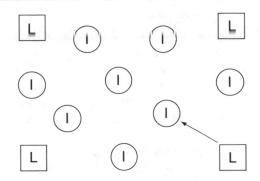

Fig. 4.11 A legitimate codeword (L) transformed by an error(s) into an illegitimate (I) codeword for an error detecting code

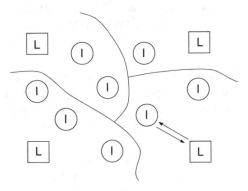

Fig. 4.12 Space of legitimate (L) and illegitimate codewords (I) for an error correcting code. A legitimate codeword is mapped by an error into an illegitimate codeword. The illegitimate codeword is mapped to the nearest (and correct) legitimate codeword

to become 1's. Recall that a 0 or 1 will be represented by a distinct waveform. Electrical, radio, or optical noise, which distorts such a waveform by adding a random-like component, may cause the receiver to make a mistake in waveform recognition, thus allowing a "0" to be received as a "1" or vice versa.

How does one then reliably transmit data in the presence of noise? How can one be sure that a transmission of financial data is accurate, for instance? The solution to this problem that people have come up with over the years is the use of error codes. Using a mathematical algorithm at the transmitter, extra check bits are added to each block of data (packet). An inverse version of the algorithm is run by the receiver on the received data and check bits.

There are two types of error codes. An error detecting code allows the presence of (certain types of) errors to be detected at the receiver. Under an error detecting code the receiver will not know which bits in the block have errors, only that there are some. In this case the receiver usually asks the transmitter to retransmit the data block that originally had errors. This protocol is called Automatic Repeat Request (ARQ). A more powerful error correcting code allows the receiver to correct (certain types of) errors on the spot without asking for a retransmission. This is particularly efficient when propagation delays are large. As an extreme example it would be very inefficient for a space probe to Saturn to ask for a retransmission if an error(s) were detected when the round trip delay is measured in hours.

Consider m bits of data with r check bits for a total of $m + r = n$ bits as in Fig. 4.10.

A basic concept involving the difference between two n bit codewords is Hamming distance. The Hamming distance between two codewords of equal size is the number of bit positions that are different. For instance, the codewords 1011 and 1101 are a distance 2 apart as two bit positions differ.

In Fig. 4.10 all 2^m data possibilities may be used but not all 2^n codeword possibilities are used. In an error detecting code, some combinations of data and check bits are legitimate and some are not. Consider a "space" of all possible codewords as in Fig. 4.11. As illustrated in the figure, an error will hopefully change a legitimate codeword in the space

(indicated by a boxed L) to an illegitimate one (indicated by a circled I) which can be recognized as such by the receiver.

Generally a code will protect against some degree of the most likely errors. As an example, a parity code can protect against single bit errors. A CRC code detects single, double, and odd numbers of bit errors as well as certain burst errors. It is always possible for an error code to be overwhelmed by too many errors. For instance, in Fig. 4.11 if an error is such that it causes a legitimate codeword to be transformed into another legitimate codeword, this will escape detection at the receiver.

A diagram similar to Fig. 4.11 for error correcting codes appears in Fig. 4.12. In this figure each legitimate codeword in the code space is surrounded by a neighborhood of illegitimate codewords. When the receiver detects an illegitimate codeword it assumes that the closest legitimate codeword is the correct one. Closeness here may in a Hamming distance sense. Naturally if a legitimate codeword is distorted too much, it may become a codeword in a different neighborhood and be mapped into a wrong but legitimate codeword.

We now discuss three block codes in detail. Note that while it is beyond the scope of this book, stream coding is also possible.

4.4.2 Parity Codes

This is a simple code that can detect single bit errors. One adds a single check bit to the data block such that the number of 1's in the entire codeword (data bits plus check bit) is even (if one wants to use "even parity") or odd (if one wants to use "odd parity"). We will always use even parity in examples in this chapter.

For instance with even parity 1011 becomes 10111 and 00101 becomes 001010 where the last bit is the appended parity/check bit. The receiver simply counts the number of 1's in the received block. If it is even the message is assumed to be correct, if it is odd there has been an error. In the case of an error a request to the transmitter for a retransmission is usually sent by the receiver. Parity codes will be used as an element of the Hamming error correcting codes of the next section.

A burst error is a series of errors affecting a number of consecutive bits. For instance a lightning strike of a certain duration may cause a burst error in a serial transmission. A trick can be used to still use parity codes in the presence of burst errors.

For this coding trick, arrange a number of codewords in a table with one codeword per row. But transmit a column, not a row, at a time over the serial channel. Then if a burst error occurs in the channel that is not too long in duration, only one bit in a number of consecutive codewords will be affected and the errored codewords will be detected.

It should be noted that errors can occur both in data and check bits. Therefore any code must be able to handle both types of errors. It can be seen that this is true of parity coding, for instance.

4.4.3 Hamming Error Correction

In this section we will take a detailed look at a 1 bit Hamming error correcting code. This code can correct 1 bit errors at the receiver with 100% accuracy. The code uses parity bits as building blocks.

Before proceeding to the actual coding mechanism, a question that needs to be answered is how many check bits are needed for a given number of data bits. Recall there are m data bits, r check bits, and $n = m+r$ total bits in a codeword. There are 2^m legal messages (i.e., all possible messages in m bits). Each such legal message is associated with itself and n possible corrupted messages that are corrupted by one bit being flipped to its opposite value. Thus 2^m times $(n + 1)$ should be less than the total number of codewords (2^n) or

$$2^m(n + 1) \leq 2^n \qquad (4.6)$$

Let $n = m + r$ and

Table 4.6 Computing number of check bits

r	Inequality	Holds?
1	$12 \leq 2$	No
2	$13 \leq 4$	No
3	$14 \leq 8$	No
4	$15 \leq 16$	Yes!

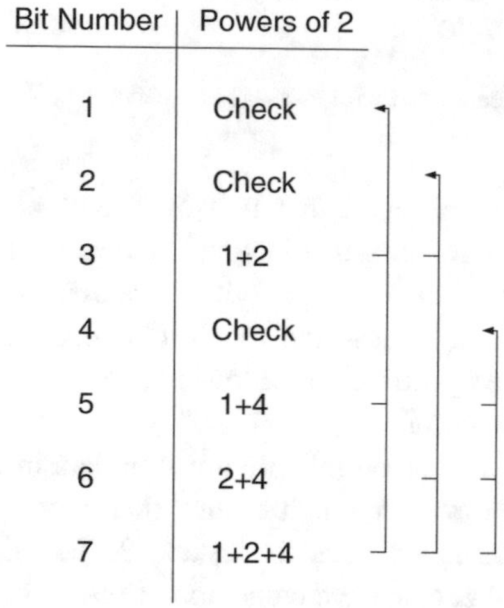

Bit Number	Powers of 2
1	Check
2	Check
3	1+2
4	Check
5	1+4
6	2+4
7	1+2+4

Fig. 4.13 Coverage of parity bits in Hamming code in seven bit codeword

$$2^m(m + r + 1) \leq 2^{m+r} \qquad (4.7)$$

Divide both sides by 2^m and one obtains

$$(m + r + 1) \leq 2^r \qquad (4.8)$$

This equation needs to be satisfied for a given number of message bits, m, by the number of check bits, r. For instance, suppose that $m = 10$ (ten message bits) and it is desired to find r. We can start with $r = 1$ and keep incrementing it until the inequality holds (Table 4.6).

Thus 4 check bits are needed for 10 data bits. As the number of data bits Increases, the percentage overhead in check bits decreases.

How does a transmitter implement Hamming error code correction? We will set up the codeword so that check bits appear in bit positions that are powers of two (i.e., 1, 2, 4, etc....). For four data bits and the required three check bits, this is illustrated in Fig. 4.14.

In the Hamming code a number of parity bits, each covering overlapping parts of the codeword, are placed in the powers of two bit positions. To see which check/parity bits are associated with which codeword bits, consider the table of Fig. 4.13. The seven bit positions are listed in the first column. Bit positions 1, 2, and 4 are listed as check bits. Each bit position number is expanded as a sum of numbers that are powers of two. For instance, 5 is $1 + 4$ and 7 is $1 + 2 + 4$.

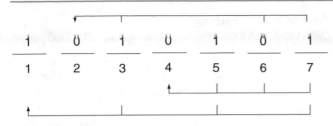

Fig. 4.14 Four bit message (1101) embedded in Hamming codeword with check bit coverage shown

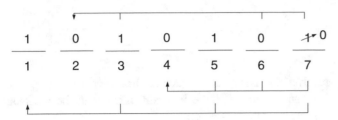

Fig. 4.15 Seven bit Hamming codeword with bit in seventh position flipped by noise from 1 to 0

Now each check bit will be (we will say) an even parity bit for itself and bits where the check bit's position number appears in the sum of powers of two. Thus, for instance, check bit 1 covers bits 1, 3, 5, and 7. The check bit in bit position 2 covers bits 2, 3, 6, and 7. Finally the check bit in bit position 4 covers bits 4,5, 6, and 7.

A transmitter implementing the 1 bit Hamming error correction code will format the codeword and insert the check bits (though check bits are often grouped together at the end of a packet after the data bits for transmission). As an example, consider Fig. 4.14. The four message bits (1101) need three check bits for a total of 7 bits in the codeword. From the figure one can see that the first check bit should be a 1 under even parity as the number of 1's then in bit positions 1, 3, 5, and 7 is even. Likewise the bit position 2 check bit is 0 so that the number of 1's in bit positions 2, 3, 6, and 7 is even. Finally, the check bit in position 4 is set to 0 so that the number of 1's in bit positions 4, 5, 6, and 7 is even. These seven bits (1010101) are then transmitted by the transmitter.

Let us see how a receiver can correct a single bit error. Instead of receiving the correct codeword, 1010101, say the seventh bit is flipped by channel noise so that 1010100 is actually received (Fig. 4.15).

The receiver checks if the check bits and the bits they cover still have even parity. For this received codeword each group of four bits associated with a check bit has odd parity. Since check bits 1, 2, and 4 are now associated with (incorrect) odd parity one adds 1+2+4=7 so that receiver knows that bit 7 is in error. Thus the receiver will change the incorrect 0 in bit position 7 to a 1 and so error correction is achieved.

One can see the receiver procedure. For instance, if bits 1 and 2 only are associated with odd parity, the error is in bit position $1 + 2 = 3$ or bit 3. If only bit 4 is associated with odd parity, only the check bit in position **4** has an error.

Why does the Hamming code procedure work? By way of an example from Fig. 4.13, if bit 5 has an error, check bits 1 and 4 will be made odd so adding the odd bit position numbers $(1 + 4 = 5)$ will indicate that bit 5 has an error. This method will work for a single bit error in any bit position.

Since check bits only appear in powers of two bit positions, the number of check bits becomes a smaller proportion of the codeword size as the codeword length increases. For instance, in a 1023 bit codeword there are only 10 check bits.

4.4.4 The CRC Code

Introduction

Error detecting codes allow a receiver to detect the presence of an error(s), though the receiver will not know which bit(s) are in error and so cannot correct the error. However, the receiver can ask for a retransmission that hopefully will be error free (otherwise the receiver will ask for a second retransmission and so on...).

Error detecting codes generally require less overhead in the form of the number of check bits than error correcting codes. This can be true even if the overhead of retransmissions is accounted for.

The cyclic redundancy check (CRC) code is a powerful error detecting code that can detect single bit errors, double bit errors, odd numbers of errors, and many types of burst errors. It is used in Ethernet.

The CRC Algorithm

Cyclic redundancy codes are based on polynomial arithmetic. This sounds complicated but really is not. For instance, the binary number 1100110 can be represented as

$$1x^6 + 1x^5 + 0x^4 + 0x^3 + 1x^2 + 1x^1 + 0x^0 \qquad (4.9)$$

Here each bit is a coefficient of a power term in x.

In a CRC code both the transmitter and the receiver agree to use some "generator polynomial," $G(x)$, as the basis of the error detection.

The division of polynomials is used in CRC codes. In short, the transmitter selects check bits to append to the message (data bits) such that the resulting codeword polynomial is exactly divisible by $G(x)$ (i.e., there is a remainder of zero). The receiver divides the polynomial of its received codeword by the same $G(x)$. If there is a zero remainder, the receiver assumes that there is no error. If the receiver's division does produce a remainder, the receiver assumes that there must be an error.

Even though the CRC code is based on binary numbers, a base 10 example can make this idea more concrete. Suppose

Fig. 4.16 Exclusive or (XOR) truth table and four bit exclusive or arithmetic example

A	B	A$\pm$B
0	0	0
0	1	1
1	0	1
1	1	0

$$\begin{array}{r} 1111 \\ -1011 \\ \hline 0100 \end{array}$$

Fig. 4.17 Cyclic redundancy check code transmitter example

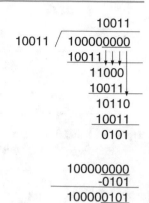

that the message is 27 and the generator number is 25. The transmitter appends a check digit to 27 to create 270. It then adds 5 to to the number so that the result, 275, is exactly divisible by the generator number. Then 275 is transmitted. The receiver divides the received number by 25. If there is a remainder, it is assumed that there is an error. If there is no remainder, it is assumed that correct reception has been achieved. In this case the message 27 is recovered from the codeword.

Recall that in doing division one needs to periodically subtract numbers. For CRC codes this is done in a special way using exclusive or arithmetic with no carries. In Fig. 4.16 the exclusive truth table is illustrated. One can see that $0 - 0$ is 0, $0 - 1$ is 1, $1 - 0$ is 1, and $1 - 1$ is 0. In the same figure next to the truth table is a four bit example. The numbers in each column are subtracted independently of what happens in other columns (there are no carries). Thus $1111 - 1011 = 0100$.

Now for the detailed procedure. Let $M(x)$ be the message (data) polynomial, $G(x)$ be the generator polynomial, and $T(x)$ be the polynomial of the transmitted codeword. We have

(a) Append r zero bits to $M(x)$ to create $x^r M(x)$. Here r is the number of bits associated with the generator polynomial minus one. The number of check bits will also be r.
(b) Divide $G(x)$ into $x^r M(x)$.
(c) Subtract the remainder from $x^r M(x)$ to create $T(x)$.

The last step is necessary so that $T(x)$ is exactly divisible by $G(x)$. An example of a CRC code transmitter division appears in Fig. 4.17. The generator is 10011 and the message is 10000. Four check bits are added (the number of generator bits minus one). The result of the division with its exclusive or subtractions is a remainder of 0101. This remainder is subtracted from $x^r M(x)$ (see bottom of Fig. 4.17). But note that the result of subtracting a remainder from the (zero) check bits is just the remainder. Thinking about this and the exclusive or truth table one can see that this is always true. Thus the receiver does not actually do the subtraction, it simply substitutes the remainder into the check bit positions.

What of the receiver? It simply divides $G(x)$ into what it receives, $R(x)$. It does not add check bits, these are already incorporated into $R(x)$. If there is a remainder, the receiver assumes that the codeword has an error and requests a retransmission. If there is a zero remainder, the receiver assumes that there is no error.

In practice CRC code division is most efficiently implemented using shift register hardware [180, 227].

CRC Code Protection

The CRC code protects against single bit errors, double bit errors, odd numbers of errors, and many types of burst errors. In this subsection it is demonstrated why this is true.

The polynomial associated with the transmitted codeword is $T(x)$. We will define an error polynomial $E(x)$ that has a 1 in every bit position where there is an error. For instance, with $E(x) = 100010$ there are errors in the first and fifth positions. However, $E(x)$ is an analytical tool. The receiver does not know which bit position is in error. What is received is $R(x) = T(x) + E(x)$ where exclusive or addition is used. From the exclusive or truth table (Fig. 4.16) one can see that if a bit in $E(x)$ is a zero (no error), adding it to the corresponding $T(x)$ bit simply leaves the $T(x)$ bit unchanged. On the other hand, if a bit in $E(x)$ is a 1 (error), the corresponding bit in $T(x)$ is inverted (1 to 0 or 0 to 1, an error).

Since $T(x)$, the codeword the transmitter transmits, is exactly divisible by the generator polynomial, $G(x)$, one has at the receiver when $R(x)$ is divided by $G(x)$

$$\frac{R(x)}{G(x)} = \frac{T(x) + E(x)}{G(x)} \rightarrow \frac{E(x)}{G(x)} \qquad (4.10)$$

We now examine each type of error that the CRC code can detect. This basically comes down to the question of selecting $G(x)$'s that give the most protection against errors. This involves a consideration of whether $E(x)/G(x)$ has a remainder. See [227] for an alternate treatment.

Single Bit Errors For a single bit error there is a single 1 in some position (say the ith position). Now $E(x)$ is a power of two ($E(x) = x^i$). If $G(x)$ has two or more non-zero terms, it will not divide $E(x)$ evenly so a remainder will be produced and the error will be detected. In base ten terms, if $E(x)$ is 32 (a power of two) and $G(x)$ is 7 $(1 + 2 + 4$, three terms), then 32/7 produces a remainder. Thus any good $G(x)$ should include two or more non-zero terms so that the receiver can detect single bit errors.

Double Bit Errors In this case $E(x)$ has two non-zero terms

$$E(x) = x^i + x^j \tag{4.11}$$

$$E(x) = x^j(x^{i-j} + 1) \tag{4.12}$$

$$E(x) = x^j(x^k + 1) \tag{4.13}$$

Now consider $E(x)/G(x)$. One should select $G(x)$ so (a) it does not divide x^j evenly (has two or more terms, see the previous subsection) and (b) $G(x)$ does not divide $x^k + 1$ evenly up to some maximum k. Since $k = i - j$ this is the maximum distance between two bit errors (i.e., maximum packet length) that can be tolerated. Small polynomials that can protect packets thousands of bits long are known by mathematicians.

Odd Number of Errors If there is an odd number of errors, $E(x)$ will have an odd number of non-zero terms. We will use the observation that there does not exist a polynomial with an odd number of terms which is divisible by $(x + 1)$. So to detect an odd number of errors, all one has to do is be sure $(x + 1)$ is factor of the $G(x)$ used.

The observation, which is not obvious, can be proven by contradiction [226]. Assume that $E(x)$ has an odd number of terms and *is* divisible by $(x + 1)$. Then certainly $(x + 1)$ can be factored out

$$E(x) = (x + 1)F(x) \tag{4.14}$$

Now let $x = 1$

$$E(1) = (1 + 1)F(1) \tag{4.15}$$

But with exclusive or arithmetic $1 + 1 = 0$ so

$$E(1) = (1 + 1)F(1) = 0 \times F(1) = 0 \tag{4.16}$$

However, a little practice with simple examples will show that if $E(x)$ has an odd number of terms, $E(1)$ always equals 1 (for instance, $E(1) = 1 + 1 + 1 = 1$). Again, we are using exclusive or arithmetic. Thus we have a contradiction in our original assumption and the observation that a polynomial

with an odd number of terms is not divisible by $(x + 1)$ is correct.

So any good CRC generator polynomial should have $(x + 1)$ as a factor to catch an odd number of errors.

Burst Errors A burst error, as has been said earlier, is an error consisting of a number of consecutive errored bits. In this case the codeword associated with $E(x)$ might look something like this:

$$0000000111100000000 \tag{4.17}$$

Or

$$E(x) = x^i(x^{k-i} + x^{k-i-1} + \ldots + x^1 + 1) \tag{4.18}$$

Here x^i is a shift of the burst. In selecting $G(x)$ one can utilize the fact that if $G(x)$ contains a term of $x^0 = 1$, then x^i is not a factor of $G(x)$ so that if the degree (i.e., highest power) of the expression in parenthesis is less than the degree of $G(x)$, there will be a remainder.

Burst errors of length less than or equal to r will be detected if there are r check bits. Thus all one needs to do to accomplish this is include a term of 1 in $G(x)$.

There is also some protection against larger bursts. For instance, if the burst length is greater than $r + 1$ or if there are multiple, smaller bursts, then the probability that the receiver does not catch an error is $(1/2)^r$. The same probability is $(1/2)^{r-1}$ if the burst length is $r + 1$. See Tanenbaum [226] for more details.

Certain generator polynomials are international standards. They all have more than two terms, they all have $(x + 1)$ as factor and they all have a term of +1. Among them are

$$x^{12} + x^{11} + x^3 + x^2 + x^1 + 1 \tag{4.19}$$

$$x^{16} + x^{12} + x^5 + 1 \tag{4.20}$$

4.5 Line Codes for Networking

A line code is a code that is used in a digital communication system for improving some aspect(s) of the system's performance. A number of codes have been developed over the years to solve a particular problem with digital receivers. Digital receivers need frequent transitions between 0's and 1's in data to properly receive digital signals (i.e., for clock recovery). They have clocking problems with very long runs of 0's and 1's in the data. But actual data may have such long runs. The following are a number of solutions, in chronological order, to this problem.

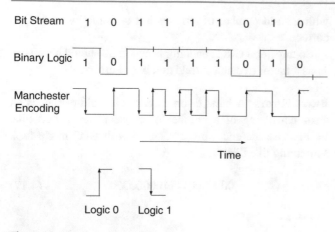

Fig. 4.18 Manchester encoding

Table 4.7 4B5B conversion table

Number	Binary	4B5B
0	0000	11110
1	0001	01001
2	0010	10100
3	0011	10101
4	0100	01010
5	0101	01011
6	0110	01110
7	0111	01111
8	1000	10010
9	1001	10011
10	1010	10110
11	1011	10111
12	1100	11010
13	1101	11011
14	1110	11100
15	1111	11101

4.5.1   Manchester Encoding

An early line code for networking was Manchester encoding. It was originally developed at the University of Manchester for storage on the old style magnetic drum of a computer (Wikipedia). It was used in the original 10 Mbps Ethernet in the early 1980s.

Referring to Fig. 4.18 [196], under Manchester encoding if a logic 0 needs to be transmitted, a transition is made from 0 to 1 (low to high voltage) in the middle of the bit waveform. If a logical 1 needs to be sent, the opposite transition is made from 1 to 0 (high to low voltage) in the middle of the bit waveform. The voltage level makes a return to its original level as necessary. For example, in the figure, from left to right, the first bit to be transmitted is a 1 so there is a downward transition. This is followed by a 0 so there is next an upward transition. This is followed by a run of four 1's resulting in four downward transitions.

Note that the signaling rate is variable. That is, the number of transitions per second is twice the data rate for long runs of a logic level and is equal to the data rate if the logic levels alternate. For this reason, Manchester encoding is said to have an efficiency of 50% or an overhead of 100%. More modern line codes achieve higher efficiencies (see below). Also Manchester encoding has no DC bias and so can be capacitively or inductively coupled for isolation purposes (Wikipedia).

Manchester encoding is also known as phase encoding. It has been widely used in consumer infrared protocols and also in RFID technology (Wikipedia). It is not suitable for use at more modern higher data rates because of the high number of transitions per second.

4.5.2   mBnB Encoding

The basic idea in mBnB codes is to add n-m extra bits to each m data bits so that one is assured of "frequent"

binary transitions. That is, only 2^m of the 2^n codewords (ones with frequent transitions) are used to transmit the original 2^m messages. There is more than one way of doing this. See Wikipedia for more detailed treatment of the following encodings.

4B5B Encoding

In 4B5B encoding a table is used to map every 4 data bits to be transmitted into 5 bits (Table 4.7). It is done in such a manner that there will be at least two transitions in every block of 5 bits. The corresponding table appears nearby (Wikipedia).

While the entry for 0 looks like it only has one transition in a long run of zeroes, there would still be two transitions per block because the first bit is 1 and the last bit is 0.

The 4B5B code became well known because of its use in FDDI (Fiber Distributed Data Interface) backbone token ring network technology in the 1980s. The name 4B5B is associated with the FDDI version.

Note that 4B5B has a 25% overhead. It is 80% efficient. Efficiency is even higher in more modern codes such as 64b66b. The 4B5B encoding was also used in 100 Mbps Fast Ethernet which is known as IEEE standard 802.3u which was standardized around 1995.

Other 4–5 bit encodings are possible and some of them have been used for magnetic recording.

8b10b Encoding

An 8b10b code is a line code that produces 10 bit blocks for 8 bit words. The code was invented and patented by IBM ([91, 239], Wikipedia). It is used to produce DC balance, bounded disparity, and sufficient transitions for digital receivers to do clock recovery. The use of 8b10b encoding became especially popular after IBM's patent expired. It has been used in Gigabit Ethernet (with the exception of twisted pair 1000BaseT), Fiber Channel, InfiniBand, and a number of other networks as well as in audio recorder and storage applications.

There is more than one way to implement 8b/10b encoding (Wikipedia). In this section the IBM implementation is largely discussed.

Since 8 data bits are mapped into 10 bits, an 8b/10b code can be said to be 80% efficient compared to Manchester encoding's 50% efficiency [196]. A code that encodes 8 input bits is very useful since 8 bits (1 byte) is a fundamental unit of computing. Would not an 8b/9b code be better? While in theory it would be more efficient, the inventors of 8b/10b found the implementation and performance of an 8b/9b encoding to be problematic [239].

The encoding scheme of 4B5B uses a single table to do the encoding. What of 8b/10b? It uses a 5b/6b code (and table) to encode the first 5 input bits and a 3b/4b code (and table) to encode the last three bits of the input ([106], Wikipedia). These are concatenated to produce a 10 bit output.

There are data symbols and special control signals. A data symbol is denoted as:

$$D.x.y$$

Here D indicates a data symbol, x has a range of 0–31 (there are 2^5 possible values), and y has a range of 0–7 (there are 2^3 possible y values). There are up to 12 control characters that can be transmitted instead of data symbols. These include characters for the start of a frame (packet) and the end of a frame.

The "bounded disparity" aspect for an (IBM) 8b/10b code provides a guarantee that the difference between the number of ones and zeros in a stream of at least 20 bits is no more than two. Also, for clocking there are not more than five ones or zeroes consecutively. Because of all this there is less need for the lower bandwidth component of the channel. That is, long runs of the same bit would lead to a frequency spectra with a significant low frequency component. A single error in the encoded stream can produce an error burst of at most 5 in the decoded stream ([239], Wikipedia).

Also a special feature of 8b/10b encoding is that it is DC free. Over long intervals the number of 1's and 0's transmitted are equal. There is no DC component. As mentioned, the difference between the number of 0's and 1's transmitted is no more than ± 2. Moreover at the end of each symbol the difference is either -1 or $+1$. This is the running disparity.

There are several advantages to having a code with DC balance (Knowledge Transfer):

1. The laser duty cycle (time on and off) in optical networks is 50% yielding good power dissipation and performance.

2. DC balance provides a restriction on the degree of offset amassing at receivers which helps in determining level detect thresholds.

3. With DC balance, there is no need for DC restoration circuitry. These are problematic to design at higher frequencies.

Actually, one can do much better than having a 25% overhead as 8b/10b does, which the next encoding, 64b/66b will demonstrate.

64b/66b Encoding

Can the line code 8b/10b, used in Gigabit Ethernet, be used in 10 Gbps Ethernet to provide clock recovery and DC balance? The problem is that in optical networks with the 25% overhead of 8b/10b one would require 12.5 Gbps lasers. At the time that 10 Gbps Ethernet was created this was a complex challenge. Instead, a 64b/66b line code was developed which mapped 64 input bits (a basic unit of computer systems) into 66 output bits. The extra two bits means that the overhead is only 3.125% and 10 Gbps lasers already developed for SONET OC-192 [198] could be used for 10 Gbps Ethernet [232, 233] (Wikipedia).

The 64b/66b code for 10 Gbps Ethernet is significant in addressing performance successfully in a statistical, rather than deterministic, sense as in 8b/10b. The 66 bits of the 64b/66b code is created by inserting two bits at the beginning of the 64 bit payload as follows (Wikipedia):

1. For 64 bits of data the prefix is 01_2.
2. For an 8 bit type field followed by 56 bits of control messages and/or data, the prefix is 10_2.
3. The prefixes 00_2 and 11_2 are not used and will produce an error when present.

These prefixes ensure a bit transition every 66 bits.

The entire 64 bit payload is "scrambled" by a scrambler function. This is done to produce a roughly even distribution of 0's and 1's in a statistical sense. This is quite different from the table lookup technique of 8b/10b. Encoding and scrambling are performed in hardware through a linear feedback register.

While it is theoretically possible for a random pattern to cause a run of 65 ones or 65 zeroes, for a 10 Gbps data rate, this occurs once about every 1900 years. Certain safeguards are in place in SONET and 64b/66b coding to prevent non-random patterns from de-synchronizing clocking (Wikipedia). DC balance for 64b/66b coding is statistical and not absolutely bounded. The scrambler's output approximates random binary bits (Wikipedia).

The Mean Time to False Packet Acceptance (MTTFPA) over 10 Gbps Ethernet is specified on the order of one billion years on a single link. At least four bit Hamming distance protection for all packet data can meet this (Wikipedia).

The line code 64b/66b has been used in 10 Gbps Ethernet, 10 Gbps Ethernet: Passive Optical Network (10 G-EPON), 100 Gbps Ethernet, some Fiber Channel versions, Infini-Band, and in other places. Other codes have been developed including 64b/67b, 128b/130b, and 128b/132b to address certain design issues (Wikipedia).

4.6   Network Coding

Network coding is a clever concept by which functions of information are transmitted on certain network links, rather than the information itself, in an effort to boost network capacity and minimize necessary network infrastructure. The concept was originated in a paper by Ahlswede et al. [7] in 2000. Significant research effort has been invested in this concept since then.

4.6.1   Bits Are Different

Let us consider a "butterfly network" example (Wikipedia). Node A in Fig. 4.19 originates data stream 1 and node B originates data stream 2. We want to deliver both streams to both nodes C and D (that is, do multicasting) over the indicated network topology.

We will look in more detail at a single bit moving from node A (b_1) and from node B (b_2) to both nodes C and D. Let us say each link has a capacity of 1 bit/s. Using direct links, node A sends bit b_1 to node C and node B sends bit b_2 to D.

If we use normal switching at node E, one has the network equivalent of a network traffic jam. One would like to send bits b_1 and b_2 over the link EF. Link EF, like the other links,

has a capacity of 1 bit/s so this EF link would be the network bottleneck. The maximum data rate for each stream would be 1/2 bit a second since b_1 and b_2 share link EF.

Suppose now that network coding is used in the example. The exclusive OR function (see Fig. 4.20) of b_1 and b_2 is sent over link EF. The value of this function is written as $b_1 + b_2$. The exclusive OR function of b_1 and b_2 is neither b_1 or b_2 but contains partial information on both bits. Node C, for instance, can recover bit b_2 from bit b_1 (which it receives directly) and the exclusive OR of b_1 and b_2 (received from node F). That is, an examination of the exclusive OR table of Fig. 4.20 will reveal that if you know any two entries in a row, the third can be uniquely determined.

Now, links AC, EF, and BD send 1 bit/s so that the network coded implementation has twice the capacity of the switched implementation previously discussed. Another switched alternative to the network of Fig. 4.19 would be to modify the topology so that there are two additional 1 bit/s links leading into nodes C and D as in Fig. 4.21. Certainly we can convey 1 bit/s of each data stream now to each node C and D but at the additional cost of what could be four long distance links (AC, EC, ED, and BD) rather than the three long distance links (AC, EF, and BD) that the network coding implementation requires.

In fact, other functions besides the exclusive OR function can be used in network coding. But Li et al. [153] have shown in 2003 that additions and multiplications are all that need to be used for networks with multicasting. Research issues for

Fig. 4.20 Exclusive OR (XOR) function truth table

A	B	A+B
0	0	0
0	1	1
1	0	1
1	1	0

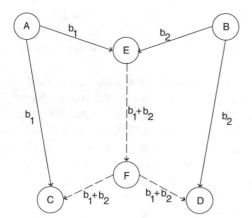

Fig. 4.19 A sample network for network coding

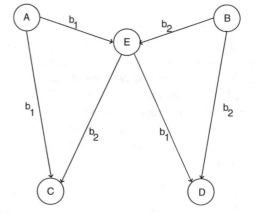

Fig. 4.21 A different sample network

network coding include removing implementation obstacles, exploring security and data integrity issues, and its use in mobile (e.g., ad hoc) networks.

As Effros, Koetter, and Médard point out in their introduction to network coding [158], networks to date rely on a transportation like paradigm. Under this (switching) paradigm bits are unchangeable objects, like cars on a road, that must transit links (roads) to reach a destination. But bits are information. So it may not be too surprising that they can be modified in transit in such a way as to improve network performance.

4.6.2 Extensions and Challenges

What has been described is an example of what is called linear coding. Linear coding reaches the performance upper bound in multicast problems with one source. However, for multi-source, multi-sink, and arbitrary demand situations there is no simple solution. An open problem is general network situations with arbitrary demands (Wikipedia).

A drawback in using linear coding over a large finite field is that it has high encoding/decoding computational complexity. Triangular network coding [186] (Wikipedia) is a solution to the problem without suffering performance loss. Pre-processing in triangular network coding is based on Gaussian elimination. Back substitution can be used to recover the original data at the receiver, because of this basis in Gaussian elimination, in $O(N^2)$ computational complexity instead of the usual $O(N^3)$ complexity. Triangular network coding has a coding rate similar to the linear network coding rate.

There are challenges in applying network coding to a wireless environment including the presence of bit errors due to fading, collisions, half duplex links, and the use of broadcasts [71, 158]. There has been some work on the use of network coded cooperation (i.e., processing multiple received signals) to make use of the spatial diversity of direct links to deal with fading channel impairments in order to increase data rates and improve the robustness of wireless networks with multiple sources [21].

Finally, it should be mentioned that network coding is related to network security issues such as homomorphic signatures and secret sharing. Network coding is susceptible to pollution attacks (i.e., transmitting garbage). Homomorphic encryption signature algorithms can be used with network coding to prevent pollution attacks. Secret sharing enables a group of participants to each hold a "share" of a secret without holding the entire secret. The secret can be reconstructed by combining some number of shares (Wikipedia).

4.7 Quantum Cryptography

4.7.1 Introduction

Quantum cryptography uses principles from quantum mechanics to assure security. Traditional cryptographic systems assure security by making a brute force attack (i.e., trying all possible keys) take so long as to be impractical. This is usually done by relating the encryption to well-known mathematical problems that are known, as long as people have worked on them, not to have a simple solution. Quantum cryptography uses physical quantum principles to prevent easy unauthorized decryption.

It should be pointed out that although current traditional cryptographic algorithms are not susceptible to cracking by today's computers, the situation may be different decades from now. Computers become faster year by year and over decades they will be significantly faster. This is a problem because information encrypted today often needs to be secure for decades. Moreover, if fully capable quantum computers (computers based on quantum computation) are developed over the next several decades, they will easily crack certain (but not all) of today's ciphers. This leads to post-quantum encryption. Post-quantum cryptography involves encryption algorithms being developed today to resist unauthorized decryption in a future world of powerful quantum computers.

4.7.2 Quantum Physics

Quantum mechanics (or quantum physics) involves the behavior of small discrete amounts of energy or matter (e.g., photons and atoms). It was the subject of much research in the last century and this century. Work started with research by Max Planck on black body radiation in 1900 and by Albert Einstein on the photo-electric effect in 1905.

In standard digital logic systems (the building blocks of today's computers) information is represented by the two logic values of "0" and "1." In quantum information processing information is represented by "qubits." Unlike the binary nature of digital logic bits, qubits exist in a simultaneous combination of states. This is often discussed in terms of a two-dimensional complex vector space.

When one attempts to measure a qubit it collapses into one of two states depending on the measurement apparatus.

This all has some implications. Quantum computers are powerful in principle because they can sort through many states (solutions) concurrently. On the other hand, the measurement of a quantum state "disturbs" the state, so perfect copying is not possible (e.g., if using photons and optical fiber, straightforward repeaters cannot be used) [159].

4.7.3 Quantum Communication

Quantum principles can be used for quantum communication in the following ways [120]:

1. Quantum key distribution (QKD) to provide secure distribution of keys to be used in conventional cryptographic systems. This will be discussed in the next subsection.
2. Quantum teleportation for moving information in quantum computers. Quantum teleportation is a means of sharing quantum information (such as the state of a photon or atom) between two locations. This requires a classical information link so information transfer does not occur at a faster than light speed. In spite of the name, quantum teleportation refers to a transfer of information (not matter as in the Star Trek television series).
3. Entanglement swapping for setting up long distance quantum networks. "Entanglement" is a condition where the properties (position, momentum, spin, and polarization) of particles at a large distance are correlated. Measurements made on one particle of an entangled pair of particles are known by the other particle of the entangled pair (Wikipedia). Quantum repeaters cannot be built because the state of quantum particles cannot be copied. But particles on either end of a network can be entangled (using a series of entangled network segments between them) to achieve a repeater effect and thus lay the foundation of a quantum network.
4. Quantum "seals" for maintaining the integrity of physical boundaries. Using quantum principles, quantum seals allow the testing of the soundness, authenticity, and physical layer security of a communication channel.

In the next section, quantum key distribution, the most developed of these techniques, will be discussed.

4.7.4 Quantum Key Distribution (QKD)

Quantum key distribution (QKD) is a quantum cryptographic technique used to distribute keys for conventional cryptographic systems. But why are keys important? The whole basis of the security of classical encryption systems is in the key. For the older symmetric key model the same key is used for both encryption and decryption. In the more recent public key cryptographic systems there are two keys: often a public one for encryption and a private one for decryption.

It is often assumed that the method of encryption/decryption is known to attackers and ciphertext (the encrypted text) is also known to attackers. The security of the cryptographic system is in the key(s). But how does one distribute the keys to users in a secure manner? This is the well-known key distribution problem. There are five basic approaches to key distribution [9].

- Classical information theoretically secure key agreement methods.
- Classical computationally secure public key cryptography.
- Classical computationally secure symmetric key cryptography.
- Quantum key distribution (QKD).
- Trusted couriers.

With the large volume of data that compact storage devices can hold the use of trusted couriers is not beyond the realm of possibility. However, in this section the emphasis is on QKD. Actual QKD systems have been in use for a number of years.

The idea for QKD can be traced back to work by Stephen Wiesner about 1969 on unforgeable bank notes. The first cryptographic application of this, QKD, was published by C.H. Bennett and G. Brassard in 1984 and the work is known as BB84 [29].

The original BB84 paper used photon polarization states to transmit information but any two pairs of conjugate states can be used. The sender (called Alice in the cryptographic literature) transmits quantum states using photons to Bob (the receiver) over a quantum communication channel. Let us think of this channel as implemented on a fiber but it could be implemented on a free space link.

As an example, one can use a rectilinear basis (i.e., photons polarized or made directional as vertical (0°) or horizontal (90°)) and also use a diagonal basis of 45° and 135°. In the tables in our example the use of a rectilinear basis is indicated by + and the use of a diagonal basis is indicated by ×.

In our example Alice and Bob use a series of rectilinear and diagonal filters. If a vertically/horizontally polarized signal is applied to a rectilinear filter, the polarization (i.e., angle) is preserved when the photon leaves the filter. The same is true when a diagonally polarized photon is applied to a diagonal filter: the angle of the diagonally polarized photon is preserved by the filter. On the other hand, if a diagonally polarized photon is applied to a rectilinear filter, then there is a 50% chance a vertically polarized photon is produced by the filter and a 50% chance a horizontally polarized photon is produced by the filter. Likewise, if a rectilinearly polarized photon is applied to a diagonal filter, then there is a 50% chance the filter produces a 45° photon and a 50% chance of producing a 135° photon.

For the example to be presented, Table 4.8 shows the two different bases and the encoding.

So, for instance, under a rectilinear basis, a "0" is represented by a vertically polarized photon. Likewise, under a

Table 4.8 Bases

Basis	0	1
+	↑	→
×	↗	↖

Table 4.9 A QKD example

Alice's bits	1	0	1	0	0	0	1	1
Alice's random basis	×	×	+	×	×	+	+	×
Alice's transmission	↖	↗	→	↗	↗	↑	→	↖
Bob's basis	×	+	+	×	×	+	×	+
Bob's measurements	↖	↑	→	↗	↗	↑	↖	↑
Basis exchange								
Shared secret key	1		1	0	0			

diagonal basis a "0" is represented by a photon polarized at 45° and a "1" by a photon polarized at 135°.

In our example (Table 4.9) Alice and Bob want to establish a shared secret key. Alice generates a random series of bits. She also randomly picks a basis (rectilinear or diagonal) to use for each bit. Now using Tables 4.8 and 4.9, in the first column of Table 4.9, to transmit a "1" with diagonal polarization, Alice transmits a diagonally polarized signal at 135°.

Bob receives the photons with his own random series of filters. He does not yet know Alice's choice of filters so some of his guesses will match Alice's and some will not.

In the first column of the accompanying table, there is a match (both choose diagonal filters) so Bob measures correctly a photon which is diagonally polarized at 135° (a "1"). In the second column Bob guesses the wrong basis so he measures (with 50% probability) a vertically polarized photon. The process continues until the last of Alice's bits.

At this point Alice and Bob exchange the bases they used over a non-secure channel. They only keep the corresponding bits where they agreed on the choice of filter. This is their shared secret key.

In the cryptographic literature "Eve" is a person trying to illegitimately acquire the message. Even if she had access to the transmission of polarized photons between Alice and Bob, she cannot read them because measuring them will change them and Bob and Alice could detect this. She does not have access to Bob's measurements so in theory she does not know the secret key.

There is also a way for Alice and Bob to check for the presence of an eavesdropper. They compare a specific subset of the key string. An eavesdropper in gaining information on the photons' polarizations will cause errors in the measurements made by Bob. However, such errors can also be produced by noise in the transmission line and by detector imperfections.

If the percentage of errors is below a certain threshold, two procedures can be used to remove the errored bits and make Eve's knowledge of the key to be an arbitrarily small value. These procedures are reconciliation and privacy amplification ([175], Wikipedia):

- Reconciliation: This procedure removes errors due to bad choices of measurement basis, errors due to channel noise, and errors due to eavesdropping. This procedure is a recursive search for errors. Alice's and Bob's key sequences are divided into blocks and the parity of blocks is compared. When parities do not match blocks are divided into smaller blocks in a binary search approach to find errors. A cascade approach (Wikipedia) is used to find multiple errors in the same block.

- Privacy Amplification: Reconciliation is done over a classical (insecure) channel. Thus an eavesdropper may gain certain information on the key by monitoring this channel as well as the quantum channel (though this later monitoring will introduce errors). Under privacy amplification Alice's and Bob's keys are used to produce a shorter key in a manner so that Eve has arbitrarily small knowledge about the key. This is done using a universal hash function (Wikipedia) which essentially maps the original keys into a shorter key with a known probability of the amount of Eve's knowledge.

Implementation

Quantum key distribution is the most developed application of quantum cryptography to date. A number of companies offer QKD systems. A number of systems have been fielded such as the DARPA quantum network which is a ten node network in Massachusetts. There are also systems in Switzerland, Tokyo, and Los Alamos. Systems have been fielded with key bit rates on the order of 10^4 to 10^6 bps (higher bit rates for shorter distances).

Absolute Security?

It should be noted that the security of a QKD system can be subverted due to threats such as hacking QKD computers and threats peculiar to quantum systems. Scarni and Kurtsiefer [206] feel that these threats can be protected against but claims of absolute security have to be put into this context.

4.8 Conclusion

In this chapter we have largely looked at fundamental network algorithms. Keep in mind though that they are usually used in very stochastic (random) environments. That is, traffic flow and routing patterns, which states a protocol executes and channel errors can all be very well modeled as being random like.

4.9 Problems

1. How many shortest paths are there in a rectangular mesh network of size 5×10 nodes between opposite diagonal corner nodes? Show the calculation.

2. Does a shortest path algorithm like the Dijkstra or Ford Fulkerson find the shortest between two nodes or is it more general?

3. How can one find the shortest paths between all pairs of nodes using the shortest path algorithm?

4. How does one use the shortest path algorithm to find the k shortest link disjoint paths between two nodes?

5. What is the difference between the "algorithm" and "routing" tables in this chapter? How are these used in table driven routing?

6. Make a "routing" table for node C in Fig. 4.3.

7. How are paths found in source routing?

8. What is flooding? How can the number of packets sent in flooding be reduced?

9. Why are paths between nodes in hierarchical networks sometimes longer than direct connections?

10. Why is the concept of switching elements useful for VLSI design?

11. What is a disadvantage, in terms of implementation, of some of the multicasting techniques mentioned in the chapter (specifically putting multiple addresses into packets and the use of spanning trees)?

12. Explain why it is more energy efficient for a packet to make several smaller hops rather than one large hop in an ad hoc network.

13. Under what condition(s) are reactive routing algorithms more efficient than proactive routing algorithms?

14. Why is protocol verification a challenging problem for large networks?

15. What is the difference between a deadlock and a livelock?

16. What is an unspecified reception?

17. What is the Hamming distance between 00001111 and 11001100?

18. In Fig. 4.12 can a legitimate codeword be mapped into another legitimate codeword through some error? What is the result?

19. Can a parity code detect an odd number of errors?

20. What happens in the base 10 example of "The CRC Algorithm" subsection of Sect. 4.4.4 if the codeword 250 is received?

21. Run the Dijkstra, Ford Fulkerson, and bottleneck bandwidth algorithm to create the algorithm tables for the network of Fig. 4.22. Let the link numbers be shortest path distances for the Dijkstra and Ford Fulkerson algorithms and let the link numbers be bandwidths (i.e., data speeds) for the bottleneck bandwidth algorithms. Here node A is

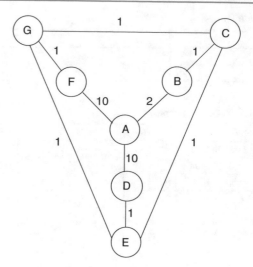

Fig. 4.22 A routing problem

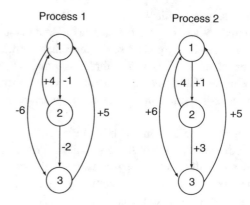

Fig. 4.23 A reachability tree problem

the root. Label the columns in each table, from left to right, as B, C, D, E, F, G.

22. Make a copy of Fig. 4.6 and indicate the paths taken by packets entering inputs 1 and 6, destined for output 011.

23. Create the reachability tree for the state machine protocol representation of Fig. 4.23. Indicate any states involved in a deadlock or unspecified reception.

24. A parity bit is appended to six bits. Even parity is used. If the received codeword is 1011000, is there an error?

25. A block of data is transmitted using a Hamming code. The data is 1011. Find the complete codeword sent by the transmitter. Use even parity.

26. Four bits of information are received using a Hamming code. The received codeword is 0010010. Assume the use of even parity. Is there an error? If so, in which bit is the error located? Check bits are in powers of two positions.

27. Consider a CRC code transmitter problem. Find the checksum to append to the message bits 1110001. Use the generator codeword 10011.

28. Consider the received CRC codeword 100010111. Is there an error? There are four check bits at the end of the codeword. The generator codeword is 10011.

29. Consider the following 4B5B encoded stream:

    01001 10100 11110

    Write the corresponding binary stream. What decimal number does it represent?

30. For network coding, in Fig. 4.19 find the sequence for b2 from the following.

    b1: 1010010110

    b1+b2: 0110010001

31. For quantum key distribution, create a table like Table 4.9 using different random choices for Alice's bits, Alice's random basis and Bob's basis.

Abstract

This chapter provides an in-depth look at the theory of scheduling computation and communication involving divisible (i.e., partitionable) loads being processed on networks of processors. This is a very useful methodology for understanding, designing, and analyzing load distribution scheduling. A literature survey begins the chapter. This is followed by a discussion of time optimal scheduling in single level tree (i.e., star) networks. Other sections cover equivalent processors, product form solutions, infinite size network performance, time-varying environments, and multi-installment scheduling. Applied scheduling problems involving monetary cost optimization and signature searching are presented. The chapter concludes by discussing mathematical programming solutions for divisible load scheduling.

5.1   Introduction

The increasing prevalence of multiple processor systems and data intensive computing creates a need for efficient scheduling of computing loads and related transmissions. An important class of such loads are divisible ones: data parallel loads that are completely partitionable among processors and links.

Over the past three decades a mathematical tool has been created to allow tractable performance analysis of systems incorporating both communication and computation issues, as in parallel and distributed processing [40, 46, 57, 78, 195, 196]. A key feature of this divisible load scheduling theory (known as "DLT") is that it uses a linear mathematical model. Thus as in other linear models such as Markovian queueing theory or electric resistive circuit theory, divisible load scheduling theory is rich in such features as easy computation, a schematic language, equivalent network element modeling, equivalent infinite network modeling, and numerous potential applications.

A "divisible" load is a computational/communication load that is for all purposes perfectly partitionable among processors and links in a network. In addition, there are no precedence relations among the data. Therefore load can be arbitrarily assigned to links and processors in a network. One can think of a divisible load as a massive data file for which we want to assign fractions of it to a number of processors tied together by an interconnection network to achieve the benefits of parallel processing. Moreover we seek a time optimal solution: the load must be distributed from a source (or sources) to processors over the network in a scheduled fashion so that all load is processed in a minimal amount of time.

Thus the theory is well suited for modeling a large class of data intensive computational problems, such as in grid and cloud computing. It also has applications for modeling and scheduling for some meta-computing applications and parallel and distributed computing problems. As a secondary benefit, it sheds light on architectural issues related to parallel and distributed computing. Moreover, the theory of divisible load scheduling is fundamentally deterministic. While stochastic features can be incorporated, the basic model has no statistical assumptions that can be the Achilles' heel of a performance evaluation model.

This section of this chapter describes reasons why divisible load scheduling theory is a tractable, flexible, and realistic modeling tool for a wide variety of potential applications. Moreover, a tutorial introduction to modeling using divisible load theory follows. It includes a discussion of single level tree modeling, equivalent processors, product form solutions, the performance of infinite size networks, multi-installment scheduling, monetary cost optimization, signature searching, time-varying modeling, and the linear programming solution of divisible load scheduling models.

Since the original work on this subject in 1988 (Agrawal [6], Cheng [66], see Anderson [12] for an antecedent), a

© Springer Nature Switzerland AG 2020
T. G. Robertazzi, L. Shi, *Networking and Computation*, https://doi.org/10.1007/978-3-030-36704-6_5

fairly large number of journal papers and books containing an expanding body of work on this subject have been published on a worldwide basis. This has been an active research area with over a hundred and forty journal papers, tutorials [46, 195], a monograph [40], and a number of recent book chapters [57, 84, 196] on the topic appearing over the years. The original motivation for some of this work was "intelligent" sensor networks doing measurements, communications, and computation. However most recently mentioned applications involve parallel and distributed computing.

A typical divisible scheduling application might consist of a credit card company that each month needs to process 10 million accounts. It could conceivably send 100,000 of the records to each of a 100 processors. Note that simply splitting the load equally among the processors does not take different computer and communication link speeds, the scheduling policy, the interconnection network, or other features into account and so is a sub-optimal policy in terms of solution time (and speedup). Divisible load scheduling theory provides the mathematical machinery to do solution time optimal processing.

Some other typical examples follow:

1. A bank, insurance company, or online service may want to process large numbers of customer records for such purposes as billing, data mining, targeted direct mail advertising or to evaluate the profitability of new policies/services. A mid size cap fund would have a need to process complex financial records of many companies in order to make the best investment decisions or evaluate new investment strategies.
2. An individual photograph typically needs to be represented by a great deal of digital information. Satellite imagery in particular can generate many thousands of such images in relatively short amount of time. It is physically/economically impossible for humans to look at all such images in their entirety and in detail. One may want to process such images for particular patterns or features for such purposes as oil/gas exploration, weather or planetary exploration. Another example of divisible image processing is searching millions of fingerprint or facial recognition records for a match.
3. Engineers and scientists working at corporations, research labs, and universities have a need to process large amounts of data for various reasons (engineering studies, looking for particular patterns, public health studies). In fact large physics collider experiments at government funded research labs generate petabytes and exabytes of data a year, all of which must be processed. Given increasing sensor and data collection capabilities, even modest engineering experiments can generate copious amounts of data.
4. With advances in sensor design and implementation and the increasing prevalence of multiple processor systems

(in everything from cars to scientific equipment) there is a need for processing and performance prediction of sensor generated loads.

Thus there are many potential situations where a tractable and accurate approach to load scheduling would be useful. Ten advantages of using divisible load scheduling theory for this purpose appear below (also see [195]).

5.1.1 Ten Reasons

Ten salient reasons to consider using divisible load theory will now be discussed.

1. A Tractable Model The key to doing optimal divisible load scheduling is what is called the optimality principle [40, 66]. That is intuitively, if one sets up a continuous variable model of scheduling and assumes that all of the processors stop computing at the same time instant, one can solve for the optimal amount of total load to assign to each processor/link using a set of linear equations or in many cases, as in queueing theory, recursive equations. The model accounts for heterogeneous computer and link speeds, interconnection topology, and scheduling policy. The model also can simply handle loads with different computation and communication intensities. Gantt chart like schematics easily display a schedule being considered.

Linear divisible load theory is a theory of proportions. As an example, consider two processors connected by a link of infinite data rate. Suppose that one processor is twice as fast as the other. Then regardless of which processor the load originates on, two thirds of the load should be assigned to the faster processor and one third of the load should be assigned to the slower processor. Generalizing this example of proportions, divisible load theory allows the amount of load as well as when the load should be placed on processors and links to be calculated. For linear divisible load theory this can be done by mathematical programming, or more relevant to this section, by linear equation solution or in many cases by algebraic recursions.

The relation between queueing theory, electric circuit theory, and divisible load theory is particularly interesting. In their basic form all three theories are linear ones. There are a number of commonalities. These include a schematic language, recursive or linear equation solution, the concept of equivalent networks, the possibility for time-varying modeling, and the possibility of results for infinite sized systems.

The tractable nature of divisible load theory is in contrast to the nature of the more traditional indivisible load problem. That is, if one has to assign atomic jobs/tasks, each of which must run on a single processor, then one has a problem in combinatorial optimization which often is NP complete [52,212]. Precedence relations provide an additional complication. It should be emphasized that divisible load theory is not applicable to all computer scheduling problems but it is applicable to an important class of such problems.

2. Interconnection Topologies Over the years divisible load modeling has been successfully applied to a wide variety of interconnections topologies. These include linear daisy chains [66, 162], trees [37, 38, 65, 67, 137], buses [23], hypercubes [48, 183] and two-, three-, and k-dimensional meshes [49,51,60,78,102] and Gaussian networks [255,256]. Figure 5.1, for instance, shows a possible diamond shaped pattern/flow of load distribution originating from a single processor in a two-dimensional mesh network. Asymptotic results for infinite sized networks have also been developed [26,98,192]. These are useful as, for sequential load distribution at least, speedup saturates as more processors are added. Therefore if one can guarantee performance close to that of an infinite sized network with a small to moderate number of processors, one has useful design information.

It should be mentioned that in finding an "optimal" schedule one is doing this in the context of a specific interconnection network and its parameters and in the context of a specific scheduling strategy. In fact, in most of this chapter it is assumed that the order of load distribution is fixed. In a single level tree with sequential distribution, for instance, this means that there is a predetermined order for children to receive load from the root. Optimization of divisible load schedules for fixed distribution orders is largely an algebraic problem. However one can also optimize the load distribution and solution reporting order using techniques such as simulated annealing, tabu search, genetic programming, or other heuristic algorithms [63,86]. In some cases analytical results for optimal ordering are possible [5,40].

3. Equivalent Networks As in other linear theories, such as Markovian queueing theory and resistive electric circuit theory, a complex network can be exactly represented by an exactly "equivalent" network element. For some network topologies such as trees, aggregation can be done recursivley, one subtree of a larger tree at a time.

The manner in which equivalent processors can be found is straightforward. One finds, using the usual divisible load scheduling techniques, an expression for a sub-network in a more complex network. For example, such a simple building block may be a pair of adjacent processors and their connecting link in a linear daisy chain network or a single level sub-tree in a multi-level tree network. One sets the computing speed of a single equivalent processor equal to this sub-network expression. One then continues this process of aggregating sub-networks of processors, including inter-

mediate equivalent processors, until a single processor is left with a computing speed equivalent to the original network. Final expressions for equivalent processor computing speed may be either closed form or iterative in nature. This concept is discussed in greater detail in Sect. 5.3.

4. Installments and Sequencing A number of applied optimization problems arise in considering divisible load scheduling. For instance, instead of a node in a tree sequentially distributing load to its children, improved performance results if load is distributed in installments (some to child 1, child 2...child M, more to child 1...) [30, 38, 213, 214]. Performance under sequential multi-installment load distribution strategies does tend to saturate as the number of installments is increased. See Sect. 5.6 for a more detailed discussion of this topic.

Some sequencing results are surprising. For instance, consider a linear daisy chain network where all processors and links have the same speed. Under one basic sequential scheduling strategy, if load originates at any interior processor, the same solution time results whether load is first distributed to the left or right parts of the network. Other results are more intuitive. For instance, distributing load over a very slow link to a relatively fast processor may degrade overall network solution time [162].

5. Scalability In early studies of divisible load scheduling it was found that if load is distributed from a root node to its children sequentially, as in a tree network, speedup saturates as more nodes are added. Overall the solution time improvement for optimal sequential load distribution over simply dividing the load equally among the processors, if link speed is of the order of processor speed, is on the order of 20–40% [142]. As mentioned above, simply increasing the number of installments also suffers from saturated performance as the number of installments is increased. However it has been found [121, 122, 124] that if a node transmits load simultaneously to all of its children in a single level tree, speedup is scalable. That is, speedup then grows linearly in the number of children. As long as a node CPU can load output buffers to all links, performance is scalable. While there was some qualitative sense that this is the case in parallel processing, divisible load theory allows a simple quantitative answer to this problem.

6. Metacomputing Accounting A devilish problem in making metacomputing (i.e., distributed computing with payment to computer owners) practical is accounting. That is, how does one take problem size and system parameters into consideration for monetary accounting? Divisible load theory can incorporate an intuitive linear model for computing and communication costs [56, 62, 63, 218, 244].

Fig. 5.1 A possible diamond
shaped load distribution flow in
2D-mesh network. From
Blazewicz and Drozdowski [49]
© Foundations of Computing
and Decision Sciences

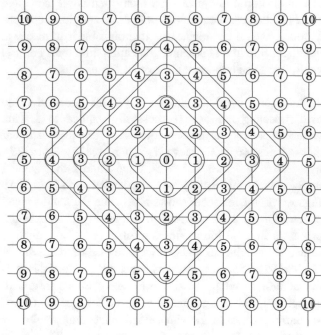

Fig. 5.2 Time-varying
background jobs and their
influence on available processor
effort. From Robertazzi [195] ©
IEEE

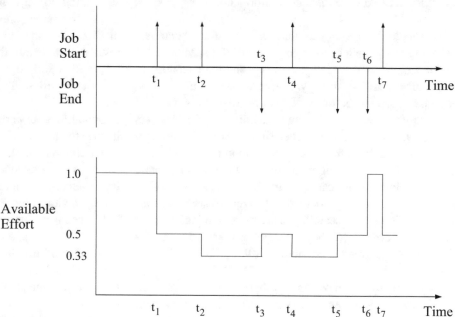

Heuristic rules of simple to moderate complexity can be developed for efficiently (in terms of both cost and performance) assigning load. This can be done in the case where only computation cost is considered or in the case of where both computation and communication costs are considered.

A problem related to metacomputing is parallel processor configuration design. That is, how does one optimally arrange links and processors with certain characteristics (speed and/or cost for instance) in a given topology? Heuristic rules that can be used for the configuration design problem are

similar in spirit to those of the metacomputing problem [62]. A more detailed discussion of this topic appears in Sect. 5.7.

7. Time-Varying Modeling In actual practice the effort that a computer can devote to a divisible job depends on the status of other background jobs. The same sharing is true of the capacity that a link can provide to transmitting part of the job because of other ongoing transmissions [34]. Figure 5.2 illustrates the time-varying effort received by a single divisible job on a single processor due to "background" jobs sporadically utilizing CPU effort. The upper part of the figure shows background jobs commencing execution (upward arrows) and terminating execution (downward arrows). The lower part of the figure shows the normalized CPU effort available for the single divisible job which runs for the duration of the diagram. Solution time optimization is possible for divisible loads if the start and end times and effort of such background jobs and messaging are known [219, 244]. Integral calculus is used to accomplish this. With less than perfect knowledge of background processes, stochastic modeling can be combined with deterministic divisible load theory. See Sect. 5.9 for a more detailed discussion of this concept.

8. Unknown System Parameters Naturally it may be difficult to obtain accurate estimates of available processor effort and link capacity, which are key inputs to divisible load scheduling models. Thus a number of "probing" strategies have been proposed [100] where some small fraction(s) of a load are sent to processors across a network of links for a sort of exploratory processing to allow the estimation of currently available processing capacity at nodes and bandwidth on links.

While an elegantly simple idea, actual implementations must take several complications into account. These include the time-varying nature of available processor effort and link capacity, the release times of processors (i.e., the times at which processors become free to accept additional load) and assuring that load is distributed on the fastest processors and links. All in all, these probing strategies are a promising approach to robust divisible load scheduling.

9. Extending Realism Efforts have been made to extend the applicability of divisible load scheduling. This includes a consideration of systems with finite buffers [47, 80, 152], finite job granularity [43], start-up costs or fixed charges [41, 44], scheduling with processor release (i.e., availability) times [39, 47], scheduling multiple divisible loads [27, 147], probabilistic computation/communications speeds [14], energy issues [83, 85], real time computing [155, 156, 160, 161] and processor priorities [97]. Moreover efforts to produce a synthesis of deterministic divisible load modeling and stochastic (queueing) modeling have been made [143, 144, 171, 219]. Specialized applications of divisible load scheduling include databases [79, 141, 142, 144], image processing, multimedia systems [17–20, 45], matrix multiplication [70, 99, 146], aligning protein sequences [182], cloud computing [129, 154, 224], Map Reduce modeling [31–33, 35], and fault tolerant scheduling [25, 246].

10. Experimental Work Experiments with actual distributed computer systems show that divisible scheduling theory can be a useful prediction tool [196].

5.1.2 Implications

The tractable and realistic nature of divisible load modeling and analysis bodes well for its widespread utility. In a sense this is due to the rich linear mathematics that underly it, as with its cousins, queueing and circuit theory. Certainly this flexible analytic structure has been a major reason that these are such large fields. A second important reason, of course, is the breadth of applications. In this context the outlook for future divisible load scheduling theory work and accomplishments is quite promising. The increasing ubiquity of sensor generated data, multiple processor systems, and data intensive computing creates a need for efficient scheduling that should drive further work on theory, applications, and software.

5.2 Some Single Level Tree (Star) Networks

Single level tree (star) networks are a good topology to use to convey an understanding of divisible load modeling and scheduling policies. In the following subsections some basic single level tree networks are modeled. For the scheduling policies described, analytical expressions for the optimal allocation of load to each processor as well as the speedup and solution time are found. This modeling and solutions are from [122].

We'll distinguish in the following between different types of distribution and the relative start of computation and communication. Under *sequential distribution* load is distributed from a root node to one child at a time. Under *simultaneous distribution* load is distributed from a root to all of its children concurrently. With *staggered start* a child node must receive all of its load before beginning to process load. With *simultaneous start* a node begins processing as soon as it begins to receive load. That is, under simultaneous start a node can receive load and process it at the same time. There are thus four scheduling scenarios with these two sets of possible features. Note that simultaneous distribution was first proposed in divisible load modeling by Piriyakumar and Murthy [183] and simultaneous start was first proposed in divisible load modeling by Kim [138].

A related type of terminology is to note that some processors have front-end sub-processors so that they may compute and communicate at the same time. If a processor does not have a front end, it can compute or communicate, but not do both at once. So one can, for example, consider a single level tree network with staggered start where the root computes as it distributes load. In this case the root must have a front end but the children do not have to.

A final piece of terminology sometimes used is to say a root is "intelligent" if it can distribute load as it computes (essentially it has and uses a front-end sub-processor).

The variables we will use in the following are

α_i: The load fraction assigned to the ith link-processor pair.
w_i: The inverse of the computing speed of the ith processor.
z_i: The inverse of the link speed of the ith link.
T_{cp}: Computing intensity constant:
  the entire load is processed in $w_i T_{cp}$ seconds by the ith processor.
T_{cm}: Communication intensity constant:
  the entire load can be transmitted in $z_i T_{cm}$ seconds over the ith link.
T_i: Finish time of the ith processor.
T_f: The finish time. Time at which the last processor ceases computation.

Note that finish time is called "makespan" in the scheduling literature. Also, the total load size is normalized to one.

Then $\alpha_i w_i T_{cp}$ is the time to process the fraction α_i of the entire load on the ith processor. Note that the units of $\alpha_i w_i T_{cp}$ are [load] × [sec/load] × [dimensionless quantity] = [seconds]. Likewise, $\alpha_i z_i T_{cm}$ is the time to transmit the fraction α_i of the entire load over the ith link. Note that the units of $\alpha_i z_i T_{cm}$ are [load] × [sec/load] × [dimensionless quantity] = [seconds]. The inclusion of T_{cp} and T_{cm} allows the relative communication and computation intensity of a job to be adjusted.

5.2.1 Sequential Load Distribution

Consider a single level tree network where load is distributed sequentially from the root to the children processors, as in Fig. 5.3. The root first transmits all of child p_1's load to it, then the root transmits child p_2's load to it, and so on. A child processor starts processing as soon as it begins to receive load. Thus it is assumed that transmission speed is fast enough relative to computation speed that no child processor "starves" for load. Thus we have a case of sequential distribution and simultaneous start. This is a completely deterministic model.

Note that if all of the link speeds are the same, then one has modeled a bus interconnection network.

The process of load distribution can be represented by Gantt chart-like timing diagrams, as illustrated in Fig. 5.4. In this Gantt chart like figure there is a graph for each processor. The horizontal axis indicates time. Communication is shown above the time axis and computation is shown below the time axis.

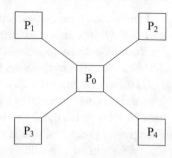

Fig. 5.3 A star (single level tree) interconnection architecture

The condition for an optimal solution is that all processors stop processing at the same time. Otherwise load could be transferred from busy to idle processors to improve the solution time [217]. One can thus write the timing equations as

$$\alpha_0 w_0 T_{cp} = \alpha_1 w_1 T_{cp} \tag{5.1}$$

$$\alpha_1 w_1 T_{cp} = \alpha_1 z_1 T_{cm} + \alpha_2 w_2 T_{cp} \tag{5.2}$$

$$\alpha_2 w_2 T_{cp} = \alpha_2 z_2 T_{cm} + \alpha_3 w_3 T_{cp} \tag{5.3}$$

$$\vdots$$

$$\alpha_{m-1} w_{m-1} T_{cp} = \alpha_{m-1} z_{m-1} T_{cm} + \alpha_m w_m T_{cp} \tag{5.4}$$

Referring to this set of equations and the figure, the first equation equates the processing time of the root to the processing time of child p_1. The second equation equates the processing time of child p_1 to the communication time from the root to child p_1 plus the computation time of child p_2 and so on.

The fundamental recursive equations of the system can be formulated as more compactly follows:

$$\alpha_0 w_0 T_{cp} = \alpha_1 w_1 T_{cp} \tag{5.5}$$

$$\alpha_{i-1} w_{i-1} T_{cp} = \alpha_{i-1} z_{i-1} T_{cm} + \alpha_i w_i T_{cp} \quad i = 2, 3, \ldots, m \tag{5.6}$$

The normalization equation for the single level tree with intelligent root is

$$\alpha_0 + \alpha_1 + \alpha_2 + \cdots + \alpha_m = 1 \tag{5.7}$$

This gives $m+1$ linear equations with $m+1$ unknowns. From (5.5)

$$\alpha_0 = \frac{w_1}{w_0} \alpha_1 = \frac{1}{k_1} \alpha_1 \qquad \text{Here, } k_1 = w_0/w_1 \tag{5.8}$$

From (5.6)

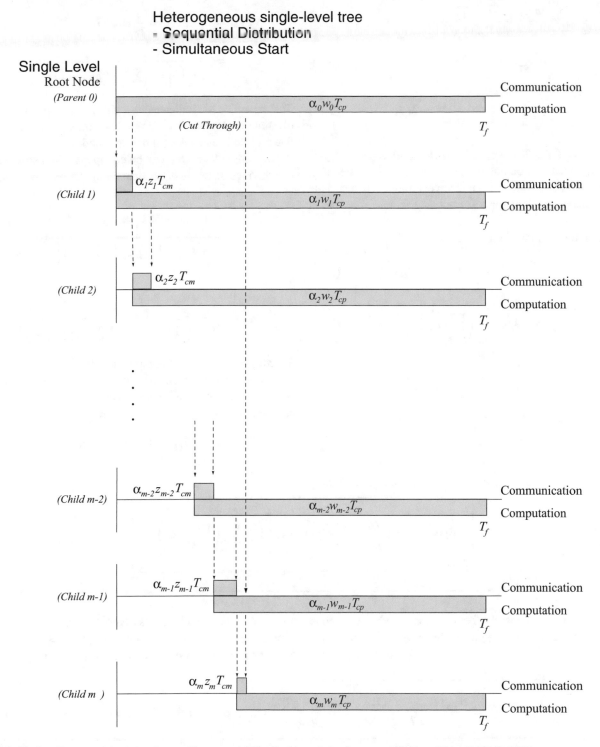

Fig. 5.4 Timing diagram of single level tree with sequential distribution and simultaneous start. From Hung [124] © IEEE

$$\alpha_i = \frac{w_{i-1}T_{cp} - z_{i-1}T_{cm}}{w_i T_{cp}}\alpha_{i-1} \qquad (5.9)$$

$$= q_i \alpha_{i-1} \qquad (5.10)$$

$$= \left(\prod_{l=2}^{i} q_l\right) \times \alpha_1 \qquad i = 2, 3, \ldots, m \qquad (5.11)$$

Here, $q_i = (w_{i-1}T_{cp} - z_{i-1}T_{cm})/w_i T_{cp}$. Also, $w_{i-1}T_{cp} > z_{i-1}T_{cm}$. That is, communication time must be faster than computation time, otherwise it is not economical to distribute load. Other conditions for best choosing w_i and z_i appear in [40].

Then, the normalization equation leads to

$$\alpha_0 + \alpha_1 + \alpha_2 + \cdots + \alpha_m = 1 \qquad (5.12)$$

$$\left[\frac{1}{k_1} + 1 + \sum_{i=2}^{m}\left(\prod_{l=2}^{i} q_l\right)\right]\alpha_1 = 1 \qquad (5.13)$$

$$\alpha_1 = \frac{1}{\frac{1}{k_1} + 1 + \sum_{i=2}^{m}\left(\prod_{l=2}^{i} q_l\right)} \qquad (5.14)$$

The finish time (makespan) is

$$T_{f,m} = \alpha_0 w_0 T_{cp} = \frac{1}{k_1}\alpha_1 w_0 T_{cp}$$

$$= \frac{1}{1 + k_1\left[1 + \sum_{i=2}^{m}\left(\prod_{l=2}^{i} q_l\right)\right]}w_0 T_{cp} \qquad (5.15)$$

The term, $T_{f,m}$, indicates the finish time for the single divisible load solved in a single level tree, which consists of one root node as well as m children nodes.

The single level tree can be collapsed into a single node, and the inverse of the equivalent computation speed w_{eq} is

$$w_{eq}T_{cp} = T_{f,m}$$

$$= \frac{1}{1 + k_1\left[1 + \sum_{i=2}^{m}\left(\prod_{l=2}^{i} q_l\right)\right]}w_0 T_{cp} \qquad (5.16)$$

Then,

$$\gamma_{eq} = \frac{w_{eq}}{w_0} = \frac{1}{1 + k_1\left[1 + \sum_{i=2}^{m}\left(\prod_{l=2}^{i} q_l\right)\right]} \qquad (5.17)$$

Speedup is the ratio of computation time on one processor to computation time on the entire tree with m children. It is a measure of parallel processing advantage. Since the computation time on a single root processor is

$$T_{f,0} = \alpha_0 w_0 T_{cp} = 1 \cdot w_0 T_{cp} \qquad \alpha_0 = 1, \qquad (5.18)$$

the speedup is

$$\boxed{\text{Speedup} = \frac{T_{f,0}}{T_{f,m}} = \frac{1}{\gamma_{eq}} = 1 + k_1\left[1 + \sum_{i=2}^{m}\left(\prod_{l=2}^{i} q_l\right)\right]} \qquad (5.19)$$

As a special case, consider the situation of a homogeneous network where all children processors have the same inverse computing speed and all links have the same inverse transmission speed (i.e., $w_i = w$ and $z_i = z$ for $i = 1, 2, \ldots, m$). Note the root w_0 can be different from w_i. Then

$$q_i = \frac{w_{i-1}T_{cp} - z_{i-1}T_{cm}}{w_i T_{cp}} = \frac{w T_{cp} - z T_{cm}}{w T_{cp}} = 1 - \sigma \qquad i = 2, 3, \ldots, m \qquad (5.20)$$

Here, $\sigma = z T_{cm}/w T_{cp}$. Consequently

$$\text{Speedup} = 1 + k_1\left[1 + \sum_{i=2}^{m}\left(\prod_{l=2}^{i} q_l\right)\right]$$

$$= 1 + \frac{w_0}{w}\left[\frac{1 - (1 - \sigma)^m}{\sigma}\right] \qquad (5.21)$$

If one plots the speedup (or solution time) of this sequential load distribution policy versus the number of children processors, one would see a speedup saturation (approach towards a constant) as the number of children increases. This makes intuitive sense as no matter how many children there are, the root can only distribute load to one child at a time. Thus adding more processors does not significantly improve performance. The calculation of the saturation level for an infinite size network appears in a later section. More scalable scheduling strategies appear in the succeeding subsections.

5.2.2 Simultaneous Distribution, Staggered Start

The structure of a single level tree network with intelligent root, $m + 1$ processors and m links is illustrated in Fig. 5.5.

If one thinks about it, the performance of the sequential scheduling in the previous section is limited as the root sends load to only one processor at a time. What if the root could send load to all of its children simultaneously? That possibility is discussed in this subsection. Simultaneous transmission of load to all children from the root is possible as long as the CPU is fast enough to continually load buffers for each of its output links.

All children processors are connected to the root processor via direct communication links. The intelligent root processor, assumed to be the only processor at which the divisible load arrives, partitions a total processing load into $m + 1$ fractions, keeps its own fraction α_0, and distributes the other fractions $\alpha_1, \alpha_2, \ldots, \alpha_m$ to the children processors respectively and concurrently.

After receiving all of its assigned fraction of load, each processor begins computing immediately (i.e., staggered start) and continues without any interruption until all of its assigned load fraction has been processed. Again, it is assumed that $z_i T_{cm} < w_i T_{cp}$ or communication speed is faster than processing speed. In order to minimize the processing finish time, all of the utilized processors in the network must finish computing at the same time [40, 217]. The process of load distribution can be represented by Gantt chart-like timing diagrams, as illustrated in Fig. 5.6. The nodes of Fig. 5.5 also contain miniature timing diagrams.

Note, again, that this is a completely deterministic model.

Since for a minimum time solution all processors must stop processing at the same time instant, one can write the fundamental timing equations, using Fig. 5.6, as

$$\alpha_0 w_0 T_{cp} = \alpha_1 z_1 T_{cm} + \alpha_1 w_1 T_{cp} \quad (5.22)$$

$$\alpha_1 z_1 T_{cm} + \alpha_1 w_1 T_{cp} = \alpha_2 z_2 T_{cm} + \alpha_2 w_2 T_{cp} \quad (5.23)$$

$$\alpha_2 z_2 T_{cm} + \alpha_2 w_2 T_{cp} = \alpha_3 z_3 T_{cm} + \alpha_3 w_3 T_{cp} \quad (5.24)$$

$$\alpha_{m-1} z_{m-1} T_{cm} \quad + \alpha_{m-1} w_{m-1} T_{cp}$$
$$= \alpha_m z_m T_{cm} + \alpha_m w_m T_{cp} \quad (5.25)$$

For example, the first equation equates the root processing time to the communication time from the root to child processor p_1 plus p_1's processing time. The second equation equates the communication time from the root to p_1 plus the computing time of p_1 to the communication time from the root to p_2 plus the computing time of p_2. The pattern can be naturally generalized.

The normalization equation for the single level tree with intelligent root is

$$\alpha_0 + \alpha_1 + \alpha_2 + \cdots + \alpha_m = 1 \quad (5.26)$$

This yields $m + 1$ linear equations with $m + 1$ unknowns. Now, one can manipulate the recursive equations to yield the solution.

$$\alpha_0 = \frac{z_1 T_{cm} + w_1 T_{cp}}{w_0 T_{cp}} \alpha_1 = \frac{1}{k_1} \alpha_1 \quad (5.27)$$

Here k_1 is defined as $w_0 T_{cp}/(w_1 T_{cp} + z_1 T_{cm})$. Also

$$\alpha_i = \frac{w_{i-1} T_{cp} + z_{i-1} T_{cm}}{w_i T_{cp} + z_i T_{cm}} \alpha_{i-1} = q_i \alpha_{i-1} \quad i = 2, 3, \ldots, m \quad (5.28)$$

Fig. 5.5 Structure of a single level tree with simultaneous distribution and staggered start

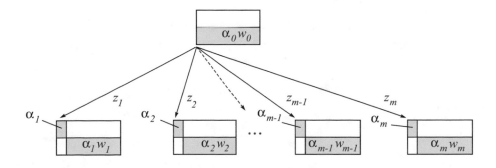

Fig. 5.6 Timing diagram of single level tree with simultaneous distribution and staggered start

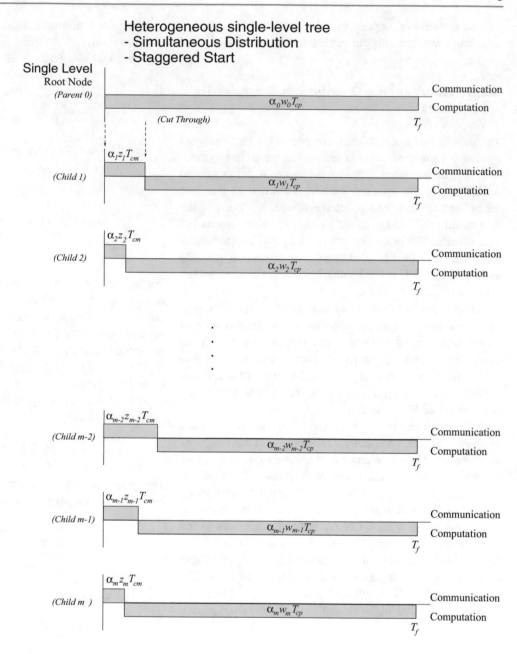

Here, we designate

$$q_i = \frac{w_{i-1}T_{cp} + z_{i-1}T_{cm}}{w_i T_{cp} + z_i T_{cm}} \quad (5.29)$$

Then, (5.28) can be represented as

$$\alpha_i = q_i\alpha_{i-1} = \left(\prod_{l=2}^{i} q_l\right)\alpha_1 \quad i = 2, 3, \ldots, m \quad (5.30)$$

Employing (5.26) and (5.28), the normalization equation becomes

$$\frac{1}{k_1}\alpha_1 + \alpha_1 + \sum_{i=2}^{m}\alpha_i = 1$$

$$\left[\frac{1}{k_1} + 1 + \sum_{i=2}^{m}\left(\prod_{l=2}^{i} q_l\right)\right]\alpha_1 = 1$$

$$\alpha_1 = \frac{1}{\left[\frac{1}{k_1} + 1 + \sum_{i=2}^{m}\left(\prod_{l=2}^{i} q_l\right)\right]} \quad (5.31)$$

From Fig. 5.6 the finish time is achieved as

$$T_{f,m} = \alpha_0 w_0 T_{cp} = \frac{1}{k_1}\alpha_1 w_0 T_{cp} \quad (5.32)$$

The term, $T_{f,m}$, indicates the finish time for the single divisible load solved in a single level tree, which consists of one root node as well as m children nodes.

Also, $T_{f,0}$ is defined as the finish time for the entire divisible load processed on the root processor. In other words, $T_{f,0}$ is the finish time of a network composed of only one root node without any children nodes. Hence

$$T_{f,0} = \alpha_0 w_0 T_{cp} = 1 \times w_0 T_{cp} = w_0 T_{cp} \qquad (5.33)$$

Now, collapsing a single level tree into a single equivalent node, one can obtain the finish time of the single level tree and the inverse of equivalent computing speed of the equivalent node as follows:

$$T_{f,m} = w_{eq} T_{cp} = \alpha_0 w_0 T_{cp} = \frac{1}{k_1} \alpha_1 w_0 T_{cp} \qquad (5.34)$$

Since $\gamma_{eq} = w_{eq}/w_0 = T_{f,m}/T_{f,0}$, one obtains the value of γ_{eq} by (5.32) dividing (5.31). That is,

$$\gamma_{eq} = \frac{1}{k_1}\alpha_1 = \frac{1}{k_1} \times \frac{1}{\left[\frac{1}{k_1} + 1 + \sum_{i=2}^{m}\left(\prod_{l=2}^{i} q_l\right)\right]}$$

$$= \frac{1}{1 + k_1\left[1 + \sum_{i=2}^{m}\left(\prod_{l=2}^{i} q_l\right)\right]} \qquad (5.35)$$

Since speedup is the ratio of job solution time on one processor to job solution time on the $m + 1$ processors one obtains the value of speedup from $T_{f,0}/T_{f,m}$, which is equal to $1/\gamma_{eq}$. Thus,

$$\boxed{\text{Speedup} = \frac{1}{\gamma_{eq}} = k_1 \times \frac{1}{\alpha_1} = 1 + k_1\left[1 + \sum_{i=2}^{m}\left(\prod_{l=2}^{i} q_l\right)\right]} \qquad (5.36)$$

Speedup is a measure of the achievable parallel processing advantage.

Two cases are discussed for the single level tree below:

1. General Case Since $\prod_{l=2}^{i} q_l$ can be simplified as $(w_1 T_{cp} + z_1 T_{cm})/(w_i T_{cp} + z_i T_{cm})$, the speedup and γ_{eq} can be derived as

$$\text{Speedup} = 1 + k_1\left[1 + \sum_{i=2}^{m}\left(\prod_{l=2}^{i} q_l\right)\right] \qquad (5.37)$$

$$\text{Speedup} = 1 + \frac{w_0 T_{cp}}{w_1 T_{cp} + z_1 T_{cm}}\left[1 + \sum_{i=2}^{m} \frac{w_1 T_{cp} + z_1 T_{cm}}{w_i T_{cp} + z_i T_{cm}}\right]$$

$$\boxed{\text{Speedup} = 1 + w_0 T_{cp} \sum_{i=1}^{m} 1/(w_i T_{cp} + z_i T_{cm})}$$

$$(5.38)$$

$$\boxed{\gamma_{eq} = 1/\left(1 + w_0 T_{cp} \sum_{i=1}^{m} \frac{1}{w_i T_{cp} + z_i T_{cm}}\right)} \qquad (5.39)$$

2. Special Case As a special case, consider the situation of a homogeneous network where all children processors have the same inverse computing speed and all links have the same inverse transmission speed. In other words, $w_i = w$ and $z_i = z$ for $i = 1, 2, \ldots, m$). Note that the root inverse computing speed, w_0, can be different from those $w_i, i = 1, 2, \ldots, m$.

Consequently

$$k_1 = \frac{w_0 T_{cp}}{w_1 T_{cp} + z_1 T_{cm}} = \frac{w_0 T_{cp}}{w T_{cp} + z T_{cm}}$$

$$q_i = \frac{w_{i-1} T_{cp} + z_{i-1} T_{cm}}{w_i T_{cp} + z_i T_{cm}} \qquad i = 2, 3, \ldots, m$$

$$= \frac{w T_{cp} + z T_{cm}}{w T_{cp} + z T_{cm}} = 1$$

$$\gamma_{eq} = \frac{1}{1 + k_1\left[1 + \sum_{i=2}^{m}\left(\prod_{l=2}^{i} q_l\right)\right]}$$

$$= \frac{1}{1 + \frac{w_0 T_{cp}}{w T_{cp} + z T_{cm}}\left[1 + \sum_{i=2}^{m}\left(\prod_{l=2}^{i} 1\right)\right]}$$

$$= \frac{1}{1 + \frac{w_0 T_{cp}}{w T_{cp} + z T_{cm}}\left[1 + (m - 1)\right]}$$

$$= \frac{1}{1 + m \times \frac{w_0 T_{cp}}{w T_{cp} + z T_{cm}}} \qquad (5.40)$$

$$\text{Speedup} = \frac{1}{\gamma_{eq}} = 1 + m \times \frac{w_0 T_{cp}}{w T_{cp} + z T_{cm}} = 1 + k_1 \times m \tag{5.41}$$

Our finding is that the computational complexity of the speedup of the single level homogeneous tree, staggered start in this case, is equal to $\Theta(m)$, which is linear in the number of children nodes. Speedup is linear as long as the root CPU can concurrently (simultaneously) transmit load to all of its children. That is, the speedup of the single level tree does not saturate (in contrast to the sequential load distribution of the previous section). Thus the scheduling policy is scalable for this type of network.

5.2.3   Simultaneous Distribution, Simultaneous Start

It would seem reasonable that performance would improve if a child processor could begin processing as soon as the load starts to arrive. Let us consider this scheduling policy.

A single level tree is illustrated in Fig. 5.7. All children processors are connected to the root processor via direct communication links. The intelligent root processor partitions a total processing load into $m + 1$ fractions, keeps its own fraction α_0, and distributes the other fractions $\alpha_1, \alpha_2, \ldots,$ α_m to the children processors respectively and concurrently. The children have front ends so load can be received while processing occurs. We assume here that $z_i T_{cm} \ll w_i T_{cp}$, so that the speed of communication in a link is faster than the speed of computation of the processor which is connected to the link. Therefore, communication ends before computation.

Each processor begins processing the received data while it receives the initial data. In order to minimize the processing finish time, all of the utilized processors in the network should finish computing at the same time.

The process of load distribution can be represented by Gantt chart-like timing diagrams, as illustrated in Fig. 5.8.

Referring to Fig. 5.8 the timing equations of this scheduling policy for a single level tree network can be written as

$$\alpha_0 w_0 T_{cp} = \alpha_1 w_1 T_{cp} \tag{5.42}$$

$$\alpha_1 w_1 T_{cp} = \alpha_2 w_2 T_{cp} \tag{5.43}$$

$$\alpha_2 w_2 T_{cp} = \alpha_3 w_3 T_{cp} \tag{5.44}$$

$$\cdot$$
$$\cdot$$

$$\alpha_{m-1} w_{m-1} T_{cp} = \alpha_m w_m T_{cp} \qquad i = 2, 3, \ldots, m \tag{5.45}$$

Here the first equation represents equating the processing time of the root processor to the processing time of child processor p_1. The second equation represents equating the processing time of p_1 to the processing time of p_2. These equivalencies can be naturally generalized.

The normalization equation for the single level tree with intelligent root is

$$\alpha_0 + \alpha_1 + \alpha_2 + \cdots + \alpha_m = 1 \tag{5.46}$$

This gives $m + 1$ linear equations with $m + 1$ unknowns.

Now, one can manipulate the recursive equations to yield the solution. From (5.42), one obtains

$$\alpha_0 = \frac{w_1 T_{cp}}{w_0 T_{cp}} \alpha_1 = \frac{1}{k_1} \alpha_1 \tag{5.47}$$

Here k_1 is defined as w_0/w_1. From (5.42) to (5.45), one obtains

Fig. 5.7 Structure of single level tree with simultaneous distribution and simultaneous start. From Hung [123] © ACTA Press

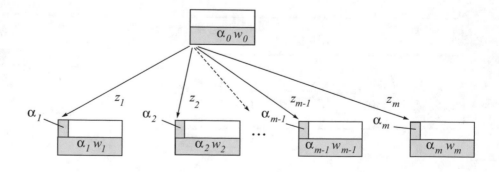

Fig. 5.8 Timing diagram of single level tree with simultaneous distribution and simultaneous start. From Hung [123] © ACTA Press

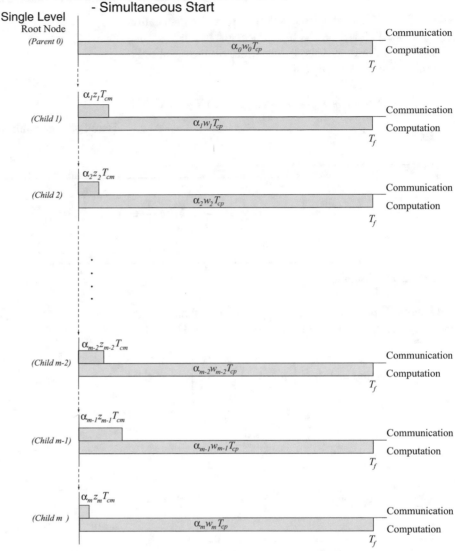

Heterogeneous Single-Level Tree
Simultaneous Distribution
- Simultaneous Start

$$\alpha_i = \frac{w_{i-1}T_{cp}}{w_i T_{cp}}\alpha_{i-1} = q_i\alpha_{i-1} \quad i = 2, 3, \ldots, m \tag{5.48}$$

Here

$$q_i = \frac{w_{i-1}}{w_i} \tag{5.49}$$

Then

$$\alpha_i = q_i\alpha_{i-1} = \left(\prod_{l=2}^{i} q_l\right)\alpha_1 \quad i = 2, 3, \ldots, m \tag{5.50}$$

Employing (5.47) and (5.50), the value of α_1 can be solved by deriving the normalization equation (5.46).

$$\frac{1}{k_1}\alpha_1 + \alpha_1 + \sum_{i=2}^{m} \alpha_i = 1$$

$$\left[\frac{1}{k_1} + 1 + \sum_{i=2}^{m}\left(\prod_{l=2}^{i} q_l\right)\right]\alpha_1 = 1$$

$$\alpha_1 = \frac{1}{\left[\frac{1}{k_1} + 1 + \sum_{i=2}^{m}\left(\prod_{l=2}^{i} q_l\right)\right]} \tag{5.51}$$

Thus, the values of α_0 and α's can be solved with respect to (5.47) and (5.50). From Fig. 5.8, the finish time is

$$T_{f,m} = \alpha_0 w_0 T_{cp} = \frac{1}{k_1}\alpha_1 w_0 T_{cp} \tag{5.52}$$

The term, $T_{f,m}$, is the finish time of a divisible job solved on the entire tree, consisting of one root node as well as m children nodes in a single level tree. It is the same as the finish time of an equivalent node for the subtree.

$$T_{f,m} = w_{eq}T_{cp} = \alpha_0 w_0 T_{cp} = \frac{1}{k_1}\alpha_1 w_0 T_{cp} \quad (5.53)$$

Also, $T_{f,0}$ is defined as the solution time for the entire divisible load solved on the root processor. In other words, it is the finish time of a single level tree, composed of only one root node without any child nodes.

$$T_{f,0} = \alpha_0 w_0 T_{cp} = 1 \times w_0 T_{cp} = w_0 T_{cp} \quad (5.54)$$

As before, γ_{eq} is equal to w_{eq}/w_0, or $T_{f,m}/T_{f,0}$. Thus, from (5.53) and (5.54)

$$\gamma_{eq} = \frac{1}{k_1}\alpha_1 = \frac{1}{k_1} \times \frac{1}{\left[\frac{1}{k_1} + 1 + \sum_{i=2}^{m}\left(\prod_{l=2}^{i}q_l\right)\right]}$$

$$= \frac{1}{1 + k_1\left[1 + \sum_{i=2}^{m}\left(\prod_{l=2}^{i}q_l\right)\right]} \quad (5.55)$$

Based on the definition of speedup

$$\boxed{\text{Speedup} = \frac{1}{\gamma_{eq}} = k_1 \times \frac{1}{\alpha_1} = 1 + k_1\left[1 + \sum_{i=2}^{m}\left(\prod_{l=2}^{i}q_l\right)\right]} \quad (5.56)$$

Two cases are discussed.

1. General Case From (5.49), $\prod_{l=2}^{i}q_l$ can be simplified as w_1/w_i. Therefore, the speedup and γ_{eq} are derived as

$$\text{Speedup} = 1 + k_1\left[1 + \sum_{i=2}^{m}\left(\prod_{l=2}^{i}q_l\right)\right] \quad (5.57)$$

$$= 1 + \frac{w_0}{w_1}\left[1 + \sum_{i=2}^{m}\frac{w_1}{w_i}\right]$$

$$\boxed{\text{Speedup} = 1 + w_0\sum_{i=1}^{m}(1/w_i)} \quad (5.58)$$

$$\boxed{\gamma_{eq} = 1/\left(1 + w_0\sum_{i=1}^{m}(1/w_i)\right)} \quad (5.59)$$

2. Special Case As a special case, consider the situation of a homogeneous network where all children processors have the same inverse computing speed and all links have the same inverse transmission speed. In other words, $w_i = w$ and $z_i = z$ for $i = 1, 2, \ldots, m$). Note that the root inverse computing speed, w_0, can be different from $w_i, i = 1, 2, \ldots, m$. This will result in

$$k_1 = \frac{w_0}{w_1} = \frac{w_0}{w} \quad (5.60)$$

$$q_i = \frac{w_{i-1}}{w_i} \qquad i = 2, 3, \ldots, m$$

$$= \frac{w}{w} = 1 \quad (5.61)$$

$$\gamma_{eq} = \frac{1}{1 + k_1\left[1 + \sum_{i=2}^{m}\left(\prod_{l=2}^{i}q_l\right)\right]}$$

$$= \frac{1}{1 + \frac{w_0}{w}\left[1 + \sum_{2}^{m}\left(\prod_{l=2}^{i}1\right)\right]}$$

$$= \frac{1}{1 + \frac{w_0}{w}[1 + (m - 1)]}$$

$$= \frac{1}{1 + m \times \frac{w_0}{w}} \quad (5.62)$$

$$\boxed{\text{Speedup} = \frac{1}{\gamma_{eq}} = 1 + m \times \frac{w_0}{w} = 1 + k_1 \times m}$$

$$(5.63)$$

Here, *speedup* is the effective processing gain in using $m + 1$ processors. Our finding is that the computational complexity of the speedup of the single level homogeneous tree is equal to $\Theta(m)$, which is proportional to the number of children, per node m. Speedup is a linear function as long as the root CPU can concurrently (simultaneously) transmit load to all of its children. That is, the speedup of the single level tree does not saturate (in contrast to the sequential load distribution as in [39, 40]). It can be seen that this speedup is larger than that of the previous section (simultaneous distribution, staggered start).

Note that one can obtain this speedup result from (5.41), the case with staggered start, if one lets $z \to 0$. In other words, the time that the root node distributes load to its children processors is negligible.

Fig. 5.9 Structure of a single level tree with simultaneous distribution, simultaneous start, and nonlinear computing time

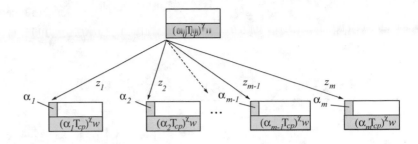

5.2.4 Nonlinear Load Processing Complexity

In the three previous subsections processing time was simply proportional to the load size. This is a linear modeling assumption. But many algorithms have a computing time that is proportional to a nonlinear function of input size. This was first considered by Drozdowksi and Wolniewicz [81] for a step function of input size. In this section we find optimal load distribution formula for a power χ dependency between computation time at a node and load size at the node in a single level tree network. All load is available at the root at $t = 0$. The structure of a single level tree network with intelligent root, $m + 1$ processors and m links is illustrated in Fig. 5.9.

All children processors are connected to the root processor via direct communication links. The intelligent root processor, assumed to be the only processor at which the divisible load arrives, partitions a total processing load into $m + 1$ fractions, keeps its own fraction α_0, and distributes the other fractions $\alpha_1, \alpha_2, \ldots, \alpha_m$ to the children processors respectively and concurrently.

While concurrently receiving its initial assigned fraction of load, each processor begins computing immediately and continues without any interruption until all of its assigned load fraction has been processed. This scheduling policy is analogous to the simultaneous distribution, simultaneous start policy of the previous subsection. In order to minimize the processing finish time, all of the utilized processors in the network must finish computing at the same time. The process of load distribution can be represented by a Gantt chart-like timing diagram, as illustrated in Fig. 5.10.

Note that this is a completely deterministic model.

From the timing diagram Fig. 5.10, the fundamental recursive equations of the system can be formulated as follows:

$$(\alpha_0 T_{cp})^\chi w_0 = (\alpha_1 T_{cp})^\chi w_1 \tag{5.64}$$

$$(\alpha_1 T_{cp})^\chi w_1 = (\alpha_2 T_{cp})^\chi w_2 \tag{5.65}$$

$$(\alpha_2 T_{cp})^\chi w_2 = (\alpha_3 T_{cp})^\chi w_3 \tag{5.66}$$

$$\vdots$$

$$(\alpha_{m-1} T_{cp})^\chi w_{m-1} = (\alpha_m T_{cp})^\chi w_m \quad i = 2, \ldots, m \tag{5.67}$$

Note that computation time is still linearly proportional to inverse computing speed. The first equation equates the processing time of the root node to the processing time of child processor p_1. The second equation equates the processing time of p_1 to the processing time of p_2 and so on.

The normalization equation for the single level tree with intelligent root is

$$\alpha_0 + \alpha_1 + \alpha_2 + \ldots + \alpha_m = 1 \tag{5.68}$$

This yields $m + 1$ linear equations with $m + 1$ unknowns. One can manipulate the recursive equations to yield a solution for the optimal allocation of load. From (5.64)

$$\alpha_0^\chi = \frac{w_1 T_{cp}^\chi}{w_0 T_{cp}^\chi} \alpha_1^\chi = \frac{w_1}{w_0} \alpha_1^\chi \tag{5.69}$$

Let $\kappa_1 = w_0/w_1$, then (5.69) becomes

$$\alpha_0^\chi = \frac{w_1}{w_0} \alpha_1^\chi = \frac{1}{\kappa_1} \alpha_1^\chi \tag{5.70}$$

One obtains

$$\alpha_0 = \sqrt[\chi]{\frac{1}{\kappa_1} \alpha_1^\chi} \left(cos\frac{2k\pi}{\chi} + \sqrt{-1}\ sin\frac{2k\pi}{\chi} \right) \tag{5.71}$$

where k takes successively the values $0, 1, 2, \ldots, n-1$

Since $\alpha_0 \geq 0$ and α_0 is *real*, therefore, we take $k = 0$ and get

$$\alpha_0 = \sqrt[\chi]{\frac{1}{\kappa_1}} \cdot \alpha_1 \quad i = 1, 2, \ldots, m \tag{5.72}$$

From (5.64) to (5.67)

$$\alpha_i^\chi = \frac{w_{i-1} T_{cp}^\chi}{w_i T_{cp}^\chi} \alpha_{i-1}^\chi = \frac{w_{i-1}}{w_i} \alpha_{i-1}^\chi \quad i = 2, \ldots, m \tag{5.73}$$

Let $\xi_i = w_{i-1}/w_i, i = 2, \ldots, m$. Then, (5.73) becomes

$$\alpha_i^\chi = \xi_i \alpha_{i-1}^\chi \qquad (5.74)$$

One obtains

where k takes successively the values $0, 1, 2, \ldots, n-1$

Since $\alpha_i \geq 0$ and α_i is *real*, therefore, we take $k = 0$ and get

$$\alpha_i = \sqrt[\chi]{\xi_i \alpha_{i-1}^\chi} \left(cos\frac{2k\pi}{\chi} + \sqrt{-1} \ sin\frac{2k\pi}{\chi} \right) \quad (5.75)$$

$$\alpha_i = \sqrt[\chi]{\xi_i} \alpha_{i-1} = \sqrt[\chi]{\prod_{l=2}^{i} \xi_l} \cdot \alpha_1 = \sqrt[\chi]{\frac{w_1}{w_i}} \cdot \alpha_1 \quad i = 2, \ldots, m \qquad (5.76)$$

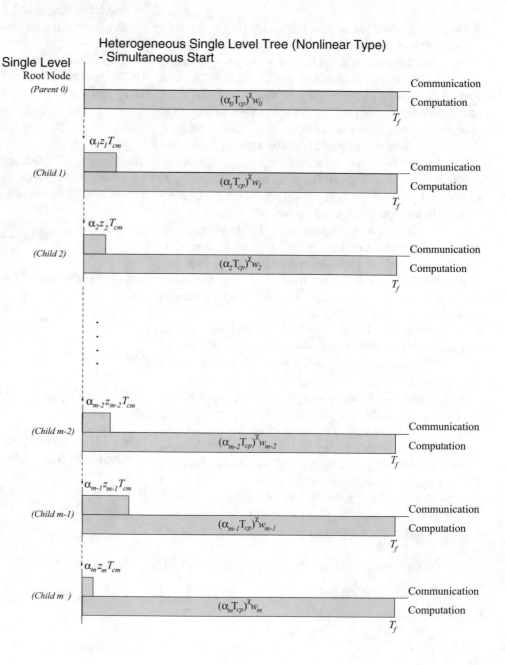

Fig. 5.10 Timing diagram of a single level tree with simultaneous distribution, simultaneous start, and nonlinear computing time

Based on the normalization equation (5.68)

$$\alpha_0 + \alpha_1 + \sum_{i=2}^{m} \alpha_i = 1 \qquad (5.77)$$

$$\left(\sqrt[x]{\frac{1}{\kappa_1}} + 1 + \sum_{i=2}^{m} \sqrt[x]{\prod_{l=2}^{i} \xi_l} \right) \alpha_1 = 1 \qquad (5.78)$$

$$\left(\sqrt[x]{\frac{w_1}{w_0}} + 1 + \sum_{i=2}^{m} \sqrt[x]{\frac{w_1}{w_i}} \right) \alpha_1 = 1 \qquad (5.79)$$

$$\left(\sum_{l=0}^{m} \sqrt[x]{\frac{w_1}{w_l}} \right) \alpha_1 = 1 \qquad (5.80)$$

$$\sqrt[x]{w_1} \cdot \left(\sum_{l=0}^{m} \sqrt[x]{\frac{1}{w_l}} \right) \alpha_1 = 1 \qquad (5.81)$$

Consequently, we obtain

$$\alpha_1 = \frac{1}{\sqrt[x]{w_1} \cdot \left(\sum_{l=0}^{m} \sqrt[x]{\frac{1}{w_l}} \right)} \qquad (5.82)$$

From (5.72)

$$\alpha_0 = \sqrt[x]{\frac{1}{\kappa_1}} \cdot \alpha_1 = \sqrt[x]{\frac{1}{\kappa_1}} \cdot \frac{1}{\sqrt[x]{w_1} \cdot \left(\sum_{l=0}^{m} \sqrt[x]{\frac{1}{w_l}} \right)}$$

$$= \sqrt[x]{\frac{w_1}{w_0}} \cdot \frac{1}{\sqrt[x]{w_1} \cdot \left(\sum_{l=0}^{m} \sqrt[x]{\frac{1}{w_l}} \right)}$$

$$= \frac{1}{\sqrt[x]{w_0} \cdot \left(\sum_{l=0}^{m} \sqrt[x]{\frac{1}{w_l}} \right)} \qquad (5.83)$$

Based on (5.76) and (5.82), one obtains

$$\alpha_i = \sqrt[x]{\frac{w_1}{w_i}} \cdot \alpha_1 = \sqrt[x]{\frac{w_1}{w_i}} \cdot \frac{1}{\sqrt[x]{w_1} \cdot \left(\sum_{l=0}^{m} \sqrt[x]{\frac{1}{w_l}} \right)} \qquad (5.84)$$

$$= \frac{1}{\sqrt[x]{w_i} \cdot \left(\sum_{l=0}^{m} \sqrt[x]{\frac{1}{w_l}} \right)} \qquad i = 2, \ldots, m \qquad (5.85)$$

From (5.82), (5.83), and (5.85), we obtain the optimal fractions of load α_i as

$$\alpha_i = \frac{1}{\sqrt[x]{w_i} \cdot \left(\sum_{l=0}^{m} \sqrt[x]{\frac{1}{w_l}} \right)} \qquad i = 0, 1, 2, \ldots, m \qquad (5.86)$$

Using Fig. 5.10, the finish time is achieved as

$$T_{f,m} = \left(\alpha_0 T_{cp} \right)^{\chi} w_0 \qquad (5.87)$$

Here, $T_{f,m}$ indicates the finish time for the single divisible load solved in a single level tree, which consists of one root node as well as m children nodes. Also, $T_{f,0}$, is defined as the finish time for the entire divisible load processed on the root processor. In other words, $T_{f,0}$ is the finish time of a network composed of only one root node without any children nodes. Hence

$$T_{f,0} = \left(\alpha_0 T_{cp} \right)^{\chi} w_0 = (1 \times T_{cp})^{\chi} w_0 = T_{cp}^{\chi} w_0 \qquad (5.88)$$

Now, collapsing a single level tree into a single equivalent node, one can obtain the finish time of the single level tree and the inverse of the equivalent computing speed of the equivalent node as follows:

$$T_{f,m} = (1 \times T_{cp})^{\chi} w_{eq} = T_{cp}^{\chi} w_{eq} = \left(\alpha_0 T_{cp} \right)^{\chi} w_0 \qquad (5.89)$$

As before, $\gamma_{eq} = w_{eq}/w_0 = T_{f,m}/T_{f,0}$ and one obtains the value of γ_{eq} by (5.88) dividing (5.89). That is,

$$\gamma_{eq} = \alpha_0^{\chi} \qquad (5.90)$$

Since speedup is the ratio of job solution time on one processor to job solution time on the $m + 1$ processors, one obtains the value of speedup from $T_{f,0}/T_{f,m}$, which is equal to $1/\gamma_{eq}$. Thus

$$\text{Speedup} = \frac{1}{\gamma_{eq}} = \left(\frac{1}{\alpha_0} \right)^{\chi} \qquad (5.91)$$

$$\boxed{\text{Speedup} = w_0 \left(\sum_{l=0}^{m} \sqrt[x]{\frac{1}{w_l}} \right)^{\chi}} \qquad (5.92)$$

Speedup is a measure of the achievable parallel processing advantage.

Special Case As a special case, consider the situation of a homogeneous network where all children processors have the same inverse computing speed and all links have the same inverse transmission speed. In other words, $w_i = w$ and $z_i = z$ for $i = 1, 2, \ldots, m$). Note that the root inverse computing speed, w_0, can be different from those $w_i, i = 1, 2, \ldots, m$. From (5.83)

$$\alpha_0 = \frac{1}{\sqrt[x]{w_0} \cdot \left(\sum_{l=0}^{m} \sqrt[x]{\frac{1}{w_l}} \right)} \qquad (5.93)$$

$$= \frac{1}{\sqrt[\chi]{w_0} \cdot \left(\sqrt[\chi]{\frac{1}{w_0}} + \sum_{l=1}^{m} \sqrt[\chi]{\frac{1}{w}} \right)} \qquad (5.94)$$

$$= \frac{1}{\sqrt[\chi]{w_0} \cdot \left(\sqrt[\chi]{\frac{1}{w_0}} + m \sqrt[\chi]{\frac{1}{w}} \right)} \qquad (5.95)$$

$$= \frac{1}{1 + m \sqrt[\chi]{\frac{w_0}{w}}} \qquad (5.96)$$

Since $\gamma_{eq} = w_{eq}/w_0 = T_{f,m}/T_{f,0}$, one obtains the value of γ_{eq}

$$\gamma_{eq} = \alpha_0^\chi = \left(\frac{1}{1 + m \sqrt[\chi]{\frac{w_0}{w}}} \right)^\chi \qquad (5.97)$$

Since speedup is the ratio of job solution time on one processor to job solution time on the $m + 1$ processors, one obtains the value of speedup from $T_{f,0}/T_{f,m}$, which is equal to $1/\gamma_{eq}$. Thus

$$\boxed{\text{Speedup} = \frac{1}{\gamma_{eq}} = \left(\frac{1}{\alpha_0} \right)^\chi = \left(1 + m \sqrt[\chi]{\frac{w_0}{w}} \right)^\chi} \qquad (5.98)$$

Again, speedup is a measure of the achievable parallel processing advantage.

If the computing capability of the root node is the same as that of children nodes for a homogeneous single level tree, i.e. $w_0 = w$, the speedup formula will become

$$\text{Speedup} = (1 + m)^\chi \qquad (5.99)$$

Again, this last result is intuitive.

Such nonlinear analysis can also be simplified by replacing a certain square root term with an approximation (based on a limited Taylor series) that is valid if communication time is much smaller than computation time which is the case in many distributed systems [222, 223].

From the above two equations it can be seen that speedup increases nonlinearly as more and more children processors are added. Are we getting something for nothing in this nonlinear increase? Not really, because as the computational complexity is nonlinear, processing fragments of the load on a number of processors is much more efficient than processing all of the load on one processor.

In fact the situation is indeed a bit more complex. For practical applications involving divisible load with nonlinear computational complexity, in some sense there is some dependency between the data and its processing. That is,

once load is divided and processed on separate processors, the individual results from each processor need to be combined through post-processing. This post-processing adds an additional computational cost. The degree to which such divisible processing for loads with nonlinear computational complexity results an overall computational cost savings is problem dependent.

Nonlinear Communication Time

There are many algorithms of practical interest where computation time is a nonlinear function of problem size. But what of communication time?

Can data moving over links between processing nodes have a communication time that is a nonlinear function of problem size? Normally one would think not. If one doubles the amount of data to be transmitted, one would think that it would take twice as long to transmit the data. This is a linear relationship.

In [237] a different way of looking at things is described. That is, what if one indexes data transmission not by time but by data structural properties?

As an example, if one transmits a square matrix of numbers and indexes data transmission by data size (i.e., number of rows/columns), the transmission time is then proportional to the square power of the number of rows. As a second example, if one transmits a binary tree of data where each node holds b bytes and indexes data transmission by the size of the tree in levels L, the transmission time is proportional to $2^L - 1$.

Some nonlinear systems have a super-linear complexity (complexity greater than a linear complexity) and some have a sub-linear complexity (complexity less than a linear complexity). One usually considers algorithm complexity in terms of data size (as in $O(N)$ where N is the data size). But there is an alternate description of data size that needs to be normalized in terms of how long it takes to communicate data.

Suppose to communicate data takes time $T_{com} = N^\chi$ and to process (compute) data takes time $T_{proc} = N^y$. Usually one says the algorithm has order $O(N^y)$ complexity. In terms of T_{com}, $N = T_{com}^{1/\chi}$ and substituting for N, $T_{proc} = T_{com}^{y/\chi}$. The actual complexity is thus $O(N^{y/\chi})$ instead of $O(N^y)$. If $y > x$, the overall complexity is super-linear. If $y < x$, the overall complexity is sub-linear.

5.3   Equivalent Processors

The basic model of divisible load scheduling, Sect. 5.2.4 notwithstanding, is a linear one. An important concept in many linear theories, such as linear electric circuits and linear queueing theory, is that of an equivalent element. An equivalent element can replace a sub-network within

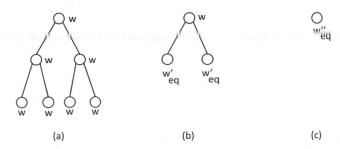

Fig. 5.11 Collapsing a dual level binary tree of processors and links into an equivalent single level tree and then into a single equivalent processor

a network or replace an entire network and provide exactly identical operating characteristics as the original sub-network or network, respectively [40, 192].

Similarly, a network of links and processors operating on a divisible load can be replaced by a single processor with identical processing capability. This is very useful for modeling and analysis. We can calculate the speedup of a large network, for instance, by replacing the network with an equivalent processor and finding the speedup of this equivalent processor.

As an example, consider finding an equivalent processor for the seven processor binary tree network in Fig. 5.11a. One can find equivalent processor w'_{eq} for the two bottom single level sub-trees in equation Fig. 5.11a using techniques appearing in [196]. Note that all of the processors in the two bottom sub-trees have inverse computation speed w. This yields the system of Fig. 5.11b.

Using similar techniques, one can find an equivalent processor for the system of Fig. 5.11b with equivalent inverse processing speed w''_{eq} which is shown in Fig. 5.11c. It that has the same effective inverse computing speed as the original network of Fig. 5.11a. Note that in Fig. 5.11b the root processor has inverse computing speed w and the two children nodes have equivalent inverse computing speed w'_{eq}. Thus the reduction from the (a) figure to the (b) figure is slightly different than the reduction from the (b) figure to the (c) figure.

In the same spirit, larger trees can be collapsed into a single equivalent processor by starting at the bottom level and working upwards. Other topologies, such as linear daisy chains of processors, can also be collapsed into a single equivalent processor. Trees are particular important because of the use of spanning trees to distribute load in other types of interconnection networks.

The following section was originally published by the first author under the title "A Product Form Solution for Tree Networks with Divisible Loads," in Parallel Processing Letters, Vol. 21, No. 1, in March 2011 and is reprinted by permission from World Scientific

5.4 Divisible Loads and Product Form Solutions

It is well known in queueing theory that there are product form solutions (solutions that are a product of terms) for the equilibrium state probabilities of both open and closed queueing networks with Markovian statistics [22, 105, 126]. Here open networks allow customers to enter and leave a network of queues and closed networks are sealed systems. Under Markovian statistics the input arrival processes for open networks are independent Poisson processes and the service times of customers at queues are independent and follow negative exponential distributions. Product form solutions in queueing networks are considered a simplifying feature. Moreover efficient numerical algorithms for solving such product form closed networks have been developed such as mean value analysis [189] and the convolution algorithm [55].

If $p_i(n_i)$ is the marginal probability that there are n_i customers in the ith queue and $p(\underline{n})$ is the joint equilibrium probability that there are n_1 customers in the first queue, n_2 customers in the 2nd queue and n_M customers in the Mth queue, then under the product form solution for open networks:

$$p(\underline{n}) = p_1(n_1)p_2(n_2)p_3(n_3)\ldots p_M(n_M) \qquad (5.100)$$

For closed queueing networks the product form solution is of the same format with the inclusion of a normalization constant factor.

For many years solutions for optimal load allocation in single layer tree networks that are essentially a one-dimensional product form solution have been known [217]. However it has been an open problem, explicitly recognized in [171], as to what type of network an M-dimensional product form solution corresponds to. This paper demonstrates that there is an M-dimensional product form solution for load distribution from a single source (root) node in a multi-level tree type interconnection network. This is significant as tree type networks are often used in an embedded fashion in other interconnection networks for load distribution [49, 51, 78, 88, 102].

In fact it has been known since 1990 [67] that optimal load distribution to processors from a root node in a multiple level tree network could be solved recursively by collapsing single layer subtrees, working from the bottom of the tree to the root, into equivalent processors and tying the solutions together. But this had not been done in explicit analytical form till 2011 [197] as is done in this section.

Fig. 5.12 A multi-level tree. From Robertazzi [197] © World Scientific

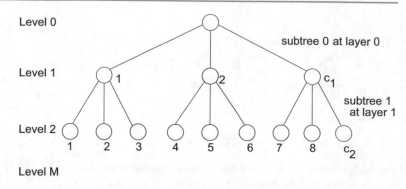

5.4.1   Model and Notation

Consider the multi-level tree of Fig. 5.12. The root is at "level" 0, the root's children are at level 1, the root's grandchildren are at level 2, and the lowest level of the finite tree is level M. Also defined are single layer "subtrees" consisting of a node and its children. The root and its children form subtree 0 at "layer" 0. A child of the root and its children are at the subtree 1 layer. Going down the tree the last subtrees are at the subtree M-1 layer. The difference between "level" and "layer" is illustrated in the figure. The nodes at the ith level, across all subtrees, are numbered from left to right as $1, 2, 3 \ldots c_i$.

It is assumed that a divisible load of volume V originates at the root at time 0. The load is distributed throughout the tree to the other nodes, each one of which can process load. There are several load distribution policies that can be used [122, 196] which are discussed below.

In terms of the processing and transmission of divisible load the following parameters are defined:

w_r: the inverse of the constant computing speed of the rth node.

z_t: the inverse of the constant transmission speed of the tth link.

T_{cp}: the computation intensity constant. The entire load can be processed on the rth processor in time $w_r T_{cp}$.

T_{cm}: the communication intensity constant. The entire load can be transmitted over the tth link in time $z_t T_{cm}$.

For the allocation of load the following variables are defined:

$\beta_{i,j}$: the fraction of load received at the root of an ith layer subtree that is sent to its jth child.

$\alpha_{i,j}$: the fraction of the total load that is processed at the ith level by the jth processor.

For the α's if the normalized total load is unity ($V = 1$):

$$\sum_{i=0}^{M} \sum_{j=1}^{c_i} \alpha_{i,j} = 1 \qquad (5.101)$$

If the total amount of load is V:

$$\sum_{i=0}^{M} \sum_{j=1}^{c_i} \alpha_{i,j} V = V \qquad (5.102)$$

5.4.2   The M Level Subtree Product Form Solution

A little thought will show that the fraction of load received at a leaf (i.e., bottom most) node in an M level tree can be written as a product of the fractions of load passed to each node above it in a path back to the root for $i = M$ and $j = 1, 2, 3 \ldots c_i$:

$$\alpha_{i,j} = \beta_{0,d} \beta_{1,e} \beta_{2,f} \ldots \beta_{i-2,h} \beta_{i-1,j} V \qquad (5.103)$$

In this equation d is the dth child in subtree 0 along the path from the root to the node of interest. Then e is the eth child in subtree 1 along the same path and so on.

This equation expresses the optimal load assigned to each (bottom level) leaf processor as a product of terms of the fractions of load received at each node along a path from the root to the leaf node of interest. These β fractions are relative. As an example, consider a three level tree. If $\beta_{0,1} = 0.5$ and $\beta_{1,1} = 0.2$, then the absolute amount of load received by the leftmost grandchild of the root is 0.10. That is, 20% of the load received by the leftmost child of the root (which is 50% of the total load) or 10% of the total load is sent to the leftmost grandchild of the root.

Note also that all processors participate in load processing. This equation allows the optimal fraction of load at any leaf node to be calculated.

One may be interested in the optimal load allocation to a node in the interior of the multi-level tree. In this case the optimal load allocated to processor i, j (i.e., $\alpha_{i,j}$) can be found from a partial product form equation similar to Eq. (5.103) for the total load delivered to node i, j minus the load node i, j gives to its children (also computed from partial product form equations similar to Eq. (5.103)). Here a "partial" product form equation is similar to (5.103) but treats an interior node as a leaf node of a partial tree, computing the load delivered to the interior node (which includes load for itself and for its descendants). However even in the case of interior node load allocation the β's must be calculated from the bottom most layer working upwards. That is, with the node of interest being a root node in a single layer subtree, the children nodes in the subtree have inverse processing speeds equivalent to the (possibly multi-level) subtrees below them that they replace. Thus one starts with the single layer subtree at the bottom of the overall multi-level tree network, collapsing them into equivalent processors and continuing this process working upwards in the multi-level tree until the β's for nodes along a path from the overall multi-level tree root to the node of interest can be solved for [40, 122].

How does one calculate the β? There is standard theory [122, 196](Sect. 5.2 this book) available in the divisible load literature for doing this. Specifically with some reuse of the i, j notation from the previous equation, for the root of a single layer subtree:

$$\beta_{i,0} = \frac{\frac{1}{k_1}}{\left[\frac{1}{k_1} + 1 + \sum_{j=2}^{m} \left(\prod_{l=2}^{j} q_l \right) \right]} \qquad (5.104)$$

For the leftmost child of a single layer subtree ($j = 1$):

$$\beta_{i,1} = \frac{1}{\left[\frac{1}{k_1} + 1 + \sum_{j=2}^{m} \left(\prod_{l=2}^{j} q_l \right) \right]} \qquad (5.105)$$

For the other children of a single level subtree ($j = 2, 3 \ldots m$):

$$\beta_{i,j} = \frac{\left(\prod_{l=2}^{j} q_l \right)}{\left[\frac{1}{k_1} + 1 + \sum_{j=2}^{m} \left(\prod_{l=2}^{j} q_l \right) \right]} \qquad (5.106)$$

Here we have written expressions for the fraction of the load that each node receives of the load given to the node's subtree's root. In this equation the j numbers the children in a specific subtree as $1, 2, 3 \ldots m$.

The k_1 and q_l are functions specific to a given load distribution policy in a single level tree network where load originates at the root node. They are usually a function of w, z, T_{cp}, and T_{cm}. They can also be functions of intermediate equivalent processing speeds of equivalent processors replacing a subnetwork consisting of a node and all the children/grandchildren... below it.

Several k and q functions appear in [122, 196]. For instance, consider a sequential load distribution where load is distributed in turn to each child once. Assume each child

commences processing as soon as load begins to be received and that the root also does load processing starting at time 0. Then:

$$k_1 = \frac{w_0}{w_1} \qquad (5.107)$$

$$q_l = \frac{w_{l-1} T_{cp} - z_{l-1} T_{cm}}{w_l T_{cp}} \qquad l = 2, 3 \ldots m \qquad (5.108)$$

In a multi-level tree for instance, the w_{l-1} and w_l are inverse equivalent processing speeds of the subtree networks that these single layer subtree children nodes replace.

As a second example, consider a policy with the simultaneous distribution of load to all children from the root. Let processing at a child start when all load for that child has been received and let the root process some of the load starting at time 0. Then:

$$k_1 = \frac{w_0 T_{cp}}{z_1 T_{cm} + w_1 T_{cp}} \qquad (5.109)$$

$$q_l = \frac{w_{l-1} T_{cp} + z_{l-1} T_{cm}}{w_l T_{cp} + z_l T_{cm}} \qquad (5.110)$$

Many policies can be substituted into the main equations. These policies differ in whether load distribution is sequential or simultaneous, whether the root does processing and whether processing at each child starts as load begins to be received or when load is completely received at a child.

If one substitutes Eq. (5.106) into Eq. (5.103) for $j = 2, \ldots m$, one has a product of terms where each term has a $\prod_{l=2}^{j} q_l$ component in its numerator. This product of products is akin to the product form solution of queueing networks. Note that the node 0 and node 1 numerator terms for β in a subtree have somewhat different values. This is because of the manner in which the β are determined in a single level tree here (i.e., using the leftmost child, node 1, as a reference processor with unnormalized allocation value of 1).

A normalization constant like term can be found from the product of the denominators of the individual β.

Thus the product form solution for the optimal load to assign to processors in a multi-level tree with divisible load distribution from the root has a finite "state" space and thus it also has a normalization constant (also called a partition function). Here the processors play a role analogous to the states of a closed queueing network Markov chain.

To summarize, to evaluate the α's for every node in a tree (a) working from the bottom to the top of the tree find the β's (b) calculate the α's for the leaf nodes using Eq. (5.103) and finally (c) working from the bottom to the top of the tree find the remaining interior node α's as described above.

5.4.3 In Summary

This section presents novel and compact analytical results for optimal load distribution in multi-level tree networks. As mentioned such trees are very general, being used as distribution networks in other interconnection topologies.

While it might be surprising that both queueing networks (a stochastic model) and divisible load scheduling (a deterministic model) admit product form solutions, the underlying reason is that, in their basic form, both are linear models. Certain patterned labelings of the elements of a linear system lead to recursive and product form solutions.

The results in this section will be of interest theoretically, for devising simple code for optimal load distribution and also as one starting point for research on very large grid and cloud networks.

5.5   Infinite Size Network Performance

5.5.1   Linear Daisy Chains

A linear daisy chain of processors where processor load is divisible and shared among the processors is discussed in this subsection. It can be shown, as in Robertazzi [196], that two or more processors can be collapsed into a single equivalent processor. This equivalence allows a characterization of the nature of the minimal time solution and closed form expressions for the equivalent processing speed of infinitely large daisy chains of processors.

The situation to be considered involves a linear daisy chain of processors, as is illustrated in Fig. 5.13. A single "problem" (or job) is solved on the network at one time. It takes time $w_i T_{cp}$ to solve the entire problem on processor i. Here w_i is inversely proportional to the speed of the ith processor and T_{cp} is the normalized computation intensity when $w_i = 1$. It takes time $z_i T_{cm}$ to transmit the entire problem representation (data) over the ith link. Here z_i is inversely proportional to the channel speed of the ith link and T_{cm} is the normalized communication intensity when $z_i = 1$.

It is assumed that the problem representation can be divided among the processors in divisible fashion. That is, fraction α_i of the total problem is assigned to the ith processor so that its computing time becomes $\alpha_i w_i T_{cp}$. It is desired to determine the optimal values of the α_i's so that the problem is solved in the minimum amount of time. The situation is non-trivial as there are communication delays incurred in transmitting fractional parts of the problem representation to each processor from the originating processor.

Two cases will be considered: processors that have front end communication sub-processors for communications offloading so that communication and computation may proceed simultaneously and processors without front-end communication sub-processors so that communication and computation must be performed at separate times.

A timing diagram for a linear daisy chain network of four processors with front-end communication sub-processor (as in Fig. 5.13) is illustrated in Fig. 5.14. It is assumed that the problem (load) originates at the left most processor.

At time 0, processor 1 can start working on its fraction, α_1, of the problem in time $\alpha_1 w_1 T_{cp}$. It also simultaneously communicates the remaining fraction of the problem to processor 2 in time $(\alpha_2 + \alpha_3 + \alpha_4) z_1 T_{cm}$. Processor 2 can then begin computation on its fraction of the problem (in time $\alpha_2 w_2 T_{cp}$) and communicates the remaining load to processor 3 in time $(\alpha_3 + \alpha_4) z_2 T_{cm}$. The process continues until all processors are working on the problem. Note that the store and forward switching method is used here but other protocols could be modeled as well.

A similar, but not identical, situation for a linear daisy chain network with processors that do not have front-end communication sub-processors is illustrated in Fig. 5.15. Here each processor must communicate the remaining load to its right neighbor before it can begin computation on its own fraction.

In [66] recursive expressions for calculating the optimal α_i's were presented. These are based on the simplifying premise that for an optimal allocation of load, all processors must stop processing at the same time. Intuitively this is because otherwise some processors would be idle while others were still busy. Analogous solutions have been developed for tree networks [67] and bus networks [23, 40].

In [196] the concept of collapsing two or more processors and associated links into a single processor with equivalent processing speed is presented. This allows a complete proof (an abridged one appears in [66]) that for the optimal, minimal time solution all processors must stop at the same time. Moreover, for the case without front-end communication sub-processors it allows a simple algorithm, to determine when it is economical to distribute load among multiple processors. Finally, the notion of equivalent processors enables the derivation of simple closed form expressions for the equivalent speed of a linear daisy chain network containing an infinite number of processors. This provides a limiting value for the performance of this network architecture and scheduling policy. This last point will now be discussed.

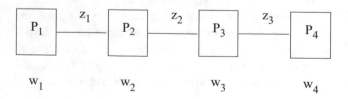

Fig. 5.13 Linear daisy chain network

Fig. 5.14 Linear network with front-end communication sub-processors

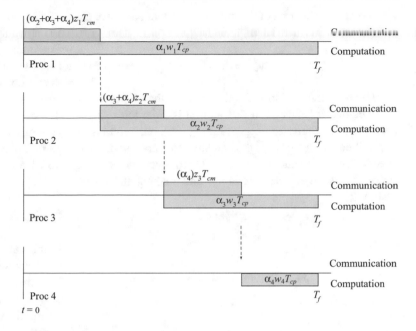

Fig. 5.15 Linear network without front-end communication sub-processors

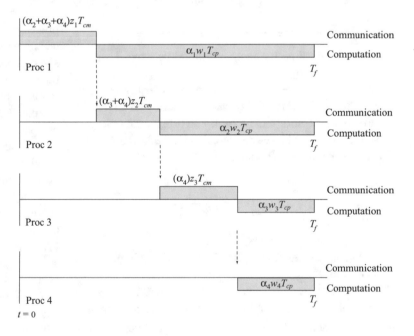

Infinite Number of Processors

A difficulty with the linear network daisy chained architecture is that as more and more processors are added to the network, the amount of improvement in the network's equivalent speed approaches a saturation limit. Intuitively, this is because of the overhead in communicating the problem representation down the linear daisy chain in what is essentially a store and forward mode of operation.

It is possible to develop simple expressions for the equivalent inverse processing speed of an infinite number of homogeneous processors and links. These provide a limiting value on the performance of this architecture. The technique is similar to that used for infinitely sized electrical networks to determine equivalent impedance.

Let the load originate at a processor at the left boundary of the network (processor 1). The basic idea is to write an expression for the speed of the single equivalent processor for processors $1, 2 \ldots \infty$. This is a function of the speed of the single equivalent processor for processors $2, 3 \ldots \infty$. However these two speeds should be equal since both involve an infinite number of processors. One can simply solve the resulting implicit equation for this speed.

Consider, first, the case where each processor has a front-end communication sub-processor. Let $w_i = w$ and $z_i = z$.

Let the network consist of P_1 and an equivalent processor for processors $2, 3 \ldots \infty$. Then

$$w_{eq}^{fe} = \hat{\alpha}_1 w \tag{5.111}$$

Here $\hat{\alpha}_i$ is the total load fraction of what the ith processor receives that it keeps (load distribution is from left to right here). Using a derivation appearing in [196], one can find an implicit equation for w_{eq}^{fe}, the equivalent inverse processing speed, for an infinite network with front-end processors. From (5.176) there with equality, and making the above assumption

$$w_{eq}^{fe} = \frac{z\rho + w_{eq}^{fe}}{w + z\rho + w_{eq}^{fe}} w \tag{5.112}$$

Here $\rho = T_{cm}/T_{cp}$. Solving explicitly for w_{eq}^{fe} results in

$$w_{eq}^{fe} = \left(-z\rho + \sqrt{(z\rho)^2 + 4wz\rho}\right)/2 \tag{5.113}$$

The solution time for such an infinite network is simply given by $T_{sol} = w_{eq}^{fe} T_{cp}$.

In a similar manner, an expression for the equivalent processing speed of a linear daisy chain network with an infinite number of processors with no front-end (nfe) communication sub-processors can be determined. Again, the load originates at processor 1 at the left boundary of the daisy chain.

$$w_{eq}^{nfe} = \sqrt{wz\rho} \tag{5.114}$$

The solution time for this infinite network is simply given by $T_{sol} = w_{eq}^{nfe} T_{cp}$.

This last expression is somewhat intuitive. Doubling w and z doubles w_{eq}^{nfe}. Doubling either w or z alone increases w_{eq}^{nfe} by a factor of $\sqrt{2}$. These results agree very closely with numerical results presented in [66]. It is straightforward to show that $w_{eq}^{fe} < w_{eq}^{nfe}$. Thus, in this limiting case, solution time is always reduced through the use of front-end processors.

It is also possible to use the above results to calculate the limiting performance of an infinite sized daisy chain when the load originates at a processor at the interior of the network (with the network having infinite extent to the left and to the right). Expressions (5.113) or (5.114) can be used to construct equivalent processors for the parts of the network to the left and right of the originating processor. The resulting three processor system then can be simply solved.

The concept of collapsing two or more processors into an equivalent processor has been shown to be useful in examining a variety of aspects related to these linear daisy chain networks of load sharing processors. Expressions for the performance of infinite chains of processors are particularly useful as if one can construct a finite sized daisy chain that approaches the performance of a hypothetical infinite system, one can feel comfortable that performance cannot be improved further for this particular architecture and load distribution sequence.

5.5.2 Tree Networks

In this subsection load distribution for networks with a tree topology with sequential load distribution is discussed [26]. This material is more general than a simple consideration of hard-wired tree networks of processors. This is because a natural way to distribute load in a processor network with cycles is through the use of an embedded spanning tree.

A homogeneous binary tree network of communicating processors will be considered in this subsection. The general technique developed here can be applied to other types of symmetrical and homogeneous tree networks. In the tree there are three types of processors: root, intermediate, and terminal processors. Each tree has one root processor that originates the load. An intermediate processor can be viewed as a parent of lower level processors with which it has a direct connection. Also it is a child of an upper level processor with which it has a direct connection. The terminal processors can only be children processors.

Every processor can only communicate with its children processors and parent processor. Each of the processors in the tree is assumed to have the same computational speed, $1/w$. The communication speed between a parent processor and each of its children is also assumed to have the same value, $1/z$.

In this section a binary tree where processors are equipped with front-end processors for communications off-loading will be discussed. Therefore communication and computation can take place in each processor at the same time.

In [67] a finite tree for the above case was discussed. The minimum processing time is achieved when all processors in the tree stop at the same time. Moreover formal proofs of optimality of single level trees are available [217]. As the size of the tree becomes larger, the share assigned to the root processor becomes smaller and so the processing time decreases. On the other hand, adding more processor (node) levels to the tree will result in more overhead time spent in communicating small fractions of load to the new processors. At some point, adding more processors will not decrease the fractions of load assigned to the root processor substantially and so there is not a considerable improvement

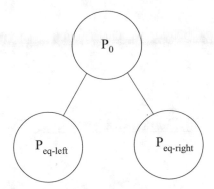

Fig. 5.16 Three node reduced tree network

in the processing time. In that case, it may be advisable not to add more processors (hardware) to the tree since the cost of doing so may not be worth the small improvement in the performance of the system.

The idea behind obtaining the asymptotic processing time (i.e., processing time for an infinite size network) for this tree is to collapse the tree into three processors as shown in Fig. 5.16. The right side of the tree below the root, P_0, has been replaced by one "equivalent" processor with equivalent processing speed w_{eq}^∞. The same is true for the left side of the tree below the root, P_0, where it was replaced with one "equivalent" processor that has an equivalent computational speed w_{eq}^∞. Naturally, as the left and right subtrees are homogeneous infinite trees in their own right, an equivalent processor for either one of them has the same computational speed as one for the entire tree.

The timing diagram for this equivalent system, that preserves the characteristics of an infinite size binary tree, is shown in Fig. 5.17. From this figure it can be seen that the computing time of the root processor, $\alpha_0 w T_{cp}$, equals the communication time between the parent processor (root processor) and the left processor, $\alpha_l z T_{cm}$, plus the computing time of the left equivalent processor, $\alpha_l w_{eq}^\infty T_{cp}$. Also the computing time of the left side equivalent processor, $\alpha_l w_{eq}^\infty T_{cp}$, equals the communication time between the root processor and the right side "equivalent" processor, $\alpha_r z T_{cm}$, plus the computing time of the right equivalent processor, $\alpha_r w_{eq}^\infty T_{cp}$. If the three processors in Fig. 5.16 are replaced with one equivalent processor, then the computing time of the root processor, $\alpha_0 w T_{cp}$, equals the computing time of the equivalent processor, $w_{eq}^\infty T_{cp}$. The three equations just explained are listed below:

$$\alpha_0 w T_{cp} = \alpha_l z T_{cm} + \alpha_l w_{eq}^\infty T_{cp} \qquad (5.115)$$

$$\alpha_l w_{eq}^\infty T_{cp} = \alpha_r z T_{cm} + \alpha_r w_{eq}^\infty T_{cp} \qquad (5.116)$$

$$\alpha_0 w T_{cp} = w_{eq}^\infty T_{cp} \qquad (5.117)$$

Also the sum of the fractions of the load equals one

$$\alpha_0 + \alpha_r + \alpha_l = 1 \qquad (5.118)$$

Now, there are four equations with four unknowns, namely $w_{eq}^\infty, \alpha_0, \alpha_r$, and α_l. With some algebra one can show that w_{eq}^∞ can be determined by iteratively solving the equation

$$w_{eq}^\infty = \frac{w(z T_{cm} + w_{eq}^\infty T_{cp})}{z T_{cm} + w_{eq}^\infty T_{cp} + w T_{cp} + \frac{w w_{eq}^\infty T_{cp}^2}{z T_{cm} + w_{eq}^\infty T_{cp}}} \qquad (5.119)$$

Solving this equation is equivalent to solving the following cubic equation

$$\left(w_{eq}^\infty\right)^3 + [2z\rho + w]\left(w_{eq}^\infty\right)^2 - [z\rho(w - z\rho)]\, w_{eq}^\infty - wz^2\rho^2 = 0 \qquad (5.120)$$

Consequently, the ultimate finish time for an infinite tree network with front-end processors, T_{fe}^∞, can now be computed by

$$T_{fe}^\infty = w_{eq}^\infty T_{cp} \qquad (5.121)$$

Naturally

$$\text{Speedup} = \frac{w}{w_{eq}^\infty} \qquad (5.122)$$

In a similar way the solution time for a homogeneous tree without front ends can be found. This technique does not apply to heterogeneous (i.e., networks with different link and processor speeds) infinite networks. It can be extended though to symmetrical networks with three or more children per node.

5.6 Multi-Installment Scheduling

Very early work on creating and analyzing divisible load schedules involved what may be called single installment strategies. For instance, if one is looking at sequential distribution in a single level tree network, the root node distributes all of child one's load in one continuous installment, then the root node distributes all of child two's load in a single continuous installment and so on, finishing with the distribution of load from the root to the last child in one continuous installment.

The inefficiency in a single installment strategy (see Fig. 5.4) is that all but the first child must wait, possibly a significant amount of time, before starting computation. This lowers the overall utilization (i.e., fraction of time useful work is done) of the processors.

In a multi-installment strategy, first described in [38], child one first receives a partial installment of its load, then child two receives a partial installment, then child 3 ... on to child N. At this point child one receives a second partial installment, child two receives a second partial installment,

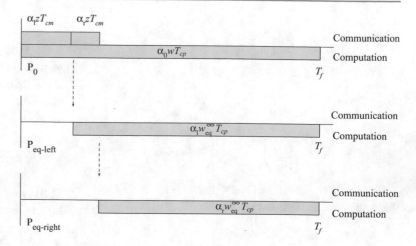

and so on. The process repeats for a fixed number of installments (or rounds as they are sometimes called). Under a multi-installment strategy the processors can start processing earlier than under a single-installment strategy. Utilization is increased and finish time (makespan) decreases.

In [38] recursive equations for single level tree networks for the general case and closed form solutions where all links have identical transmission speeds are found. Tradeoffs between the number of installments and the number of processors are discussed. This work assumes linear costs. That is, computation time and communication time is proportional to load size.

In [249] an affine cost model is assumed. Here besides time delay terms proportional to load size, there are also fixed constant latencies. That is,

$$T_{comp,i} = l_{comp,i} + \frac{chunk_i}{S_i} \qquad (5.123)$$

$$T_{com,i} = l_{com,i} + \frac{chunk_i}{B_i} \qquad (5.124)$$

Here $T_{comp,i}$ and $T_{com,i}$ are the computation and communication time of the ith processor/link, respectively. The fixed latencies are $l_{comp,i}$ and $l_{com,i}$ for computation and communication, respectively. The installment/round data size is $chunk_i$. Also S_i is the processing speed and B_i is the data transmission speed.

The fixed latencies are sometimes called start-up costs (which has nothing to do with monetary cost)[41, 50, 221]. Affine costs can also be used for single load distribution strategies.

Yang goes on to propose a uniform multi-round (UMR) algorithm—uniform because it assumes that chunk sizes are fixed within each round. The near optimal number of rounds is calculated. The size of the chunks grows exponentially in each round,

In [247, 248] UMR is modified to handle concurrent transmission for root to child under certain conditions (the

parallel transferable uniform multi-round (PTUMR) algorithm). The algorithm optimizes the size of the chunks, the number of rounds, and the number of children to which data is transmitted in parallel.

Multi-installment scheduling in systems with limited memory is discussed in [30, 82]. An exact branch and bound algorithm, genetic algorithm, and heuristic algorithms are examined in these works.

A feature one sees in models of divisible load scheduling is a consideration of "release times." The release time [39, 47, 231] is the time at which a processor becomes available to process load. This feature has been studied with multi-installment scheduling processing with arbitrary release times [39] and in multi-installment systems with limited memory [47]. Note that closed form solutions can be found when the processor release times are identical and heuristic algorithms can be used when the processor release times are arbitrary.

Other works on multi-installment scheduling include [28, 118, 213, 214, 238].

In some of these works it is shown that the amount of improvement that results in adding installments is greatest for the first few installments and is smaller for latter installments (i.e., "diminishing returns").

5.7 Computing/Communication Monetary Costs

For many years there has been a vision of "computer utilities" that would provide computing services in exchange for some monetary payment. This has come to pass in some sense when one considers the many online services such as social networks that charge for service. It is perhaps even better realized in cloud computing services such as Amazon Web Services and Microsoft Azure that make available processing power at some monetary charge. In the future, owners of costly high performance computing machines using now exotic technologies such as cryogenics or quantum computing may charge for their use.

In this spirit this section looks at some representative results and algorithms involving optimizing divisible loads when there are monetary charges for computing and communication. This section is based on work from [62, 63, 218].

5.7.1 A Computing Costs Only Example

Consider a bus network [218] with

- A root control processor, P_0, on the bus distributing divisible load to N children processors, $P_1, P_2 \ldots P_N$. Thus there are a total of $N + 1$ processors. Note that such a bus network with $N + 1$ nodes is equivalent to a single level tree network under sequential distribution, where all the links have the same communication speed (homogeneous) but the children speeds may be different from each other (heterogeneous).
- The root processor has a front-end sub-processor so that it can distribute single load installments over the bus to each child processor in order $1, 2 \ldots N$ at the same time it processes some of the load.
- The time to communicate the single installment of load from the root processor to the nth child processor is $\alpha_n z T_{cm}$ where we use the same variable definitions as earlier in the chapter. The time to process the nth load installment on the nth processor is $\alpha_n w_n T_{cp}$.
- Under the optimality principle assume that all processors stop at the same time. It is assumed in this entire section that for any given load distribution sequence or arrangement profile that the α's are determined using the methodology of Sect. 5.2 Other approaches are possible but this approach is a natural one.
- A linear cost model is assumed. Only computing is charged here. That is,

$$C_{total} = \sum_{n=0}^{N} \alpha_n c_n w_n T_{cp} \qquad (5.125)$$

Here c_n is the cost per second of processing load on the nth processor.

With the bus interconnection network, the root processor can only distribute load to one processor at a time. If one assumes that load can be distributed to children nodes in any sequence (order) with only one installment per child, then with some algebra it can be shown [218] that the optimal monetary cost sequence is such that

$$c_0 w_0 \leq c_1 w_1 \leq c_2 w_2 \leq \ldots \leq c_N w_N \qquad (5.126)$$

Here load is processed at the root processor with the smallest value of $c_n w_n$, the first child to receive load has the next smallest value of $c_n w_n$ and so on.

In terms of dimensioning one has

$$\alpha_n c_n w_n T_{cp} \longleftrightarrow [load][cost/sec][sec/load]$$

$$[dimensionless] \qquad (5.127)$$

Thus the overall unit is monetary cost. Recall that w represents inverse computing speed. Also, the units of $c_n w_n$ are $[cost/load]$.

Thus the optimization criteria above holds that processing should be done in non-decreasing order of cost per load or $c_n w_n$. This is an intuitively pleasing result.

5.7.2 Two Quantities to Optimize

Two additional optimization problems are possible. This is because we seek to minimize both computing cost and finish time (i.e., makespan). One problem is to minimize monetary cost with a bound on finish time and the other is to minimize finish time with a bound on monetary cost.

First consider minimizing cost with a bound on finish time. Assume the faster a processor, the more expensive it is to use. One can reduce the total cost below what was achieved in the previous section but finish time, T_f, will increase. That is, one uses cheaper but slower processors more than in the previous section.

The minimal cost is achieved if all load is processed by the cheapest processor only:

$$C_{total}^{min} = \min_{n=0,1,2\ldots N} c_n w_n T_{cp} \qquad (5.128)$$

With a bound on T_f, more than one processor may be used. Call the bound T_f^{bound}. A heuristic algorithm to do this minimization is presented in [218]. It solves:

$$\min_{T_f \leq T_f^{bound}} Cost = \min_{T_f \leq T_f^{bound}} \sum_{n=0}^{N} \alpha_n c_n w_n T_{cp} \qquad (5.129)$$

As a second optimization problem, one can also minimize finish time with a bound on monetary cost.

$$\min_{C \leq C^{bound}} T_f \qquad (5.130)$$

A heuristic algorithm based on binary search for this problem appears in Sohn [219].

5.7.3 Computation and Communication Monetary Cost

Suppose one adopts a linear model for communication monetary cost in addition to a linear model for computation monetary cost. Consider a single level tree with root control processor P_0 and children processors $P_1, P_2 \ldots P_N$. As in the previous part of this section, also assume that the control processor communicates sequentially a single installment of load to each child. Each child starts computing when it finishes receiving its load from the root processor. All processors stop computing at the same time T_f. Link speeds can be different (heterogeneous) as well as processor speeds.

Let c_n^l be the cost per second of communicating the nth installment load on the nth link. Let c_n^p be the cost per second of processing the nth load on the nth processor. Then the cost model becomes the following

$$C_0 = c_o^p w_0 T_{cp} \tag{5.131}$$

$$C_n = c_n^l z_n T_{cm} + c_n^p w_n T_{cp} \quad n = 1, 2, \ldots N \tag{5.132}$$

where

$c_n^p w_n$: computing cost per unit load of utilizing the nth processor.

$c_n^l z_n$: communication cost per unit load of utilizing the nth link.

$c_n^p w_n + c_n^l z_n$: processing cost per unit load of utilizing the nth link processor pair.

Now $\alpha_n C_n$ is the cost of processing the assigned fraction of load (α_n) on the nth processor and the total cost is

$$C_{total} = \alpha_0 C_0 + \sum_{n=1}^{N} \alpha_n C_n \tag{5.133}$$

Two types of optimization problems are discussed in [62, 63]. In the "arrangement problem" (or "computer configuration" problem), one is given a set of N links and N children processors and one is free to put them together in the best single level tree network arrangement for optimal distribution in a monetary cost sense. That is, initially links are not assigned to particular processors. Under the "sequencing problem" one is given link-processor pairs (i.e., each link fixed to a child processor) and one must find the monetary cost optimal sequence of load distribution from the root to the children processors.

The algebra becomes substantive in these cases. For the arrangement problem, heuristic adjacent pairwise swapping of processors is considered. A number of (not exhaustive) algebraic conditions when such a swapping decision decreases the total cost is described. However a basic algorithm has a fairly large tendency to converge to locally optimal

solutions. Adding more heuristic features achieves much better results [62].

For the sequencing problem a basic greedy algorithm can be developed [63] that has very good probabilistic convergence results though it is not optimal (tabu search and simulated annealing can do better).

5.7.4 In Summary

Through the work cited here we have learned that these monetary cost optimization problems lead to fairly complex exercises in algebra with no simple solutions when both communication and computation costs are considered.

The problem is susceptible to combinatorial optimization algorithms to find good but not necessarily globally optimal solutions.

A limitation was doing cases with more than about 20 children processors. It was hard numerically to distinguish between solutions that were very close to each other in order to do an optimization.

5.7.5 Extensions

The underlying mathematics of monetary cost and time based divisible load scheduling optimization may be of use in other applied areas. For instance, [40] speaks of a truck(s) hauling a divisible physical quantity (such as sand) from a source to processing plants in a sequential or concurrent manner. Another possibility is pumping crude oil from an oil field concurrently over several pipelines to refineries in a simultaneous distribution manner. Certainly specific aspects of the divisible load paradigm will be more or less valid in other applied contexts than in the cases discussed in this chapter.

5.8   Signature Searching

A "signature" is a data pattern of interest in a very large data file. Signature searching involves searching through massive amounts of data that can be produced with today's technology looking for (possibly multiple) signatures. Potential applications include DNA sequence detection, biometrics, speech recognition, aerospace applications such as radar, large scientific experiments, and many more.

In this section we consider the statistics of finding signatures in flat files. That is one can view a data set as a huge, linear (flat) file possibly indexed by time. The file should be optimally partitioned among processing nodes in a network so that processing (search) time is minimized. Flat, linear, files are a natural choice for early database implementation though at a later point more sophisticated database models are often used.

Signature searching is similar in spirit to "string searching." String searching generally involves finding a pattern of length m in a text of length of n over an alphabet. Both signature searching and string searching are special cases of pattern searching.

5.8.1 Uniformly Distributed Signatures

Single Signature

Consider a uniformly distributed single signature. It will be assumed that the network topology is either a linear daisy chain network or a single level tree (star) network, each of $(M + 1)$ nodes. If one looks at a single level tree network, sequential load distribution will be assumed. Assume also single installment computing.

Let

α_m: load fraction assigned to mth processor. Here $m = 0, 1, 2 \ldots M$.

$T_f(M)$: optimal finish time of network with $M + 1$ processors.

S_m : time when the mth processor begins processing load.

$\mathbf{Y}$: a random variable that indicates the amount of time until a signature is found.

Then

$$E[\mathbf{Y}] = \sum_{m=0}^{M} \alpha_m \frac{T_f(M) + S_m}{2} \quad (5.134)$$

Intuitively, the average signature search time is the weighted sum of the midpoint of each processing segment weighted by the size of each segment. In [144] a more general derivation of this result is given.

Multiple Signatures

The solution for the expected time of searching for multiple signatures becomes more involved since the last signature to appear in the data set is not always the last signature to be detected. This occurs because the processors sift through the data concurrently, modifying the order in which detection is made. However the expected time of finding all of the signatures can be obtained by finding the expected time to find the last signature position. This is an example of "order statistics," the statistics of a numerical ordering of random variables.

Using the concept of an equivalent processor, the expected time of finding all of L signatures in a data set under independent and uniformly distributed data locations can be found as [144]

$$E[\mathbf{Y_L}] = ? \frac{L}{L+1} \left[T_f(M) - \sum_{m=0}^{M} \alpha_m \frac{T_f(M) + S_m}{2} \right] \quad (5.135)$$

$$+ \left[\sum_{m=0}^{M} \alpha_m (T_f(M) + S_m) - T_f(M) \right] \quad (5.136)$$

Here L is the number of (known) signatures and $\mathbf{Y_L}$ is the random variable indicating the position of the last signature.

What of finding the lth, $1 \leq l \leq L$, signature if the signatures are uniformly distributed? A derivation that results in an expression for these for linear daisy chain and single level tree networks appears in [148]. The derivation is based on the kth ordered sample of a uniform distribution, known as the beta distribution.

5.8.2 Arbitrary Distributions of Signature Location

Suppose that a single signature in a massive flat file is distributed according to an arbitrary but known probability distribution under sequential processing. Then one can reduce the expected time to find the signature by processing the data record in decreasing order of the probability (likelihood) of the signature appearing in the data in a "greedy" manner [148].

Consider Fig. 5.18. It shows divisible load computation only, in a single level tree network with single installment sequential distribution. Here the root processor P_0 does computation as well as load distribution communication though the data load distribution is not shown in the diagram.

Here also S_i is the ith start time of processing by the ith processor.

Consider now Fig. 5.19. Here the part of the data record with the highest probability of containing a single signature is processed by P_0 from S_0 to S_1. The data with the next highest probability of containing a signature is processed by $P0$ and $P1$ between S_1 and S_2. The process continues with the data with the next highest probability of containing a signature processed by P_0, P_1, and P_2 between S_2 and S_3 and so on. The process is described in detail in [148].

5.8.3 Searching in a Networked File System

A particular example of questions that can be addressed in the context of signature searching appears in [251, 252]. Here flat file data records are placed in nodes in an interconnection network sometime prior to searching. Thus load distribution is not taken into account. This assumption is sometimes jus-

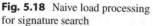

Fig. 5.18 Naive load processing for signature search

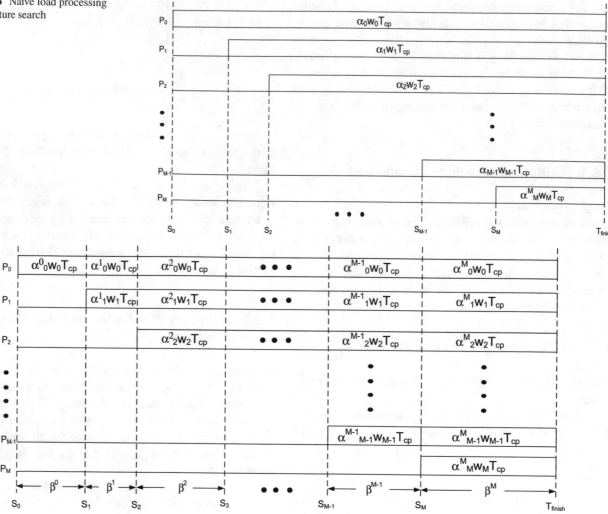

Fig. 5.19 Greedy load processing for signature search

tified as load is often spontaneously distributed to processors without being monitored, scheduled, or timed. One can pose a number of interesting problems involving finding expected search time and speedup under different search protocols that differ in a number of assumptions.

- Whether the number of signatures in a file at a node is known or unknown a priori.
- Whether the data file stored at a node contains one or more than one signature.
- If no signatures are found in a file(s) at a node in a tree interconnection network, whether this implies there are no signatures in files in the children nodes of the node.
- Whether search, particularly in a tree of processors, is sequential or concurrent.
- Which type of interconnection network is considered.

Analyses for these cases can be found in [251, 252].

5.8.4 In Summary

Signature searching occurs in many potential applications so problems of this type will be of interest into the future.

5.9 Time-Varying Environments

All of the previous sections investigated divisible loads under the assumption that a processor can compute only a single job at a time. Under this assumption, the next job can be served only after the processor finishes the computation of the currently running job. However, most practical time-sharing computer systems can handle more than one job at a time. It is therefore natural to study divisible loads in multiprogrammed and multiprocessor environments [219, 244].

In the previous sections processor and link speed were constant. Let us consider situations where they may vary with time. The processors here can be assumed to be multiprogrammed so that there are a number of jobs running in the *background* in addition to a divisible load of interest. These background jobs consume processor and link resources so that the divisible load of interest may see time-varying processor and link speed. It is immaterial for our purposes whether the background jobs are divisible or indivisible. The processor speed and the channel speed depend on the number of jobs which is currently served under a processor or transmitted through a channel. When there are a large number of jobs running in a processor, the processor speed for a specific job of interest becomes slower than when it has fewer jobs. The channel speed also becomes slower when there are a large number of background job related transmissions passing through a link than when there are fewer transmissions using the link.

The purpose of this section is to show how to determine the optimal fraction of an entire divisible workload to be distributed to each processor to achieve the minimal processing time when the processor speeds of the processors are time-varying variables. To determine the optimal fraction of the workload deterministically, the processor speed over the duration of the divisible load computation must be known in advance before the load originating processor starts distributing the workload to each processor. If the exact arrival times and departure times of the background jobs are known, one can determine the exact time-varying processor speed and the channel speed. This is suitable for *production jobs* that are performed in a system repeatedly for a known period. If the arrival and the departure times of the background jobs are not known, but the stochastic arrival process and the stochastic departure process of the jobs can be assumed to be Markovian, the optimal fraction of the workload can still be found by a stochastic analysis which makes use of well-known Markovian queueing theory. In this section a deterministic numerical method to find the optimal allocation of the entire workload in terms of minimal processing time is presented when the background jobs' arrival and departure times are known.

5.9.1 Time-Varying Processor Speed

The distributed computing system to be considered here consists of a control processor for distributing the workload and N processors attached to a linear bus as in Fig. 5.20. New arriving measurement data is distributed to each processor under the supervision of the control processor. The control processor distributes the workload among the N processors interconnected through a bus type communication medium in order to obtain the benefits of parallel processing. Note that the control processor is a network processor which does no processing itself and only distributes the workload.

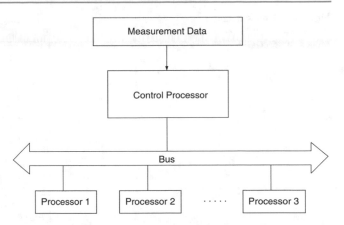

Fig. 5.20 Bus network with load origination at control processor

Each processor is a multiprogrammed processor that can simultaneously process multiple jobs. Thus the processor speed varies with time and it depends on the amount of workload. The processor speed varies under the following processor sharing rule: The processor devotes all its computational power evenly to each job. That is, if there are m jobs running under a certain processor, each job receives $1/m$ of the full computational power of the processor. This behavior is similar to a fair resource scheduling policy as used in UNIX systems. It is assumed here that there is no limitation of the number of jobs to be simultaneously processed in a single processor, even though the processor speed for a specific job will be very slow if there are a large number of jobs running simultaneously under the processor.

In this section we assume that background jobs start and terminate simultaneously across all processors and that negligible bus communication is needed to support their running. The technique can be extended to situations with different background job start/stop times on different processors. The main problem in this section is to find the optimal fraction of a divisible load which is distributed to each of N processors to minimize the processing finish time when the communication delay is nonnegligible.

The timing diagram for the bus network with load origination at a control processor is depicted in Fig. 5.21. In this timing diagram, communication time appears above the horizontal time axis and computation time appears below the axis. In this section, the channel speed is assumed to be a constant while the computing speed of each processor is assumed to be time-varying, as described above.

At any time, the processor effort available for the divisible loads of interest varies because of background jobs which consume processor effort. These background jobs can arrive at or terminate on the processors at any time during the computation of the divisible load that the control processor is going to distribute. The arrival and departure times of the background jobs over intervals during which the divisible

Fig. 5.21 Timing diagram for
bus network with time-varying
processor speed

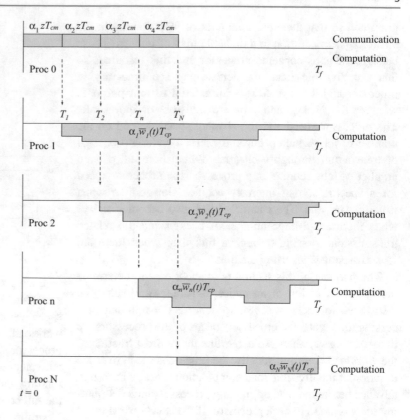

load is processed, however, should be exactly known. This is
the reason that this section represents deterministic models
of the load sharing problem. When the arrival and departure
times are unknown and the statistics of the arrival and
departure process of the jobs are known to be Markovian,
then this load sharing problem can be stochastically analyzed
as in [219, 244].

Referring to Fig. 5.21, at time $t = 0$, the originating
processor (the control processor in this case) transmits the
first fraction of the workload to P_1 in time $\alpha_1 z T_{cm}$. The
control processor then sequentially transmits the second
fraction of the workload to P_2 in time $\alpha_2 z T_{cm}$, and so on.
After P_1 completes receiving its workload from the control
processor (an amount of α_1 of the entire load), P_1 can start
computing immediately and it will take a time of $T_f - T_1$
to finish. Here $T_1 = \alpha_1 z T_{cm}$. The second processor P_2 also
completes receiving the workload from the control processor
at time $T_2 = (\alpha_1 + \alpha_2) z T_{cm}$ and it will start computing for
a duration of $T_f - T_2$ of time. This procedure continues
until the last processor. For optimality, all the processors
must finish computing at the same time. Intuitively, this is
because otherwise the processing time could be improved by
transferring the load from busy processors to idle ones.

Now let us represent those intervals of the computation
time as $T_f - T_1, T_f - T_2, \ldots, T_f - T_N$. The interval $T_f - T_n$
for P_n to compute the nth fraction of the entire load can be
expressed as

$$T_f - T_n = \alpha_n \overline{w}_n(t) T_{cp} \qquad n = 1, 2, \ldots, N \tag{5.137}$$

where $\overline{w}_n(t)$ is defined as the inverse of the time average of
the applied computing speed of P_n in the interval (T_n, T_f).
Since $w_n(t)$ is defined as *the inverse* of the computing speed,
to calculate the time average of $w_n(t)$ one must invert $w_n(t)$
first to make it proportional to the actual computing speed
and take the time average, and then invert it again. That is,

$$\overline{w}_n(t) = \left(E \left\{ \frac{1}{w_n(t)} \right\} \right)^{-1}$$

$$= \frac{T_f - T_n}{\int_{T_n}^{T_f} \frac{1}{w_n(t)} dt} \tag{5.138}$$

Explanatory diagrams for the computing speed of P_n are
depicted Fig. 5.22a, b, and c. Consider Fig. 5.22a, b, and c
in reverse order. Figure 5.22c shows the process which is
proportional to the computing speed of P_n which is available
for the single divisible job of interest. The divisible job
arrives at time 0. When the processor is idle in the interval
(t_0, t_1), the divisible load that is delivered from the control
processor will receive the full computational power of P_n.
Therefore, the computing speed of P_n in the interval (t_0, t_1)
for the load from the control processor is $1/w_n$ where w_n
is the *inverse* of the maximum computational power of P_n.
When there is one background job running in the processor
in the interval (t_1, t_2) due to the arrival of one background

Fig. 5.22 (a) Derivative of timing process which is inversely proportional to computing speed (b) Timing process which is inversely proportional to computing speed. (c) Timing process which is proportional to computing speed

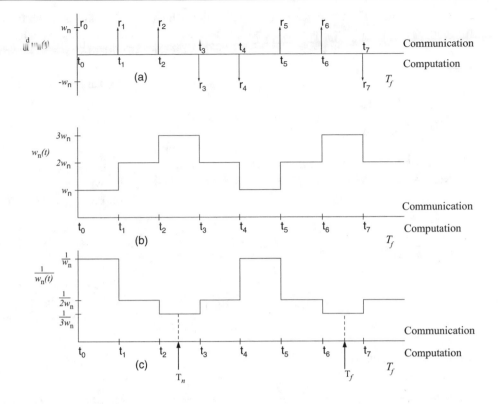

job at time $t = t_1$, the computational power of P_n is equally divided by two so that each job, one background job and the divisible load from the control processor, can receive half of the full computational power of P_n. That is, the computing speed of P_n in the interval (t_1, t_2) for each job is $1/2w_n$.

Likewise, when there are two background jobs running in the processor in the interval (t_2, t_3) due to the additional arrival of a background job at time $t = t_2$, the computational power of P_n is equally divided by three so that each job, two background jobs and the divisible load from the control processor, can receive one third of the full computational power of P_n. The computing speed of P_n in the interval (t_2, t_3) for each job is $1/3w_n$. When the processor finishes the computation of one of the background jobs at time $t = t_3$, the computing speed of the P_n for each job (at this time, there are two jobs running in the processor, one a background job and the other a divisible load fragment from the control processor) speeds up back to $1/2w_n$.

Note that the integral in the denominator of Eq. (5.138) is the area under the curve of Fig. 5.22c between times T_n and T_f.

Figure 5.22b shows the process which is *inversely proportional* to the computing speed of P_n. In other words, Fig. 5.22b is just the inverse of Fig. 5.22c. Figure 5.22a is the derivative of Fig. 5.22b. This represents the arrival and departure time of the jobs. The upright impulse r_0 indicates the arrival of a divisible job that is present in the system for the duration of the timing diagram. The upright impulses (r_1, r_2, r_5, r_6) represent the arrival of each background job

and the upside down impulses (r_3, r_4, r_7) represent the departure or service completion of each background job.

What is deterministic in this section is that the time of each arrival and departure of the background jobs is deterministically known. That is, the time $t_1, t_2, \ldots$, etc. should be all known at time $t = 0$. This condition can be true of a production system repetitively running the same jobs. The height of the each impulse is $+w_n$ for the ones which corresponds to the arrivals and $-w_n$ for the ones which corresponds to the departure of the background job. This is because one arrival of a background job causes the computing speed to change from $1/mw_n$ to $1/(m+1)w_n$ in Fig. 5.22c so the inverse speed changes from mw_n to $(m+1)w_n$ in Fig. 5.22b for any integer m. A similar explanation can applied to the departure of background jobs.

Let us now find the expressions for Fig. 5.22a, b, and c. The expression for Fig. 5.22a is

$$\frac{d}{dt}w_n(t) = \sum_{k=0}^{\infty} r_k \delta(t - t_k)w_n \qquad (5.139)$$

where

$$r_k = \begin{cases} +1, & \text{for arrival} \\ -1, & \text{for departure} \end{cases}$$

The following equation represents Fig. 5.22b.

$$w_n(t) = \sum_{k=0}^{\infty} r_k u(t - t_k)w_n \qquad (5.140)$$

Here $u(t)$ is the unit step function (i.e., a function which has the value of 1 for positive time and has a value of 0 for negative time). A little thought yields an expression for Fig. 5.22c

$$\frac{1}{w_n(t)} = \sum_{k=0}^{\infty} \left(\sum_{j=0}^{k} r_j \right)^{-1} [u(t - t_k) - u(t - t_{k+1})] \frac{1}{w_n} \tag{5.141}$$

The next step is to find the time average of $w_n(t)$ in the interval (T_n, T_f). To find $\overline{w}_n(t)$, it is necessary to find $\int_{T_n}^{T_f} \frac{1}{w_n(t)} dt$ from Eq. (5.138).

$$\int_{T_n}^{T_f} \frac{1}{w_n(t)} dt = \frac{T_f}{w_n(T_f)} - \frac{T_n}{w_n(T_n)}$$
$$- \sum_{k=x_n+1}^{x_f} \left(\frac{1}{w_n(t_k)} - \frac{1}{w_n(t_{k-1})} \right) t_k \tag{5.142}$$

See [219] for details. Therefore,

$$\overline{w}_n(t) = \frac{T_f - T_n}{\frac{T_f}{w_n(T_f)} - \frac{T_n}{w_n(T_n)} - \sum_{k=x_n+1}^{x_f} \left(\frac{1}{w_n(t_k)} - \frac{1}{w_n(t_{k-1})} \right) t_k} \tag{5.143}$$

Using Eq. (5.137), one can also find the expression for α_n.

$$T_f - T_n = \alpha_n \overline{w}_n(t) T_{cp}$$
$$= \alpha_n T_{cp} \frac{T_f - T_n}{\int_{T_n}^{T_f} \frac{1}{w_n(t)} dt} \tag{5.144}$$

Thus

$$\alpha_n = \frac{1}{T_{cp}} \int_{T_n}^{T_f} \frac{1}{w_n(t)} dt$$
$$= \frac{1}{T_{cp}} \left[\frac{T_f}{w_n(T_f)} - \frac{T_n}{w_n(T_n)} \right.$$
$$\left. - \sum_{k=x_n+1}^{x_f} \left(\frac{1}{w_n(t_k)} - \frac{1}{w_n(t_{k-1})} \right) t_k \right] \tag{5.145}$$

Here Eqs. (5.142), (5.143), and (5.145) are functions of T_n and T_f. That is, if T_n and T_f are known, the fraction of the workload for P_n as well as the integral of the applied computing speed of the nth processor and the inverse of the average applied computing speed of P_n in the interval (T_n, T_f) can be found. This problem can be solved by a simple recursive method that can express every α_n as a function of T_f. Let us introduce an algorithm

to find the optimal fraction of workload that the control processor must calculate before distributing the load to each processor.

1. Express α_N as a function of T_f from

$$\alpha_N = \frac{1}{T_{cp}} \int_{T_N}^{T_f} \frac{1}{w_N(t)} dt$$

Since $T_N = (\alpha_1 + \alpha_2 + \cdots + \alpha_N) Z T_{cm} = Z T_{cm}$, T_N is known.

2. Express α_{N-1} as a function of T_f from

$$\alpha_{N-1} = \frac{1}{T_{cp}} \int_{T_{N-1}}^{T_f} \frac{1}{w_{N-1}(t)} dt$$

Since $T_{N-1} = (1 - \alpha_N) Z T_{cm}$, T_{N-1} is function of α_N and is also function of T_f.

3. Express α_{N-2} as a function of T_f from

$$\alpha_{N-2} = \frac{1}{T_{cp}} \int_{T_{N-2}}^{T_f} \frac{1}{w_{N-2}(t)} dt$$

Since $T_{N-2} = (1 - \alpha_N - \alpha_{N-1}) Z T_{cm}$, T_{N-2} is function of α_N and α_{N-1}, and is also function of T_f.

4. This procedure can be continued up to α_1. Then, one can express every α_n as a function of T_f. Finally, by using the normalization equation which states that $\sum_{n=1}^{N} \alpha_n = 1$, all of the α_n, as well as the actual T_f, can be found.

Note that the algorithm starts from time 0 when the initial processor speeds are known as they are a function of past arrivals and departures.

5.10 Linear Programming and Divisible Load Modeling

In this chapter divisible load scheduling models have been solved by linear equation solution and related recursions. An alternate solution technique is to use linear programming. Linear programming [115, 193] is an optimization technique for constrained linear models developed originally by George Dantzig in 1947. Generally one has a number of continuous variables, a linear objective function of the variables that must be minimized or maximized along with a set of linear constraint equations in the variables.

The simplex algorithm of George Dantzig is an effective means of solving such linear models. It can be shown that the feasible solutions of a linear programming problem lie on the extreme (corner) points of a convex polyhedron in the solution space. One might intuitively and loosely think of the convex polyhedron as a faceted diamond. The simple algorithm moves from extreme point to adjacent extreme point until the extreme point corresponding to the optimal solution is reached. A different interior point method was developed by Narendra Karmarkar in 1984. Interior point methods start at a point inside the convex polyhedron and move outward until the extreme point corresponding to the optimal solution is reached. Agrawal and Jagadish in 1988 were the first to show that linear programming could be used to solve divisible load models.

As an example, consider a divisible load distribution model for a single level tree with simultaneous distribution and staggered start, as in Sect. 5.2.2. From Fig. 5.6, one can write expressions for the finish time, T, as

$$\alpha_0 w_0 T_{cp} \leq T \tag{5.146}$$

$$\alpha_1 z_1 T_{cm} + \alpha_1 w_1 T_{cp} \leq T \tag{5.147}$$

$$\alpha_2 z_2 T_{cm} + \alpha_2 w_2 T_{cp} \leq T \tag{5.148}$$

$$\alpha_i z_i T_{cm} + \alpha_i w_i T_{cp} \leq T \tag{5.149}$$

$$\alpha_m z_m T_{cm} + \alpha_m w_m T_{cp} \leq T \tag{5.150}$$

These are the constraints (i.e., the time that each processor finishes communication and computation is less than or equal to the system finish time, T).

The objective function is

$$\min T \tag{5.151}$$

But

$$T = \alpha_0 w_0 T_{cp} \tag{5.152}$$

So the complete linear program is

$$\min \alpha_0 w_0 T_{cp} \tag{5.153}$$

$$\alpha_1 z_1 T_{cm} + \alpha_1 w_1 T_{cp} - \alpha_0 w_0 T_{cp} \leq 0 \tag{5.154}$$

$$\alpha_2 z_2 T_{cm} + \alpha_2 w_2 T_{cp} - \alpha_0 w_0 T_{cp} \leq 0 \tag{5.155}$$

$$\alpha_i z_i T_{cm} + \alpha_i w_i T_{cp} - \alpha_0 w_0 T_{cp} \leq 0 \tag{5.156}$$

$$\alpha_m z_m T_{cm} + \alpha_m w_m T_{cp} - \alpha_0 w_0 T_{cp} \leq 0 \tag{5.157}$$

$$\alpha_0 + \alpha_1 + \alpha_2 \ldots + \alpha_m - 1 = 0 \tag{5.158}$$

$$\alpha_0, \alpha_1, \alpha_2 \ldots, \alpha_m \geq 0 \tag{5.159}$$

Linear programming is more computation intensive than an analytical solution. The advantage of using it (or using linear equation solution to some extent) is that it takes less analytical effort to simply write the linear program (or set of linear equations) and solve it with a library function than to solve the model analytically. This is particularly advantageous for a complex model. On the other hand, an analytical solution often gives more intuition into the nature of the solution and is faster to solve, once found. Thus the two approaches are complementary.

5.11 Conclusion

Parallel computing is likely to be of interest for a considerable amount of time. Thus there will be increasing opportunities to analyze scheduling for parallel processors, as is done in this chapter for divisible loads.

5.12 Problems

1. What is a divisible load?
2. What is it about divisible load scheduling modeling that makes solutions tractable? What other types of modeling is it related to?
3. What is the difference between sequential and simultaneous distribution?
4. Intuitively, why is simultaneous distribution scalable?
5. What is the difference between staggered start and simultaneous start?
6. What does a front-end processor in a processor allow one to do? What should lead to a faster solution: the inclusion of a front-end processor or its absence? Why?
7. What is speedup? What does it measure?
8. Explain the Gantt chart like diagrams used in this chapter.
9. Why does speedup increase nonlinearly for nonlinear models? Is one getting something for nothing?
10. Explain the concept of equivalent processors. How does it help in finding overall network performance?
11. Intuitively, why does speedup saturate as the size of a linear daisy chain is increased (up to infinite size chains)? Why does speedup saturate in tree networks as the number of children and/or levels is increased?
12. Why is it sometimes not worth distributing load to a neighboring node in a linear daisy chain?
13. Why is the environment in which divisible load is processed often time varying?
14. What is an indivisible load?
15. Give an example of a load that is not divisible.
16. Is linear programming useful in solving divisible load scheduling problems?

17. Verify Eq. (5.21).
18. Do divisible load scheduling problems with nonlinear loads allow one to get something for nothing?
19. What is an equivalent processor?
20. Why is the existence of product form solutions in some models a good thing?
21. What is a practical advantage of a knowledge of infinite network performance?
22. Give an application example of a situation where the concept of monetary cost optimization is important.
23. Give an example of a "signature."
24. Consider a linear daisy chain with N processors where load originates at the left most processor. Store and forward switching and staggered start is used—a node receives all load for itself and its right neighbors before commencing processing.

    (a) Suppose each processor receives the entire measurement load for processing (perhaps each processor processes the load with different algorithms). Find the optimal number of processors that minimize the finish time. Note: Using too many processors leads to excessive communication delays and using too few processors leads to insufficient parallelism.

    (b) Suppose that the load is divided into fragments of equal size with one fragment assigned to each processor. Find the optimal number of processors that minimize the finish time.

    (c) Suppose for the situation of (b) that there is non-negligible solution reporting time. Starting with the right most processor, each processor reports back its solution in time T_s. Solutions are reported back in the opposite order in which load is distributed. Thus the left most processor eventually receives solutions of duration $(N - 1)T_s$. Find a means of calculating the optimal number of processors.

25. Consider a three children node homogeneous single level tree network. Let $z = 2$, $w = 6$, and $T_{cp} = T_{cm} = 1$. Find the optimal load fractions, α_i's, the equivalent (inverse) processing speed, w_{eq} and speedup for scheduling policies with:

    (a) Sequential distribution.
    (b) Simultaneous distribution and staggered start.
    (c) Simultaneous distribution and simultaneous start.

26. Consider a single level tree network of a control (root) processor and n children processors. Sequential distribution and staggered start is used. The control processor distributes load to its children but does no processing of its own. Draw the Gantt chart timing diagram.

    (a) Write the timing equations.
    (b) Solve for the optimal fraction of load to allocate to the ith processor, α_i.
    (c) Find the minimum finish time and speedup.

27. Demonstrate for a linear daisy chain network that for a minimal finish time solution, all processors must stop processing at the same time. There is no solution reporting time. Load originates at the left most processor. Do the (partial) proof in the context of the two right most processors. Assume that processor $N - 1$ keeps a fraction α of the data that it has received and transmits the remaining fraction $1 - \alpha$ to the Nth processor. There are two possibilities, either the $N - 1$st or the Nth processor will stop computing first.

28. Consider two processors without front-end processors and the link connecting them. Processor 1 processes load fraction α_1 and processor 2 processes load fraction $(1 - \alpha_1)$. Draw the Gantt chart timing diagram.

    (a) Write the timing equations of the system.
    (b) Show that for parallel processing to save time, $w_1 T_{cp} > z T_{cm}$. That is, the link must be faster than computing at processor 1.

29. (a) Show for a single level tree network that the speedup equation for staggered start reduces to that for simultaneous start if link speed goes to infinity. Intuitively, why is this so?

    (b) Show for a single level tree network that the speedup of simultaneous start is always larger than that for staggered start.

30. Consider a linear daisy chain network with front-end processors that is infinite in size in both directions where load originates at an interior processor. The load originating processor first distributes load to its left side and then to its right side. Having a front end, the originating processor computes as it distributes load. Each processor receiving load receives it for itself and for the processors beyond it. Staggered start is used for nodes receiving load. Draw a Gantt chart diagram of the originating processor and its two neighbors.

    (a) Write the timing equation of the originating processor and its two neighbors. Let β_c be the amount of total load kept by the originating processor, let β_l the amount of total load sent to the originating processor's left neighbor, and let β_r be the amount of total load sent to the originating processor's right neighbor. Naturally $\beta_c + \beta_l + \beta_r = 1$.

    (b) Write an implicit equation for w_{eqs}^∞, the equivalent inverse speed constant of the three processors.

    (c) Suggest a numerical solution technique using the result for load distribution from the boundary of a linear daisy chain of infinite size network as in Sect. 5.4.1.

31. Repeat the steps of the previous problem for load origination at an interior processor in an infinite size linear daisy chain network *without* front-end processors. That is, everything is the same as in the previous problem except that the load originating processor does not start

computing until it has distributed load to its left and right neighbors.

32. Consider an infinite size binary tree network as in Sect. 5.5.2 except that processors do not have front-end processors. That is, a node first distributes load to its left child, then to its right child and only then processes its own fraction of load. Staggered start is used for the children. Draw the Gantt chart of the root node and its children.

    (a) Write the timing equations.
    (b) Use the result of (a) to write an implicit equation for w_{eq}^{∞}, the equivalent inverse processing speed of the infinite size network.
    (c) Using (b), find a polynomial equation that can be solved for w_{eq}^{∞}.

33. Consider a system as in Sect. 5.9 but with time-varying channel speed, rather than processor speeds. The channel (bus) is shared with the divisible job of interest and other background transmissions in a "processor sharing" service discipline manner. That is, each of n transmissions on the channel receives $1/n$ of the effort.

    (a) Write a similar set of equations to those in Sect. 5.9 for this situation.
    (b) Outline a solution algorithm similar to that in Sect. refch5:sec9.

34. Write a set of equations and algorithm for a system as in Sect. 5.9 and the previous problem with both time-varying processor and channel speed. Note: Summarizing previous results, only two equations and some explanation are needed.

35. Phrase the scheduling model for sequential distribution in a single level tree (Sect. 5.2.1) as a linear programming problem.

36. Draw the Gantt chart for a single level tree of four children nodes with simultaneous distribution and staggered start. After computation is finished the ith child reports back a solution in time $\alpha_i z_i T_{cm-out}$ (i.e., solution reporting time for a processor is proportional to the load fragment size assigned to the processor). The root does no processing, it only does load distribution to the children nodes.

    (a) Solve the model for the optimal amount of load to assign to each processor.
    (b) Find the equivalent processing speed of the network as well as its speedup.

37. Draw a Gantt chart for a single level tree with m children with simultaneous distribution and simultaneous start. Here the load must be completely received by the root from an out of network node over a link with inverse transmission speed z_0 before it begins to distribute load to its children. But the root commences processing as it begins to receive load.

    (a) Solve the model for the optimal amount of load to assign to each processor.
    (b) Find the equivalent processing speed of the network as well as its speedup.

38. Consider N source nodes and M sink nodes. Source nodes distribute divisible load to the sink nodes which do the actual processing. There is a link of inverse speed $z_{i,j}$ between each source node i and each sink node j. The jth sink node has inverse processing speed w_j. A load in the amount L_i is distributed by the ith source node. Also $\sum_{i=1}^{N} L_i = L$. Finally, $\alpha_{i,j}$ is the amount of load that sink j receives from source i and α_j is the fraction of load L that sink j will receive from all of the sources ($\sum_{j=1}^{M} \alpha_j = 1$).

    (a) Draw a network diagram and Gantt chart of this situation.
    (b) Solve for the optimal amount of load to allocate to each sink and the finish (solution) time.
    (c) Does this problem have a unique solution? Comment on this.

Abstract

A general look at Amdahl's law, which expresses the parallel processing advantage of a job with serial and parallel components. Variants of Amdahl's law such as Gustafson's law are discussed. The design of different multicore architectures is examined. A CPU/GPU example is considered. Delay and energy objective functions are presented. Finally, the role of Amdahl's law in the context of local versus cloud processing tradeoffs is studied.

6.1 Introduction

In Chap. 5 we have seen that purely divisible computation and communication load in parallel systems can be mathematically modeled and optimal load distribution policies found. An older modeling paradigm for parallel systems that is still of much interest involves the ultimate performance of such computational systems. In this chapter, basic relationships concerning parallel processor performance, Amdahl's law [10] and Gustafson's law [109] are examined. These laws speak to the dependence of performance on the relative sequential and parallel components of a computational problem. Applications of these relationships (as well as generalizations) are applied to multicore architectures and cloud computing in this chapter.

6.2 Amdahl's Law

In 1967 Gene Amdahl was asked by his employer, IBM, to present a talk at a conference "to compare the computing potential of a super uni-processor to that of a unique quasi-parallel computer, the ILLIAC IV (Illinois Integrator and Automatic Computer)" [11]. This was a question of some practical interest and would remain so for some time: for better performance is it preferable to use a high performing single processor or a multiplicity of, possibly not as individually high performing, processors?

Amdahl came up with an interesting argument and related mathematical expression now called "Amdahl's law" whose persistence in computing thought surprised even him [11].

Amdahl's argument was that a typical computing problem has two parts: a sequential part where code has to be executed serially on a single processor and a parallel part whose execution time can be decreased arbitrarily by using an increasing number of parallel processors. Note that the sequential component of a program is due to system software including synchronization and coordination overheads and the sequential part of the specific problem that the code is written to solve [11].

Amdahl argued that even if one could solve the parallel part of a program in near zero time due to the use of a large number of parallel processors, the bottleneck was the sequential part of the program which could only be processed on a single processor.

The performance metric called "speedup," S, is a basic way of expressing parallel processing time advantage. It is defined as the ratio of solution time on one processor, $T(1)$, to solution time on p processors, $T(p)$:

$$S = \frac{T(1)}{T(p)} \qquad (6.1)$$

A speedup of 5, for instance, means that the effective processing is as if done by 5 processors. Say that one has a program that is 10% sequential and 90% parallel. No matter how many processors are used on the parallel part of the program, the maximum speedup is (1.0/0.1) or 10.

© Springer Nature Switzerland AG 2020

T. G. Robertazzi, L. Shi, *Networking and Computation*, https://doi.org/10.1007/978-3-030-36704-6_6

To write this mathematically, let:

f: Workload fraction that is parallelizable.
$1 - f$: Workload fraction that is serial.
p: Number of homogeneous (i.e., identical) processors.
$T(1)$: Time to solve the workload on one processor.
$T(p)$: Time to solve the workload on p processors.
T_s: Serial execution time for the entire program.

Then:

$$T(1) = T_s \qquad (6.2)$$

$$T(p) = \underbrace{(1 - f)T_s}_{serial} + \underbrace{f\frac{T_s}{p}}_{parallel} \qquad (6.3)$$

Here $T(p)$ is a weighted sum of serial and parallel execution time. The parallel execution time is fT_s/p, the parallel workload, divided by p, the number of processors used. Here also it is assumed that there is no time overlap between the serial and the parallel execution.

So one has in terms of speedup:

$$S^{Amdahl} = \frac{T(1)}{T(p)} = \frac{T_s}{(1 - f)T_s + f\frac{T_s}{p}} \qquad (6.4)$$

$$S^{Amdahl} = \frac{T(1)}{T(p)} = \frac{1}{(1 - f) + \frac{f}{p}} \qquad (6.5)$$

For an infinite number of processors:

$$\lim_{p \to \infty} S^{Amdahl} = \frac{1}{1 - f} \qquad (6.6)$$

Note that Amdahl's law assumes that problem size/workload is fixed. This point will be returned to below in discussing Gustafson's law. One can say [177]:

$$S^{Amdahl} = \frac{T(1)}{T(p)} = \frac{1}{(1 - f) + \frac{f}{p}} \leq \min\left(p, \frac{1}{1 - f}\right) \qquad (6.7)$$

Here the speedup is upper bounded by either the number of processors, p, or the asymptotic speedup, $\frac{1}{1-f}$, whichever is smaller.

Amdahl's law is an expression of a basic performance relationship that speaks to what extent a component improvement improves an entire system. Let us look at an example [177].

Suppose a program is distributed as:

- 30% of time → floating point addition.
- 25% of time → floating point multiplication.
- 10% of time → floating point division.

Which of the three following improvements would be the best to make?

- addition → 2 times as fast.
- multiplication → 3 times as fast.
- division → 10 times as fast.

The percentage of the load that is NOT improved in each situation is

1. addition → $1.0 - 0.3 = 0.7$
2. multiplication → $1.0 - 0.25 = 0.75$
3. division → $1.0 - 0.1 = 0.9$.

The improvement of speedup for each decision is

$$S^{add}_{improvement} = \frac{1}{0.7 + \frac{0.3}{2}} = 1.18 \qquad (6.8)$$

$$S^{mult}_{improvement} = \frac{1}{0.75 + \frac{0.25}{3}} = 1.2 \qquad (6.9)$$

$$S^{div}_{improvement} = \frac{1}{0.9 + \frac{0.1}{10}} = 1.1 \qquad (6.10)$$

Thus, improving the multiplier by making it operate three times faster yields the best improvement in speedup.

This specific example can be generalized [59]. Let:

k: number of computational components.
F_k: fraction of workload for the kth component.
S_k: speedup of kth component if $S_k \geq 1$. If $S_k < 1$, one has the slowdown of the kth component.
S: The system speedup.

Then for components $0, 1, 2, \ldots k - 1$:

$$\sum_{j=0}^{k-1} F_j = 1 \qquad (6.11)$$

The system speedup is

$$S = \frac{1}{\frac{F_0}{S_0} + \frac{F_1}{S_1} + \frac{F_2}{S_2} + \cdots + \frac{F_{k-1}}{S_{k-1}}} \qquad (6.12)$$

This can be written as:

$$S = \frac{1}{\sum_{j=0}^{k-1} \frac{F_j}{S_j}} \qquad (6.13)$$

This generalizes the earlier example of Amdahl's law.

Finally, one could generically normalize speedup to be between 0 and 1 by letting:

$$E(p) = \frac{S(p)}{p} \tag{6.14}$$

Here $E(p)$ is the "efficiency" for p processors and $S(p)$ is the speedup for p processors.

6.3 Gustafson's Law

In a 1988 paper J.L. Gustafson made an argument that the Amdahl's law assumption of constant problem size is usually never the case [109]. More cores are normally used to solve larger and more complicated problems. Thus one would be justified in having a parallel fraction that grows linearly in problem size (i.e., using fp instead of a single f). One could write [127]:

$$S^{Gustafson} = \frac{T(1)}{T(p)} = \frac{(1-f)T_s + fpT_s}{(1-f)T_s + \frac{fpT_s}{p}} \tag{6.15}$$

Canceling terms:

$$S^{Gustafson} = \frac{(1-f) + fp}{(1-f) + f} \tag{6.16}$$

Putting the above equation another way, assume that the normalized run time on p cores is $(1-f) + f = 1$. Then the execution time on one core is $(1-f) + pf$ (the serial run time is the same but the parallel run time is p times longer). Simplifying:

$$\boxed{S^{Gustafson} = (1-f) + pf \tag{6.17}}$$

Figure 6.1a shows a pictorial representation of Amdahl's law and Fig. 6.1b shows a pictorial representation of Gustafson's law [109]. The serial run time of $1 - f$ in all cases is followed by the parallel run time.

6.4 A General Law

Juurlink suggests one could have a parallel fraction growth factor, $scale(p)$, that is between a constant (Amdahl's law) and linear growth (Gustafson's law) (see [127]). It is not the only possibility but he suggests a square root function, $scale(p) = \sqrt{p}$. One then has in terms of speedup:

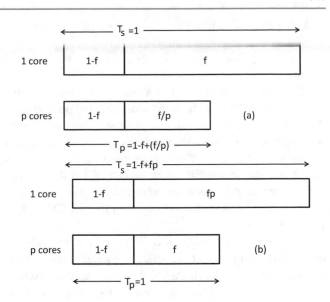

Fig. 6.1 Pictorial representations of (**a**) Amdahl's law. (**b**) Gustafson's law

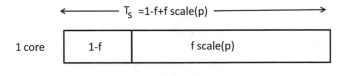

Fig. 6.2 Pictorial representation of a general law

$$S^{General} = \frac{T(1)}{T(p)} = \frac{(1-f)T_s + fscale(p)T_s}{(1-f)T_s + \frac{fscale(p)T_s}{p}} \tag{6.18}$$

Or:

$$S^{General} = \frac{(1-f) + fscale(p)}{(1-f) + \frac{fscale(p)}{p}} \tag{6.19}$$

In [127] this general speedup is referred to as GSEE (generalized scaled speedup equation).

Figure 6.2 is a pictorial representation of the general law.

6.5 Symmetric Multicore Design

Multicore chips use a multiplicity of cores (i.e., processors) on a single chip to improve chip processing throughput. But there are some generic questions that arise in designing a multicore chip that one should consider:

- Use many small and simple cores?
- Use a few hi-performing cores?

- Use a mixture of both?

These questions are addressed in this and the next few sections.

Consider first in this section a symmetric multicore chip design. That is all cores have the same size and performance. As in Marty and Hill [113], it is assumed here that a multicore chip of a certain size that is implemented in a particular technology contains n "base core equivalents (BCE) and each BCE has a normalized performance of 1.0. A core with a chip area of r BCEs has a performance $perf(r)$ where $perf(r) \leq r$. A common assumption [53, 113, 127, 184] is $perf(r) = \sqrt{r}$.

So let us assume we have n BCE's, r BCEs/core, and n/r cores. It is assumed that the serial and parallel parts of the program are executed at different times, not concurrently. One core with r BCEs is assigned to the serial part of the program. Also, n/r cores of r BCEs each are assigned to work on the parallel part of the program. Then:

$$S_{symmetric}^{Amdahl} = \frac{T(1)}{T(f,n,r)}$$

$$= \frac{T_s}{\left(\frac{(1-f)}{perf(r)}\right)T_s + \left(\frac{f}{perf(r)\cdot(n/r)}\right)T_s} \quad (6.20)$$

Normalizing by letting $T(1) = T_s = 1$:

$$S_{symmetric}^{Amdahl} = \frac{1}{\frac{1-f}{perf(r)} + \frac{f}{perf(r)\cdot(n/r)}} \quad (6.21)$$

Note that the bigger that $perf(r)$ is, the smaller $T(f,n,r)$ is and the larger is the speedup. Or:

$$S_{symmetric}^{Amdahl} = \frac{1}{\frac{1-f}{perf(r)} + \frac{fr}{perf(r)\cdot n}} \quad (6.22)$$

For Gustafson's law, again fn is used instead of f for the parallel component of workload growth. This is because the speedup is in terms of one BCE [127].

$$S_{symmetric}^{Gustafson} = \frac{T(1)}{T(f,n,r)} = \frac{(1-f)T_s + fnT_s}{\frac{1-f}{perf(r)}T_s + \frac{fn}{perf(r)\cdot\left(\frac{n}{r}\right)}T_s} \quad (6.23)$$

Canceling terms:

$$S_{symmetric}^{Gustafson} = \frac{(1-f) + fn}{\frac{1-f}{perf(r)} + \frac{fr}{perf(r)}} \quad (6.24)$$

Again, the bigger $perf(r)$ is, the smaller the execution time with n BCEs, $T(f,n,r)$ is and the larger that speedup is.

Multiplying the top and bottom by $perf(r)$ yields:

$$S_{symmetric}^{Gustafson} = \frac{((1-f) + fn)perf(r)}{1 - f + (fr)} \quad (6.25)$$

For the general case, as before, $1 \leq scale(n) \leq n$. In the general case:

$$S_{symmetric}^{General} = \frac{T(1)}{T(f,n,r)} = \frac{(1-f)T_s + f \cdot scale(n) \cdot T_s}{\left(\frac{1-f}{perf(r)}\right)T_s + \left(\frac{f\cdot scale(n)}{perf(r)\cdot\frac{n}{r}}\right)T_s} \quad (6.26)$$

Canceling and multiplying top and bottom by $perf(r)$ gives

$$S_{symmetric}^{General} = \frac{((1-f) + f \cdot scale(n))perf(r)}{(1-f) + \left(\frac{fr\cdot scale(n)}{n}\right)} \quad (6.27)$$

If $scale(n)$ is constant $(= 1)$, one has Amdahl's law. If $scale(n)$ is linear $(= n)$, one has Gustafson's law.

6.6   Asymmetric Multicore Design

In an asymmetric multicore design one or several cores are more powerful than others. This type of technology may be called performance heterogeneous multicores [113, 127].

Amdahl's law suggests it may be useful to have a large core for the serial part of a program and many small cores for the parallel part of the program. This case is now considered.

Again, there are n total BCEs. So suppose one has one large core of r BCEs for the serial part of the program. Since it is assumed that the serial and parallel parts of the program do not execute concurrently, the large core of r BCEs is also used for the parallel part of the program along with $n - r$ small cores. The large core has a performance of $perf(r)$.

For an Amdahl model:

$$S_{asymmetric}^{Amdahl}(f,n,r) = \frac{T(1)}{T(f,n,r)} = \frac{T_s}{\frac{(1-f)T_s}{perf(r)} + \frac{fT_s}{perf(r)+n-r}} \quad (6.28)$$

Canceling T_s terms:

$$S_{asymmetric}^{Amdahl}(f,n,r) = \frac{T(1)}{T(f,n,r)} = \frac{1}{\frac{(1-f)}{perf(r)} + \frac{f}{perf(r)+n-r}} \quad (6.29)$$

Likewise, for a Gustafson model:

$$S_{asymmetric}^{Gustafson}(f,n,r) = \frac{(1-f)T_s + fnT_s}{\frac{(1-f)T_s}{perf(r)} + \frac{fnT_s}{perf(r)+n-r}} \quad (6.30)$$

Canceling T_s:

$$S_{asymmetric}^{Gustafson}(f, n, r) = \frac{(1-f) + fn}{\frac{(1-f)}{perf(r)} + \frac{fn}{perf(r)+n-r}} \quad (6.31)$$

Finally, for a general model:

$$S_{asymmetric}^{General}(f, n, r) = \frac{(1-f)T_s + fscale(n)T_s}{\frac{(1-f)T_s}{perf(r)} + \frac{fscale(n)T_s}{perf(r)+n-r}} \quad (6.32)$$

Canceling:

$$S_{asymmetric}^{General}(f, n, r) = \frac{(1-f) + fscale(n)}{\frac{(1-f)}{perf(r)} + \frac{fscale(n)}{perf(r)+n-r}} \quad (6.33)$$

6.7 Dynamic Multicore Design

Under the "dynamic multicore" case we consider r BCEs that can be temporarily combined to work on the sequential part of a program. When processing the parallel part of the program, n single BCE cores are used (there are n BCEs on the chip in total).

So for the three models we have:

$$S_{dynamic}^{Amdahl}(f, n, r) = \frac{T_s}{\frac{(1-f)T_s}{perf(r)} + \frac{fT_s}{n}} \quad (6.34)$$

$$= \frac{1}{\frac{(1-f)}{perf(r)} + \frac{f}{n}} \quad (6.35)$$

$$S_{dynamic}^{Gustafson}(f, n, r) = \frac{(1-f)T_s + fnT_s}{\frac{(1-f)T_s}{perf(r)} + \frac{fnT_s}{n}} \quad (6.36)$$

$$= \frac{(1-f) + fn}{\frac{(1-f)}{perf(r)} + f} \quad (6.37)$$

$$S_{dynamic}^{General}(f, n, r) = \frac{(1-f)T_s + fscale(n)T_s}{\frac{(1-f)T_s}{perf(r)} + \frac{fscale(n)T_s}{n}} \quad (6.38)$$

$$= \frac{(1-f) + fscale(n)}{\frac{(1-f)}{perf(r)} + \frac{fscale(n)}{n}} \quad (6.39)$$

Using numerical parameters, design conclusions can be drawn from the performance equations of these sections (see [113, 114, 127] for a discussion).

6.8 A CPU/GPU Example

6.8.1 Speedup

As an example [163, 164], consider a chip with c CPUs and g GPUs. Let:

$1 - f$: The sequential component of the overall program.

f: The parallel component of the overall program.

α: The fraction of the parallel execution done on CPUs.

$1 - \alpha$: The fraction of the parallel execution done on GPUs.

β: The GPU performance with respect to a CPU performance of 1.0.

T_s: Serial execution time for the entire program.

Then using Amdahl's law:

$$S_{cpu-gpu}^{Amdahl} = \frac{T(1)}{T(c, g, f, \alpha)}$$

$$= \frac{T_s}{(1-f)T_s + \frac{\alpha f T_s}{c} + \frac{(1-\alpha)f T_s}{g\beta}} \quad (6.40)$$

$$= \frac{1}{(1-f) + \frac{\alpha f}{c} + \frac{(1-\alpha)f}{g\beta}} \quad (6.41)$$

Here it is assumed that the program execution time consists of three non-overlapping phases:

1. A sequential phase with one active core.
2. A CPU phase with c CPUs working on an α fraction of the parallel part of the program.
3. A GPU phase with g GPUs working on $(1 - \alpha)$ fraction of the parallel part of the program.

However, concurrent CPU/GPU computing can be modeled as we do in Sect. 6.8.3 [163, 164].

6.8.2 Average Power

To determine the average power consumed [163, 164], let the average power of each phase be P_s, P_c, and P_g (sequential, CPU and GPU phases, respectively). Let the individual power consumption of elements be

- CPU → power consumption of one CPU is 1.0.
- GPU → power consumption of one GPU is w_g.

Then the sequential phase un-normalized power is

$$P_s = (1 - f)\left(1 + (c - 1)k_c + gw_gk_g\right) \quad (6.42)$$

Here one CPU is active on the sequential component of a program. Also $(c - 1)$ CPUs are inactive with relative idle power consumption factor k_c and g GPUs are inactive with individual GPU power consumption w_g and idle GPU power consumption factor k_g.

Also in the CPU phase, c CPUs are active (with power consumption $c \cdot 1 = c$) and g GPUs are inactive (with idle power consumption gw_gk_g):

$$P_c = \frac{\alpha f}{c}\left(c + g w_g k_g\right) \quad (6.43)$$

Finally, in the GPU phase, g GPUs are active (power consumption $g w_g$) and c CPUs are inactive (with idle power consumption $c k_c$:

$$P_g = \frac{(1-\alpha)f}{g\beta}\left(g w_g + c k_c\right) \quad (6.44)$$

In the above the power consumed is multiplied by the execution time of each phase. Then the average normalized power consumption is

$$W_a = \frac{P_s + P_c + P_g}{(1-f) + \frac{\alpha f}{c} + \frac{(1-\alpha)f}{g\beta}} \quad (6.45)$$

If one examines the expressions for P_s, P_c, and P_g in the numerator in conjunction with the denominator, one can see that each of the individual powers is weighted by a number between zero and one so we have a normalized weighted sum of the individual powers which yields the average power, W_a. Also:

$$\frac{Performance}{Avg.Power} = \frac{S^{Amdahl}_{cpu-gpu}}{W_a} = \frac{1}{P_s + P_c + P_g} \quad (6.46)$$

This equation yields the ratio of performance to average power.

6.8.3 Concurrent Asymmetric Performance

The previous subsections assumed that the sequential, CPU and GPU parts of the processing occur at different times. In some applications though the CPU and GPU processing on the parallel component of a program can proceed at the same time (i.e., concurrently or simultaneously) [163, 164]. Let α be the fraction of the parallel load processed by the CPUs and let $(1-\alpha)$ be the fraction of the parallel load processed by the GPUs. Again let there c CPUs and let there be g GPUs and let β be the GPU performance with respect to a CPU performance of 1.0. Let f be as before. The optimal (minimal) processing time will occur if the CPU execution time, $\alpha f/c$, equals the GPU execution time, $(1-\alpha)f/(g\beta)$. If they were different, the execution time could be improved by transferring load from the busy processors to the partially idle ones (see Chap. 5).

The optimal value of α, α', can be found from:

$$\frac{\alpha f}{c} = \frac{(1-\alpha)f}{g\beta} \quad (6.47)$$

$$\alpha' = \frac{c}{g\beta + c} \quad (6.48)$$

The speedup is

$$S^{concurrent-Amdahl}_{cpu-gpu} = \frac{T(1)}{T(c,g,f,\alpha')} = \frac{T_s}{(1-f)T_s + \frac{\alpha' f T_s}{c}} \quad (6.49)$$

Here we are using the CPU execution time (it equals the concurrent GPU execution time). Also in this we assume one CPU core (with a perfromance of 1.0) is used for the sequential component of the computing. So we have

$$S^{concurrent-Amdahl}_{cpu-gpu} = \frac{1}{(1-f) + \frac{\alpha' f}{c}} = \frac{1}{(1-f) + \frac{f}{g\beta+c}} \quad (6.50)$$

The average power consumption, $W^{concurrent}_a$, can be computed. One core is used for the sequential part of a program with a power consumption of 1.0. During this time the idle power consumption for the remaining $c-1$ CPUs is $(c-1)k_c$ and for the g GPUs it is $g w_g k_g$. So as before the average power consumption for the serial part of the processing is

$$P_s = (1-f)\left(1 + (c-1)k_c + g w_g k_g\right) \quad (6.51)$$

The average power consumption for the parallel CPU and GPU computing is

$$P_{cpu-gpu} = \frac{\alpha' f}{c}\left(c + g w_g\right) \quad (6.52)$$

The average normalized power consumption is

$$W^{concurrent\,Amdahl}_{cpu-gpu} = \frac{P_s + P_{cpu-gpu}}{(1-f) + \frac{\alpha' f}{c}} \quad (6.53)$$

One can see if one examines the expressions for P_s and $P_{cpu-gpu}$ in the numerator in conjunction with the denominator that each of the two powers is weighted by a number between zero and one (i.e., in a normalizing fashion) so overall one has a normalized weighted sum of the individual powers.

Also:

$$\frac{Performance}{Avg.Power} = \frac{S^{concurrent\,Amdahl}_{cpu-gpu}}{W^{concurrent\,Amdahl}_{cpu-gpu}} = \frac{1}{P_s + P_{cpu-gpu}} \quad (6.54)$$

6.9 Delay and Energy Objective Functions

In this section the goal is to write objective functions for average delay and average energy use using Amdahl's law for use in optimization. We follow the excellent development of [59] in this.

6.9.1 Preliminaries

From the earlier section on Amdahl's law we had

K: Number of computational components (levels) of parallelism.

M: Number of classes of instructions.

F_k: Fraction of workload for the kth component.

S_k: Speedup of the kth component.

S: System speedup.

Then we had in (6.12) and (6.13):

$$S = \frac{1}{\frac{F_0}{S_0} + \frac{F_1}{S_1} + \frac{F_2}{S_2} + \cdots + \frac{F_{K-1}}{S_{K-1}}} = \frac{1}{\sum_{j=0}^{K-1} \frac{F_j}{S_j}} \quad (6.55)$$

Here:

$$\sum_{j=0}^{K-1} F_j = 1.0 \quad (6.56)$$

Let Q_j be the number of instructions associated with the jth speedup and let $\sum_j Q_j$ be the entire number of instructions. Then the fraction of workload of the jth component, F_j, is

$$F_j = \frac{Q_j}{\sum_j Q_j} \quad (6.57)$$

6.9.2 Delay as Cost (Single Processor)

Suppose that the ith class of instructions takes d_i seconds to run. Let $p(d_i)$ be the probability distribution of the delay of the M classes. From first principles, the expected delay can be written as:

$$E[D] = \sum_{i=0}^{M-1} d_i p(d_i) \quad (6.58)$$

Let:

$$p(d_i) = \frac{Q_i}{\sum_{i=0}^{M-1} Q_i} = G_i \quad (6.59)$$

Here again Q_i is the number of instructions with the ith delay, d_i. Letting $D_i = d_i$ one has a cost function:

$$\boxed{J_D = \sum_{i=0}^{M-1} G_i D_i} \quad (6.60)$$

6.9.3 Delay as Cost (with Parallelism)

Suppose that there are N parallel processors in an ideal parallel processing situation. Then:

$$E[D] = \sum_{i=0}^{M-1} \frac{d_i}{N} p(d_i) = \frac{1}{N} \sum_{i=0}^{M-1} d_i p(d_i) \quad (6.61)$$

$$\boxed{J_D = \frac{1}{N} \sum_{i=0}^{M-1} G_i D_i} \quad (6.62)$$

Now let us use Amdahl's methodology. Let F_p be the parallel component of a program and let F_s be the serial component of the same program. Naturally $F_s + F_p = 1$. Then a cost function in terms of time (delay) is

$$J_D = \underbrace{\frac{F_s}{1} \sum_{i=0}^{M-1} G_{s_i} D_{s_i}}_{serial} + \underbrace{\frac{F_p}{N} \sum_{i=0}^{M-1} G_{p_i} D_{p_i}}_{parallel} \quad (6.63)$$

It is assumed here that M, the number of classes of instructions, is the same for both the serial and parallel components. Here the "s" and "p" denote the serial and parallel variables, respectively. That is G_{s_i} and G_{p_i} are the class instruction distributions and D_{s_i} and D_{p_i} are the delays for the serial and parallel fractions of a program, respectively.

Let G_{ij}/D_{ij} be the instruction distribution/delay for the ith class and the jth component. Then for an arbitrary K levels (components) of parallelism and M instruction classes one can write

$$J_D = \sum_{j=0}^{K-1} \frac{F_j}{N_j} \sum_{i=0}^{M-1} G_{ij} D_{ij} \quad (6.64)$$

Here N_j is the number of parallel processors for the j level. For the serial component $N_j = 1$.

6.9.4 Energy as Cost

Let a class of instructions need e_i units of energy (e.g., Joules). Then, with some notational abuse, the expected energy use of a program, $E[E]$, is

$$E[E] = \sum_{i=0}^{M-1} e_i\, p(e_i) \qquad (6.65)$$

Here $p(e_i)$ is defined in a manner similar to what was done earlier. We can either have $p(e_i) = p(d_i)$ if energy and delay are strongly correlated or $p(e_i) \neq p(d_i)$ if energy and delay are independent. Let $E_i = e_i$. Then:

$$J_E = \sum_{i=0}^{M-1} G_i E_i \qquad (6.66)$$

We most likely have different energy usage for active (A) and idle (I) processors. Let:

$$N_{jh} \in \{N_{jA}, N_{jI}\} \qquad (6.67)$$

Then for K levels of parallelism where the third subscript "h" refers to the active/inactive status:

$$J_E = \sum_{j=0}^{K-1} \frac{F_j}{N_j} \sum_{h \in \{A, I\}} N_{jh} \sum_{i=0}^{M-1} G_{ijh} E_{ijh} \qquad (6.68)$$

Here if the number of active processors N_{jA} is N_j which cancels the outer N_j. The number of idle processors N_{jI} is $N - N_j$. For the serial component $N_0 = 1$.

6.9.5   Joint Delay and Energy Optimization

One common metric in design when there are two objective functions is the product of the two functions:

$$J_{ED} = \left[\sum_{j=0}^{K-1} \frac{F_j}{N_j} \sum_{i=0}^{M-1} G_{ij} D_{ij} \right] \qquad (6.69)$$

$$\times \left[\sum_{j=0}^{K-1} \frac{F_j}{N_j} \sum_{h \in \{A,I\}} N_{jh} \sum_{i=0}^{M-1} G_{ijh} E_{ijh} \right]^{\gamma}$$

The exponent γ allows different weighting of the two objective functions. A common choice is $\gamma = 2$.

Another metric [58, 59] is the energy dot product:

$$J_{E \cdot D} = \sum_{j=0}^{K-1} \frac{F_j}{N_j} \sum_{i=0}^{M-1} G_{ij} D_{ij} \left(N_j E_{ij} \right)^{\gamma} \qquad (6.70)$$

6.10   Amdahl's Law and Cloud Computing

Under the cloud computing paradigm a user or an organization can utilize computing on a provider's data center(s) resources to host applications, software, and services [198]. Some apps are very suitable to be off-loaded from a local device to a cloud and others are not. A performance question of interest is whether it is "better" for a particular app to be off-loaded to the cloud or is it better to process it locally. The latter case fits into the paradigms of fog computing and edge computing. In building models to address this question we follow the work of [72].

6.10.1   Local Processing

We wish to find an expression for t_{local}, the local program execution time. In general the local uni-processor execution time can be expressed as:

$$N_I \times CPI \times T \qquad (6.71)$$

Here:

N_I: The number of program instructions.
CPI: The average number of clock cycles per instruction.
T: The clock period.

Also let:

n_c: The number of cores.
f: The fraction of load that is parallelizable.
$1 - f$:   The fraction of load that is serial.
S_p: The parallel system speedup.
S_S: The overall system speedup.

Then of course from (6.5) Amdahl's law states that

$$S_S = \frac{1}{(1-f) + \frac{f}{S_p}} \qquad (6.72)$$

So in terms of local processing on a local device one has

$$t_{local} = \underbrace{(1-f) N_I \times CPI_{local} \times T_{local}}_{serial}$$

$$+ f \underbrace{\frac{N_I \times CPI_{local} \times T_{local}}{n_{c,local}}}_{parallel} \qquad (6.73)$$

$$t_{local} = \left[(1-f) + \frac{f}{n_{c,local}} \right] N_I \times CPI_{local} \times T_{local} \quad (6.74)$$

6.10.2 Cloud Processing

If an app is processed on a cloud, the steps are:

1. Data and possibly code is sent into the cloud through the Internet.
2. The transmitted app is executed on the cloud.
3. The results are sent back to the local device.

The steps are illustrated in Fig. 6.3. Here $N_{D,input}$ is the data sent to the cloud. Also, $N_{D,output}$ is the result data transmitted after the cloud execution from the cloud to the local device.

A number of assumptions may be made to simplify the model [72]:

- Code size is minimal with respect to data size.
- Data is transmitted at a constant rate (loosely called bandwidth BW). Startup latency is negligible.
- Internal cloud overhead times can be neglected. Díaz-del-Río assumes that the majority of overhead times takes place in parallel with other times.

It can be said that

$$N_{Data} = N_{D,input} + N_{D,output} \quad (6.75)$$

Here N_{Data} is the total data exchanged between the local device and the cloud.

Now communication and computation times may or may not overlap. If they are disjoint, the cloud execution time, t_{cloud}, has two additive components: communication and computation:

$$t_{cloud} = \underbrace{\frac{N_{Data}}{BW}}_{communication} \quad (6.76)$$

$$+ \underbrace{\left[(1-f) + \frac{f}{n_{c,cloud}} \right] N_I \times CPI_{cloud} \times T_{cloud}}_{computation}$$

If the parallelizable component can be parallelized to such an extent that the parallel execution time goes to zero, one has the simplification:

$$t_{cloud} = \frac{N_{Data}}{BW} + (1-f)N_I \times CPI_{cloud} \times T_{cloud} \quad (6.77)$$

If there is complete overlap, communication time is hidden and with the parallelizable execution time going to zero, one has

$$t_{cloud} = (1-f)N_I \times CPI_{cloud} \times T_{cloud} \quad (6.78)$$

Assume that $CPI_{local} \approx CPI_{cloud}$ and that $T_{local} \approx T_{cloud}$. In the worst case with no overlap between computation and communication:

$$S_t = \frac{t_{local}}{t_{cloud}} = \frac{\left[(1-f) + \frac{f}{n_{c,local}} \right] N_I \times CPI \times T}{\frac{N_{Data}}{BW} + (1-f)N_I \times CPI \times T} \quad (6.79)$$

Let:

$$\mu = \frac{\frac{1}{CPI \times T}}{BW} = \frac{1}{CPI \times T \times BW} \quad (6.80)$$

$$D_I = \frac{N_I}{N_{Data}} \quad (6.81)$$

Here μ is the ratio of the local machine capacity to execute instructions per second (and per core) and its data rate (i.e., bits per second). Also D_I is the app's "computing density" [72], the average number of instructions executed per data bit transmission.

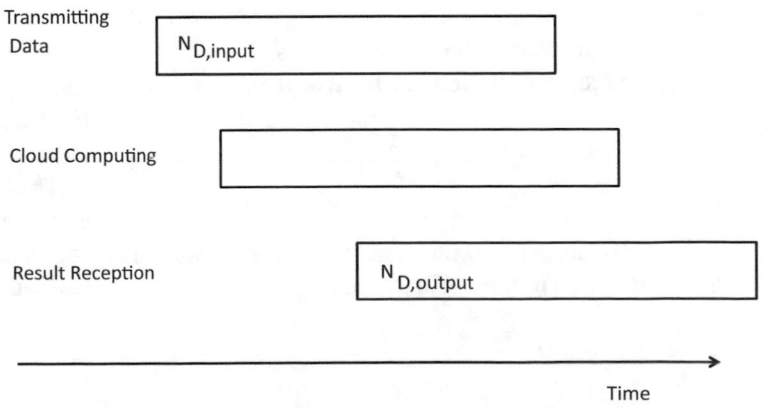

Fig. 6.3 Timing diagram of off-loaded to cloud computation

Transmitting Data — $N_{D,input}$

Cloud Computing

Result Reception — $N_{D,output}$

Time

Note that a dimensional analysis yields

$$\mu = \frac{\frac{1}{CPI \times T}}{BW} \rightarrow \frac{\frac{1}{\left[\frac{clock-cycles}{instruction} \times \frac{sec}{clock-cycles}\right]}}{[bps]} \quad (6.82)$$

$$= \frac{\left[\frac{1}{sec/instruction}\right]}{[bps]} \quad (6.83)$$

$$= \frac{[instructions/sec]}{[bps]} \quad (6.84)$$

Then dividing top and bottom of S_t by $CPI \times T$:

$$S_t = \frac{\left[(1-f) + \frac{f}{n_{c,local}}\right] N_I}{\frac{N_{Data}}{CPI \times T \times BW} + (1-f)N_I} \quad (6.85)$$

Next, dividing top and bottom by N_I and identifying μ:

$$S_t = \frac{\left[(1-f) + \frac{f}{n_{c,local}}\right]}{\frac{\mu N_{Data}}{N_I} + (1-f)} \quad (6.86)$$

Then identifying D_I and re-arranging one has

$$S_t = \frac{\frac{f}{n_{c,local}} + (1-f)}{\frac{\mu}{D_I} + (1-f)} \quad (6.87)$$

Díaz-del-Río presents a number of performance curves for speedup and comes to three conclusions:

1. The amount of overlap between communication and computation strongly influences cloud execution time.
2. The ratio μ/D_I has a strong influence on speedup. As of 2016 CPI is about one for most programs, T is on the order of 1 ns and BW is on the order of 1 Gps. So μ is a few units. As time goes μ will decrease as BW will increase at a rapid rate. Also $CPI \times T$ has plateaued because of technological limitations. So since μ/D_I will decrease in all likelihood, the advantage of cloud off-loading will increase as times goes by.
3. If a single processor is used in a local device,

$$S_t = \frac{1}{\frac{\mu}{D_I} + (1-f)} \quad (6.88)$$

If μ/D_I decreases with time, the use of simplified device hardware may be advantageous.

6.10.3 Energy Performance

Power is the rate of energy consumption per (divided by) time. Thus energy use is the sum of time intervals multiplied by the energy use during those intervals [72]:

$$\sum_i P_i t_i \quad (6.89)$$

The question to be addressed is what is the energy consumption when an app is processed locally and what is the local energy consumption when an app is processed on the cloud? Local app execution should have two components: a sequential and parallel component. For the sequential time period a single processor is active with power consumption P_1 and $(n_{c,local} - 1)$ processors are idle each with power consumption factor k_{idle}. The fraction of the time period that is sequential is of course $(1 - f)$.

For the parallel component with component f of parallel execution time, all cores are involved so the power consumption is $f n_{c,local} P_1$ divided by the speedup of using $n_{c,local}$ cores or divided by $n_{c,local}$:

$$P_{local} = (1 - f)\left(k_{idle} P_1 (n_{c,local} - 1) + P_1\right)$$
$$+ \frac{f n_{c,local} P_1}{n_{c,local}} \quad (6.90)$$

This can be simplified:

$$P_{local} = (1 - f)\left(k_{idle} P_1 (n_{c,local} - 1)\right)$$
$$+ (1 - f)P_1 + f P_1 \quad (6.91)$$
$$= (1 - f)\left(k_{idle} P_1 (n_{c,local} - 1)\right) + P_1 \quad (6.92)$$
$$= \left[(1 - f)(n_{c,local} - 1)k_{idle} + 1\right] P_1 \quad (6.93)$$

Multiplying this by the time involved yields the energy consumed:

$$E_{local} = P_1 \left[(1 - f)(n_{c,local} - 1)k_{idle} + 1\right]$$
$$\times [N_I \times CPI \times T] \quad (6.94)$$

Now, E_{cloud} is considered. This is not the energy consumed in the cloud but rather the energy consumed by the local device when the app execution is transferred to the cloud.

There is a communication component and a computation component. The worst case, no overlap of communication and computation, is considered. In addition to E_{cloud}^{proc} one must add the energy due to data transmission:

$$E_{cloud}^{xmsssn} = \frac{N_D}{BW} P_T \tag{6.95}$$

Here P_T is the data transmission power per bit.

The computation component consists of the time during which energy is consumed during the serial component plus the time of data transmission:

$$E_{cloud}^{proc} = \left[(1-f)(N_I \times CPI \times T) + \frac{N_D}{BW} \right]$$
$$\times n_{c,local} k_{off} P_1 \tag{6.96}$$

Here it is assumed that k_{off} is the off state local power consumption factor as the local device need not be running during the wait for a cloud response. For instance, the network interface card (NIC) can transmit data directly by accessing the local memory so that all local cores would be in the off state.

Let us assume that the number of (cloud) virtual processors be so large that no parallel component is necessary in the equation below. Thus adding the communication and computation (processing) components one has [72]:

$$E_{cloud} = \frac{N_D}{BW} P_t + \left[(1-f)(N_I \times CPI \times T) + \frac{N_D}{BW} \right]$$
$$\times n_{c,local} k_{off} P_1 \tag{6.97}$$

One can compute energy efficiency as the performance that can be realized with identical energy (i.e., battery life cycle). The speedup under this condition is

$$S_{t \times E} = \frac{E_{local} t_{local}}{E_{cloud} t_{cloud}} = S_E S_t \tag{6.98}$$

Substituting with $k_{off} = 0$ and simplifying one has from (6.87):

$$S_{t \times E} = \frac{\left[(1-f)(n_{c,local} - 1)k_{idle} + 1 \right] P_1}{\frac{\mu}{D_I} P_t}$$
$$\times \frac{\frac{f}{n_{c,local}} + (1-f)}{\frac{\mu}{D_I} + (1-f)} \tag{6.99}$$

For scientific applications where often $f \to 1$ or $n_{c,local} = 1$,

$$\lim_{f \to 1} S_{t \times E} \propto \left(\frac{D_I}{\mu} \right)^2 \tag{6.100}$$

So for basic devices ($n_{c,local} = 1$) or very parallel apps one can will reach energy efficiency earlier for cloud migration than for timing speedup (note the square in the expression).

Another surmise we can make is the critical role of $D_I = N_t / N_{data}$. Using 2016 values for the technology parameters ($\mu = 1$, $k_{idle} = 0.3$, and $P_1 = P_T = 1$ w), cloud migration is favored for moderate D_I. If the device is a basic one ($n_{c,local} = 1$, $D_I > 1.3$ and if $f \geq 0.5$), migration is beneficial as the speedup $S_{t \times E}$ is greater than one. Further related conclusions can be drawn from this analysis [72]. A final surmise from Díaz-del-Río is that if μ continues to reduce in size with continued technological evolution, the bounds will decrease proportionately. Thus future technology will favor cloud migration.

6.11 Conclusion

Amdahl's law and its variations provide much to think about when evaluating the performance of parallel systems. Parallel systems are increasingly prevalent. Thus these issues are likely to be of interest for a considerable amount of time.

6.12 Problems

1. Calculate S_{Amdahl}, $S_{Gustafson}$, and $S_{General}$ if the number of processors is 10 and a square root function is used in the general law for f= 0.9, 0.95, and 0.98.
2. Calculate S_{Amdahl}, $S_{Gustafson}$, and $S_{General}$ if f is 0.9 and a square root function is used in the general law for 5, 10, and 15 processors.
3. Why is there a "p" in the denominator of the parallel time expression in Eq. (6.3)?
4. What basic performance relationship/question does Amdahl's law address (see text after Eq. (6.9))?
5. Describe the difference between the assumptions behind Amdahl's versus Gustafson's law.
6. What is better, a higher speedup or a lower speedup?
7. Why are the speedup expressions for different multi-core architectures useful (symmetric, asymmetric, and dynamic)?
8. Why is it important to consider (average) power consumption and performance per average power (see CPU/GPU Sect. 6.8)?
9. What is the difference between concurrent and sequential processing (see Sect. 6.8)?
10. How can delay and energy objective functions be used?
11. What is the basic question Sect. 6.10 addresses? In the first paragraph of Sect. 6.10 what is meant by "better" – in what sense?
12. How does the analysis of Sect. 6.10 relate to doing experimental work on the same topic? Is analysis a replacement for experimental work?

Abstract

With a significant improvement in performance and efficiency recently, machine learning techniques have been widely applied in the areas of computer vision, natural language processing, and pattern recognition. While machine learning techniques have shown their superior ability in solving many complex problems, their applications to the networking area is still at an early stage. This chapter reviews state-of-the-art machine learning applications in the networking area, with the purpose of providing some insights on existing solutions and future opportunities. An overview of machine learning begins the chapter. This is followed by discussions of the applications of machine learning techniques to traffic classification, traffic routing, and resource management, respectively.

7.1 Introduction

After decades of development, networking systems have evolved from small local area networks interconnecting a limited number of computers to a very complicated and varied systems: Data center networks interconnect thousands and more machines, manage them using software, and provide large scale parallel processing which is a key part of the cloud computing technique—a novel but already phenomenal technique. Global wide area networks interconnect multiple local area networks across the world and enable real-time communication between people living in different countries or continents. Wireless sensor networks interconnect spatially scattered sensors to monitor and record specific data, which have been applied in many areas like smart city, health care monitoring, environmental sensing, and industrial monitoring. Vehicular ad-hoc networks interconnect moving vehicles and have enabled more intelligent road service and transportation systems.

To establish a networking system, many key functionalities are needed, like traffic classification, traffic predication, traffic routing, resource management, fault management, and network security. While each of those key functionalities seems to solve a unique set of problems, in modern networking systems, more and more cross-functionality problems appear to be solved, to improve the efficiency of the network and provide better quality of service to users. For example, while naive traffic routing focuses on finding paths with a single optimization criterion like minimizing distance or maximizing the minimum bandwidth, more advanced traffic routing may need to consider future traffic demand on critical links for time-sensitive network traffic and consider link failure probabilities for latency-sensitive network traffic. What makes those networking problems even more complicated is the unique properties of different types of networks. Still taking traffic routing as an example, in a data center network, one may need to consider the volume and variety of the network traffic; in a wireless sensor network, one may need to focus more on the energy consumption; and in a vehicular network, one may need to deal with a fast-changing dynamic network topology.

Conventional methods to solve the networking problems include heuristics, meta-heuristics, and optimization techniques like linear and nonlinear programming. Along with the increasing complexity of the networking problems, the solution space of those problems is greatly expanded and those conventional methods appear to have more and more limitations. The performance of heuristics relies heavily on human insights on the problems. When dealing with complicated problems, such human insights can be limited and therefore the proposed heuristics may converge to some local optimum that is far from the global optimum. While meta-heuristics and linear/nonlinear programming techniques can

T. G. Robertazzi, L. Shi, *Networking and Computation*, https://doi.org/10.1007/978-3-030-36704-6_7

avoid the caveat of heuristics to some degree, they usually suffer from long execution time when the solution space is large, which is the most common factor that prevents them from being applied in practice. Whereas, with the huge improvement in both performance and efficiency, machine learning techniques have recently shown their unique advantages in solving networking problems.

Machine learning techniques are algorithms that can automatically learn the solution space from historical data and perform a new solution search based on the learned experience. The theoretical fundament of these techniques had been well studied decades ago, yet they were not widely applied in practice until we entered the "big data era" recently in which the availability of rich data dramatically improves the performance of these techniques.

Applying machine learning techniques on the networking problems is straightforward. The networking problems can be broadly divided into three types:

(i) **The identification problems** are the problems of recognizing the property of some given network entities. Examples include identifying the category of a given network flow in order to provide corresponding level of service and labeling abnormal network traffic for network attack detection.

(ii) **The predication problems** are the problems of predicating the future value of some network metrics. Examples include predicating the traffic demand between a given pair of network nodes and predicating the probability of link failures.

(iii) **The decision making problems** are the problems of making certain decisions for some requests based on current network state. For example, the traffic routing problem is a typical decision making problem.

For each of these types of problems, there exists a specific type of machine learning techniques to solve it. Besides, most of the machine learning techniques complete new solution search very fast while taking a large amount of time for learning from historical data. This is a critical and highly desired feature since the requirement on execution speed of many network services is very high. In addition, the volume of networking data collected has dramatically increased recently as the scale of networks increases significantly, which makes the application of machine learning techniques more promising since the performance of most of the machine learning techniques closely depends on the volume of available historical data.

Observing the increasing trend of applying machine techniques to solve the networking problems, this chapter focuses on reviewing the state-of-the-art applications of machine learning techniques in the networking area, with the purpose of providing some insights on existing solutions and future opportunities. In Sect. 7.2, an overview of machine learning is provided. In Sects. 7.3–7.5, the applications of machine learning techniques to traffic classification, traffic routing, and resource management are discussed, respectively. The above three domains of the networking problems are specifically selected as they are the most closely related to the previous chapters of this book.

7.2   An Overview of Machine Learning

Machine learning is a field of study concerned with enabling computers to build models or programs that perform specific tasks and improve themselves without being explicitly programmed. This view was initially given by Arthur Samuel. In 1959, he defined machine learning as a "field of study that gives computers the ability to learn without being explicitly programmed." [205] While Samuel's definition has been agreed upon by many successors, it can be somewhat ambiguous. Tom Michell, another well-known researcher in the machine learning area, provides a more precise formalism in his book: "A computer program is said to learn from experience E with respect to some class of tasks T and performance measure P if its performance at tasks in T, as measured by P, improves with experience E." [170] To clarify this definition, let us use a spam email detection system as an example. The purpose of this system is to identify spam emails from all emails in an inbox and move them to a spam folder. In this case, task T of this system is to flag emails as spam or not; experience E is sets of existing emails including spam ones (e.g., flagged by users) and regular ones; performance measure P could be the percentage of correctly identified emails. The system learns if this percentage increases with consuming more experience E over time.

While the formalism given by Tom Michell covers a broad range of applications, today, machine learning techniques have been well developed to more specifically learn hidden properties or patterns of existing data and predict output based on learned properties. There are various objectives, such as

(i) predicting the output of new input data based on historical data that contains both inputs and corresponding outputs,

(ii) dividing existing data into multiple groups based on hidden patterns described by a set of features of those data,

(iii) building some models that interact with the outside environment and take a series of actions based on the feedback of the environment to achieve a certain long-term goal.

Fig. 7.1 Examples of supervised learning problems. (**a**) Classification. (**b**) Regression

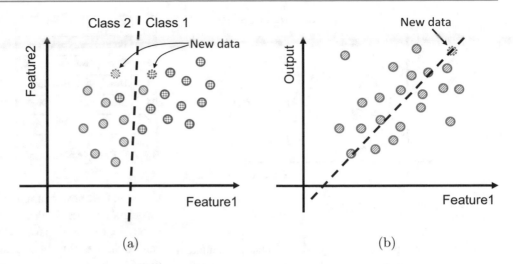

(a)   (b)

Machine learning techniques have been applied to many domains. A classic and closely related one is data mining—the process of discovering patterns in large data sets. Data mining has used many machine learning techniques. While typical data mining problems focus more on discovering previously unknown knowledge, existing machine learning techniques have gone beyond the scope of data mining to many other fields with different objectives as introduced above.

Naturally, many networking problems can be solved by machine learning techniques. For example, a typical network management problem is identifying the types of network flows and therefore providing different category of services, which can be solved by the classification ML technique. Another example is the traffic routing problem on which the reinforcement learning technique can be applied.

The following sections introduce more about machine learning categories and practical solutions.

7.2.1 Categories and Practical Solutions

Machine learning has been applied to solve a broad range of problems. Most of those problems can be grouped into several categories based on the algorithms used to solve them. The most popular ones are *supervised learning*, *unsupervised learning*, and *reinforcement learning*.

Supervised Learning

In supervised learning, a mathematical model is learned/-trained from a dataset in which each data sample is composed of inputs described by a set of variables and expected output. Such data with both inputs and expected output is usually called labeled training data and the set of variables used to describe inputs are called features. The objective in supervised learning is to map new inputs to some output. If the outputs are restricted to a limited set of values, the problems

are called *classification*; if the output is a continuous value, the problems are called *regression*. Figure 7.1 shows the pattern of classification and regression, respectively. Note that although both figures show a two-dimensional plane, in Fig. 7.1a of the classification problem, both the horizontal and the vertical axis are features and output is illustrated using different filling patterns of those data nodes; whereas, in Fig. 7.1b of regression problem, X axis is input feature and Y axis is output value. The example of identifying spam emails introduced previously is a typical classification problem. In this problem, the output is restricted to two values, i.e., spam or not, and the objective is to classify each incoming email to one of those two values.

Typical classification problems in the networking area are traffic classification problems in which the goal is to associate network traffic with some predefined classes like HTTP, FTP, WWW, or P2P. An example of regression problems is predicting home sales price. In this case, training data is a set of historical house sell records. Input features may include house size, lot size, property type, built year, zip code, and etc. Of course, output is the final sold price. There are many networking problems in the category of regression, for instance, the problem of predicting future egress network traffic on a border router using some previously observed traffic volume.

There exist many supervised learning models. Examples of the classification models include Naive Bayes, Supported Vector Machine (SVM) and K Nearest Neighbors (KNN). Examples of the regression models include Linear Regression, Lasso Regression, and ElasticNet Regression. There are also models that can be used to solve both of the classification problem and the regression problem, for example, Decision Trees and Neural Network models.

Unsupervised Learning

In unsupervised learning, a mathematical model is learned from an unlabeled training dataset, i.e., a dataset in which

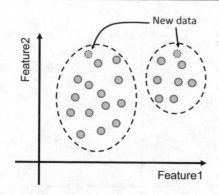

Fig. 7.2 Example of clustering problems

TID	Items
1	Bread, Ham, Coke
2	Diaper, Beer, Bread, Milk
3	Milk, Diaper, Beer, Eggs
4	Bread, Ham, Beer, Diaper
5	Bread, Ham, Milk, Coke

Association rules

(Diaper) → (Beer)

(Bread, Ham) → (Coke)

(Break, Beer) → (Milk)

Fig. 7.3 Example of association rule problems

each data sample is described by a set of features. Unlike a labeled dataset, there is no expected output in an unlabeled dataset. The objective is to discover some patterns hidden behind this given unlabeled dataset. The discovered patterns can be used to identify new data or can provide critical information in some other decision making process. There are two typical types of unsupervised learning problems: *clustering* and *association rule*. Figure 7.2 shows patterns of clustering. As shown in Fig. 7.2, the goal of clustering problems is to discover inherent groupings in the training data and assign any new data into one of those clusters. An example of cluster problems is discovering groups of similar customers given their information and activities like viewing web content, adding items to cart, making purchase, and etc. Such grouping can be used later for targeting or recommendation. Many problems in the networking area can be formatted as clustering problems. For instance, when detecting network intrusions, one can build an ML model that groups historical network behaviors into two groups, one group includes the majority of those historical behaviors while the other contains far fewer behaviors. Essentially, the major group is a group of normal network behaviors while the minor group is a group of abnormal behaviors, i.e., possible network intrusions. Any new network behaviors assigned to the minor group are examined more closely, which increases the chances of detecting an intrusion. Examples of existing popular ML techniques to solve clustering problems include K-Means clustering, Mean-Shift clustering, Density-Based Spatial Clustering of Applications with Noise (DBSCAN), and Agglomerative Hierarchical clustering.

The association rule problem is another major type of unsupervised learning problems, whose goal is to discover the relationship between various items, elements, or variables in a given large dataset. It is also referred to as market basket analysis, as this was the initial area of association rule machine learning. Figure 7.3 shows such an example. Given a set of grocery transactions, the association rule learning problem is to find the sets of items that are frequently purchased together. Note that the first frequent item set

shown in the above example is the famous Beer and Diapers association which indicates that men who go to the store to buy diapers will also tend to buy beer. Such association can be very useful when making marketing decisions and therefore boost one's sales profile. In the above example, with knowledge of the association between diapers and beer, a grocery store can place them together on the same shelf or provide promotions on just one out of the two items, which should lead to better sales than applying the same marketing methods on items without any associations. Besides market basket analysis, association rule learning has been employed in many other areas including web usage mining, intrusion detection, recommendation system, and etc. In the networking area, association rule learning can be used in network traffic analysis to discover frequent patterns of network events. These discovered patterns may provide critical information when detecting anomalies or depicting communication structures. Common association rule learning algorithms include the Apriori algorithm, the Eclat algorithm, and the FP-Growth algorithm.

Reinforcement learning is another major category of machine learning techniques that build an agent to learn how to take the best actions in an interactive environment based on feedback of the actions it takes [310]. Unlike the case of supervised and unsupervised learning in which the training dataset that contains inputs and possibly expected outputs is given in advance, the output of the agent is some actions to take in the next step which will generate feedback and new inputs. Feedback of actions taken previously is in the form of rewards or punishment. The objective is to find a suitable action model which maximizes the long-term accumulated reward.

Figure 7.4 shows the pattern of reinforcement learning. A very popular application of reinforcement learning technique is to build AI for playing computer games. AlphaGo Zero—the famous AI that beats a world champion in the game of Go—is built using reinforcement learning [305]. After conquering the game of Go, DeepMind lab, the owner of AlphaGo Zero, is currently building AI to challenge the StarCraft II which is a much more complex game than Go [313]. Many problems in the computer networking area can be solved by using reinforcement learning techniques. An

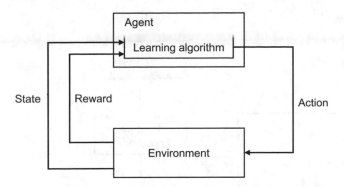

Fig. 7.4 Example of reinforcement learning problems

example is the bandwidth allocation problem in which we are required to allocate available bandwidth dynamically to all existing network flows based on many criteria—some of them can be associated with the type of flows while some others may be associated with real-time bandwidth utilization at different levels of scale. Common goals include minimizing long-term cost of the overall network or maximizing the number of flows that can be accommodated in a certain period. Due to the huge solution space created by the complexity of this problem, popular practical solutions are usually heuristic-based algorithms which have very limited ability on searching for global optimum. Reinforcement learning techniques are very suitable to solve this problem, because of their ability to automatically explore the solution space and converge on the best path towards a global optimum.

7.2.2 General Procedure of Building Machine Learning Solutions

As introduced in the previous section, there are several categories of machine learning techniques. Machine learning models falling into those categories can be very different from each other. However, there is a general procedure to build machine learning solutions in practice that is compatible with most of the machine learning models. Figure 7.5 shows this general procedure. It contains two phases: development and production.

In the development phase, the procedure begins with the *problem formulation*. This includes analyzing properties of the problem and determining an ML model to be used. In practice, problems can usually be solved by multiple ML models depending on how they are formulated. However, applying different models may result in different efficiency. Therefore, proper problem formulation is critical to the success of the solution. Once an ML model is selected, one usually determines the initial set of features—the set of variables used to describe inputs—at the same time. Following

is *data collection* and *model learning*. In the data collection step, according to the initial set of features, historical data are fetched or assembled from data storage. *Model learning* essentially is a loop of *feature engineering*, *model training*, *evaluation*, and *model tuning*. In the feature engineering step, collected historical data are transformed to well-formatted datasets. While some features can be mapped directly from existing fields of historical data, some other features may require a sort of transformations on historical data. Another common operation included in feature engineering is dimensionality reduction, which is the process of reducing the number of features input into the ML model. This process reduces the computation cost of training the model and can also increase accuracy. After feature engineering, the well-formatted dataset is divided into training, validation, and test datasets. Subsequently, the training dataset is used to train the model in the training step, while the validation and test datasets are used to evaluate the trained model. Based on evaluation results, the structure of the model or the feature configuration is updated accordingly. This is the model tuning step. This loop of preparing dataset, training, evaluating, and tuning may be repeated several times until evaluation results are good enough. Once evaluation is passed, the development phase is finished and one is ready to put the trained model into production.

In the production phase, the trained model takes new data and generates the outcome. Of course, all new data need to be transformed to the formatted dataset in the feature engineering step first. The outcome of the trained model is sent to the outside world and used as needed. At the same time, we keep collecting new results from the outside world. These new results are then used as new historical data to update the model via model learning loop. Such updates keep the trained model up-to-date to avoid bias.

More details of data collection, feature engineering, and evaluation are introduced in Sects. 7.2.3, 7.2.5, and 7.2.7.

7.2.3 Data Collection

Building an ML model effectively requires a large amount of representative data. Therefore, data collection is an important step. It usually consists of two phases: offline and online [314].

In the offline data collection phase, a large enough volume of historical data is gathered based on features of the proposed ML model. This is usually the first step of ML model development. It is very common that the dimension of the completed historical data is much larger than required. Building data pipelines that can effectively extract required data from the whole historical data pool is critical. On the other hand, data can be gathered from many public data repositories also. There are comprehensive data repositories

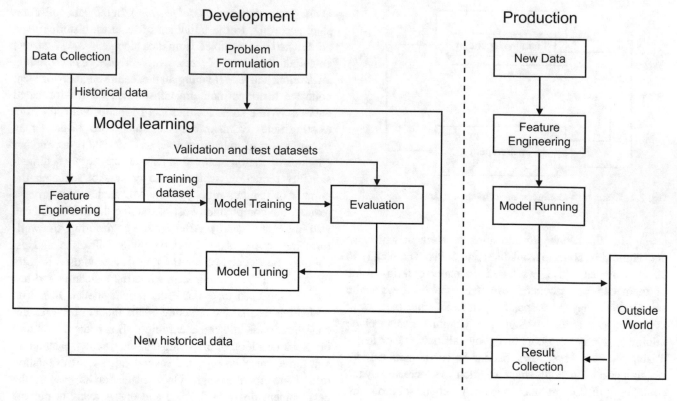

Fig. 7.5 General procedure of building a Machine Learning solution

that cover data in many domains. For example, Data.gov [264] is a comprehensive data repository providing data from multiple US government agencies, which has more than 200 thousand datasets covering 14 domains, such as agriculture, consumer, energy, health, and etc. Eurostat [265], a high-quality stats repository provided by the EU, is another example of such a large and comprehensive data repository. At the same time, there are also datasets specially focusing on a single area. For example, in computer vision area, Google's Open Images dataset [266], a dataset consisting of nine million URLs to images that have been annotated with labels spanning over 6000 categories. In the natural language processing area, Amazon Reviews [263] is a dataset that contains over 30 million reviews from Amazon. Data include product information, user information, and plain text of reviews.

In the online data collection phase, newly generated data is gathered in a continuous way. These new data can be fed back to the model as the input data of model re-training. Online data collection together with model update is an important procedure which improves model performance and keeps the model capturing properties of up-to-date data. In many cases, we can leverage monitoring tools or build real-time data pipelines to collect those new data. In the spam email detection case, after moving emails identified by the model to the spam folder, users may manually move them out, which essentially is a feedback of false positive (i.e.,

not spam but labeled as spam). Such feedback can help us evaluate accuracy of the model better. There are also emails manually labeled by users as spam. Those spam emails can be served as an input of re-training.

In the networking area, there are many monitoring and measurement tools that can be used to collect data from a real network. To gather raw network data, these tools commonly use network monitoring protocols, like Simple Network Management Protocol [306] and IP Flow Information Export [69]. Meanwhile, they provide fine control in various aspects of data gathering, such as monitoring duration, location of measurement points, data sampling rate, and etc. Generally, there are two types of networking monitoring: passive and proactive. Proactive monitoring, also known as synthetic monitoring, is a type of monitoring that actively injects synthetic traffic into network to emulate network behavior of certain end-users or applications and collects corresponding network measurement data. Due to the fact that proactive monitoring relies on synthetic traffic only, it is always available with a short measurement cycle and does not require certain applications or service to be online in advance. However, relying on synthetic traffic is also a drawback of proactive monitoring, because some actual network behaviors are very hard to simulate comprehensively. In addition, injected synthetic traffic also comes with an additional overhead which can be costly.

Passive monitoring techniques rely on actual network traffic to take measurements. They usually use specific probes or built-in data capture functionality on network devices. While passive monitoring doesn't have the drawbacks of proactive monitoring like the overhead of injecting traffic, it does have its own caveats. When the network becomes more dynamic and scaled-up, passive monitoring may not be able to capture enough traffic to comprehensively cover all representative network behaviors, or it simply requires capturing a very large amount of network traffic data which comes with a high cost. In practice, passive and proactive monitoring are more complementary rather than competitive now. There are also data repositories relevant to the networking area, for example, Measurement and Analysis on the WIDE Internet (MAWI) Working Group Traffic Archive [261], Waikato Internet Traffic Storage (WITS) datasets [262], UCI Knowledge Discovery In Databases Archive [269], and Canadian Institute for Cybersecurity datasets [260].

7.2.4 Ground Truth Collection

Network traffic data collected via measurement tools are usually sufficient for unsupervised learning models. However, to solve a supervised learning problem, it requires ground truth in addition to collected traffic data that are attributed as features. For example, in traffic classification problem, training data is a set of network traffic, each of which comprises traffic data as inputs/features and traffic category as output. Establishing ground truth, i.e., finding out the category of each network traffic collected, is a critical yet challenging step. Generally, methods of establishing such ground truth include traditional methods and active measuring methods [318].

Traditional methods include manual labeling, port-based methods, and deep packet inspection (DPI) techniques. Manual labeling requires network experts to spend vast amount of time on analyzing traffic traces, which is very costly to scale. Considering the data volume needed to properly train an ML model is relatively large, manual labeling is almost impossible for ground truth collection. Port-based methods label traffic by the port numbers. They were widely used in the early stage of the Internet. At that time, types of network traffic were limited and port numbers were good identifiers. However, with the increasing complexity of network traffic, it is much harder to label network traffic based on ports today. Some studies [293] have shown that reliability of port-based methods is questionable now. DPI technique is a type of data processing that examines the content of network packets flowing through certain checkpoints. It has been widely used in network measurement and security areas. Labeling network traffic is one of its important applications. Examples of DPI tools include L7-filter [259] and OpenDPI [311].

Notice that both port-based and DPI approaches leverage the passive mode network measurement. They analyze actual network traffic captured by passive network measurement tools.

Along with the increasing usage of active network measurement tools, many active measuring methods for ground truth collection have been proposed [277]. As shown in Fig. 7.6, the general collection procedure of these methods includes:

1. Running some applications on network hosts in a controllable environment. These applications then actively collect information of applications that send out network traffic. Collected application information is ground truth of traffic sent out by those applications.
2. Deploying measurement tools on network routers to collect network packet-level or flow-level network traffic data.
3. Performing post-processing to join application information with traffic data, i.e., attach ground truth to inputs, to form completed traffic data.

Since these active measuring methods collect application-level network data on hosts, the ground truth generated are highly reliable. This is the main reason for the increasing usage of such methods. However, due to using active measurement tools, network traffic captured by those methods are usually emulated traffic which may not be able to reflect the variety and complexity of real network traffic. In other words, the main concern of using such methods is that data collected are not representative enough.

7.2.5 Feature Engineering

The collected raw data usually cannot be directly used to train the ML model, because of many possible problems. For example, these data may spread over multiple datasets or tables, or they can contain invalid values, or only part of them is important to the problem while other data is just noise. Feature engineering is a process that transforms raw data into a well-formatted dataset in which historical data is described by features that better represent the problem to be solved. Good feature engineering leads to significant improvement in model prediction accuracy.

Essentially, the goal of featuring engineering is to retain only meaningful data from all raw data. This is critical because for most of the problems to be solved by ML techniques today, we are able to collect vast amount of raw data. The collected raw data inevitably contains a great deal of noise which would hurt model performance if this noise is included in the training data. In the networking area, there are many choices of features. They can be cat-

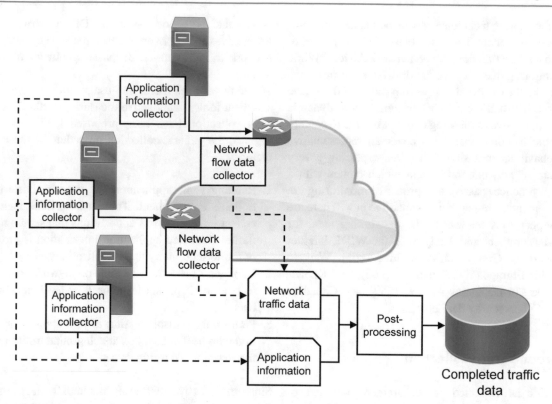

Fig. 7.6 General procedure of active measuring ground truth collection method

egorized into packet-level, flow-level, and connection-level features. Packet-level features are information extracted or derived from network packets. Examples include statistics of packet size, like mean and variance, and time series information such as the Hurst exponent. Flow-level features are statistics of network flows, for example, mean flow duration, mean number of packets per flow, mean number of bytes per flow, and etc. Connection-level features are mostly transport-layer information such as throughput and advertised window size in TCP connection headers. While there are many possible features, different features are relevant to different networking problems. The goal of feature engineering is to select the best set of features for a given problem.

Feature engineering contains two steps: data pre-processing and data transformation. In the data pre-processing step, raw data is processed to generate a well-formatted dataset that can be digested by the ML model. In the data transformation step, important/meaningful features are identified or extracted and only the corresponding part of the well-formatted dataset is retained. This step reduces the number of features and increases feature quality, which reduces the complexity of the model and improves model accuracy.

The procedure of data pre-processing commonly includes removing invalid data or outliers, joining relevant data across multiple data sources/tables, flattening fields with repeated

values into multiple columns, and aggregating data based on certain keys. The output of data pre-processing is a single table that contains all data. Figure 7.7 shows an example of such a data pre-processing on a grocery store's sales records. The goal is to predict purchase of a future customer on a specific category like meat or seafood. In this example, raw data spread over 3 tables:

- The "purchase records" table contains customers' purchase composed of user id, item, and purchase amount.
- The "item metadata" table contains category of each specific item.
- The "customer profile" table contains customer information including age, gender, and user interests.

These three tables are joined together and purchases made by the same user on the same category are aggregated together. In addition, abnormal purchase records, like that User C purchases $2323 worth of lamb, are removed as outliers; and the field "user interests" with repeated values is expanded into 3 columns such that each column in the output table contains a single value. With all these transformations, a single table is generated from raw data. Subsequently, the value of each column in this table is replaced with all numerical values. Categorized values are usually substituted by index numbers and boolean values are substituted by binary numbers. After obtaining a well-formatted

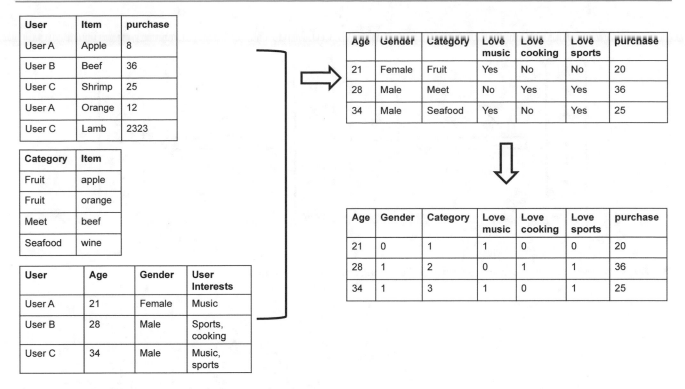

User	Item	purchase
User A	Apple	8
User B	Beef	36
User C	Shrimp	25
User A	Orange	12
User C	Lamb	2323

Category	Item
Fruit	apple
Fruit	orange
Meet	beef
Seafood	wine

User	Age	Gender	User Interests
User A	21	Female	Music
User B	28	Male	Sports, cooking
User C	34	Male	Music, sports

Age	Gender	Category	Love music	Love cooking	Love sports	purchase
21	Female	Fruit	Yes	No	No	20
28	Male	Meet	No	Yes	Yes	36
34	Male	Seafood	Yes	No	Yes	25

Age	Gender	Category	Love music	Love cooking	Love sports	purchase
21	0	1	1	0	0	20
28	1	2	0	1	1	36
34	1	3	1	0	1	25

Fig. 7.7 Example of data pre-processing in feature engineering

numerical table, one can further apply more advanced data processing technique such as data normalization and standardization.

Data transformation is the next step after data pre-processing. The general purpose is to retain only important features as training data for the ML model. Feature selection and feature extraction are two commonly used techniques which achieve the goal via different ways. Figure 7.8 shows the transformations performed in feature selection and feature extraction, respectively, and Fig. 7.9 shows examples of these two techniques.

As shown in Fig. 7.8a, feature selection is a process that identifies and removes irrelevant or redundant features. There features are unneeded for the ML model and may actually damage the performance of the model. There are three major classes of feature selection algorithms: filter method, embedded method, and wrapper method. Filter methods calculate a score for each feature via applying some statistical measure. They rank all features based on the score and keeps the top-K features. Filter methods are usually univariate and consider features independently. Examples of such methods include Chi squared test, correlation coefficient scores, and information gain. Wrapper methods employ an iterative approach. They start with training the model with an initial set of features and evaluating the performance via some objective function. In each of the following iterations, it tries to improve the performance by updating the set of selected features and repeats the training

and evaluation process. Essentially, wrapper methods treat feature selection as an optimization problem whose goal is to find the optimal set of features. Many typical optimization algorithms can be applied here, including metahueristics like genetic algorithms and simulated annealing, heuristics like recursive feature elimination, and search methods like breadth-first search. Embedded methods identify important features and enlarge their impact on the ML model within the model training procedure. The most common embedded methods are regularization methods, like LASSO, Elastic Net, Ridge Regression, and etc.

Unlike feature selection which reduces the dimensionality of training data by determining and keeping a subset of original features, feature extraction achieves this goal by creating a reduced set of features from raw data in a way that the new features can still accurately describe the original data. In the example shown in Fig. 7.9, feature extraction generates a set of three new features from the original set of six features. Note that the newly generated features are usually treated as a feature vector without a specific name for each feature and the value of those new features in data samples may have different scale than that of the original features. Feature extraction is particularly useful when the raw data is too large to be processed or to directly perform feature selection. Such cases are commonly seen in the area of imaging processing and natural language processing in which analog observations are stored in digital format. Examples of feature extraction techniques include Principal

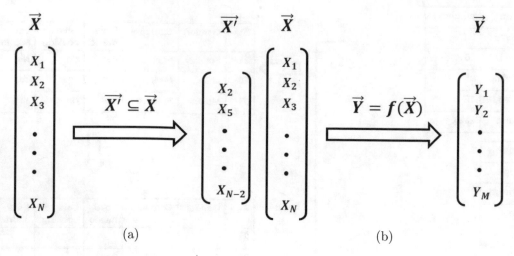

Fig. 7.8 Data transformation in feature engineering. (**a**) Feature selection. (**b**) Feature extraction

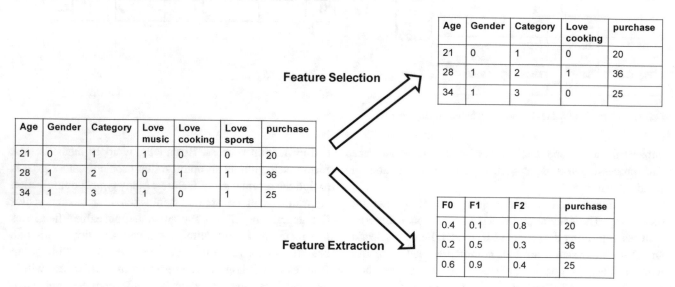

Fig. 7.9 Example of data transformation in feature engineering

Component Analysis (PCA) for tabular data and line or edge detection for image data.

7.2.6 Data Splitting

After feature engineering, the well-formatted historical data is split into training dataset, validation dataset, and test dataset. Each of these datasets serves a specific purpose.

Figure 7.10 shows the usage of training, validation, and test datasets in the ML solution. The training dataset is used to fit the model, i.e., to find the optimal values of model parameters like weights of connections between layers of neurons in a Neural Network. The validation dataset is used to evaluate the performance of the model fitted using the training dataset when tuning model hyper-parameters like the number of hidden layers and the number of neurons in each

hidden layer in a Neural Network. It may be used in some other model preparation step such as feature selection. The validation dataset can also play a role in early stopping to avoid overfitting when training the model: During the training process, evaluate the per-example error once in a while and stop training if the error increases [297]. Note that if the proposed ML model needs little hyper-parameter tuning, it may not need a validation dataset. Finally, the test dataset is used to evaluate the performance of the final model, i.e., to calculate performance metrics such as accuracy, sensitivity, F-measure, and so on. If multiple types of ML models are selected as candidate solutions, the optimal one is the one with the best final performance metrics.

One way to split historical data into these three datasets is fixed-ratio splitting. A common ratio is 60%/20%/20% among training, validation, and test datasets. If validation dataset is not needed, a common ratio between training and

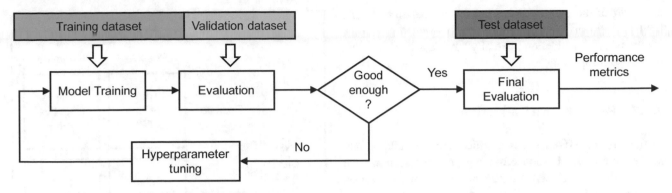

Fig. 7.10 Usage of training, validation, and testing datasets

Each round of training, validation and tuning

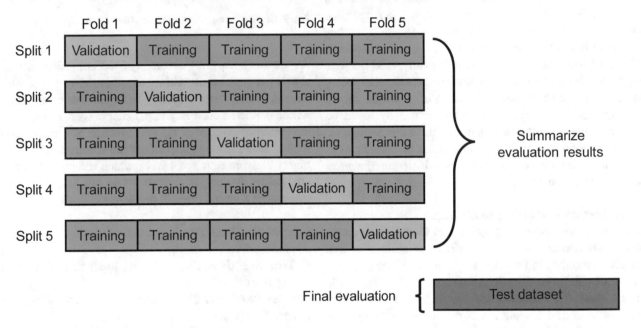

Fig. 7.11 Example of fivefold cross validation

test datasets is 70%/30%. Besides these common ratios, the actual splitting ratio can be adjusted case by case. A general rule to follow is that the more hyper-parameters that the ML model has, the larger the validation dataset that is needed. On the other hand, in the context of big data, the volume of historical data can be extremely large, with millions or billions of records. In these cases, a small portion of input data, like 1 or 2%, may be already enough for validation and testing purpose. However, a critical rule here is that validation and testing dataset must follow the same probability distribution of training dataset to avoid potential bias.

While fixed-ratio data splitting is easy to implement, it can be hard to determine a suitable ratio. In contrast, cross validation is another way of data splitting that avoids such a pre-determination of splitting ratio. Cross validation, also known as k-fold cross validation, is a technique that can be

used whenever we need two datasets, one to train/tune an ML model and the other to evaluate model performance. The general procedure of cross validation starts with randomly splitting the input dataset into k groups. Subsequently, the training plus evaluation process are repeated k times. For each time, the ith group is selected as validation dataset for evaluation and all other groups are combined together as a training dataset. After obtaining k sets of performance metrics, the final step is combining all sets of metrics together as the final model performance. Cross validation can be used in either the loop of training and hyper-parameter tuning or the final evaluation. In the former case, it is split between the training and the validation datasets; in the latter case, it is split between the training and the test datasets. Figure 7.11 shows an example in which a test dataset is pre-split out from an input dataset and fivefold cross validation is used in each round of training and hyperparameter tuning. Since each part of the input data is used for evaluation once and final performance is a summary of all evaluation results, results of cross validation generally have less bias than that of fixed-ratio splitting.

7.2.7   Performance Evaluation

Evaluating the ML model is a critical part of building any ML solutions. As introduced in the previous sections, we evaluate the model performance in each round of hyper-parameter tuning using validation datasets and at the final step using test datasets. The purpose is to provide quantified metrics so that we are able to compare two configuration of hyper-parameters or two models. However, one important thing to notice is that it is very hard to simply conclude that one model is the best based on those given metrics, as they measure different aspects of the model, such as accuracy, reliability, robustness, and so on. It is totally valid that one model has better accuracy but less robustness when being compared to another model. Selection of a specific model or a specific set of hyper-parameters is usually based on evaluation metrics and practical demands. In this section, we introduce performance evaluation metrics of classification, regression, and clustering problems which are the most common ML problems.

Classification Performance Metrics
The most straightforward evaluation metric of a classification problem is accuracy which is the ratio of number of true predictions to the total number of predictions. Let us define the set of classes as $\mathcal{C} = \{C_1, C_2, \ldots, C_N\}$ in which C_i is the ith class. We further define T_{C_i} as the number of true predictions in class C_i, and P_{C_i} as the number of predictions in class C_i. Accuracy can be then formulated as

$$\text{Accuracy} = \frac{\sum_{i=1}^{N} T_{C_i}}{\sum_{i=1}^{N} P_{C_i}}. \qquad (7.1)$$

For example, with expanding the spam email classification problem to classify input emails into three folders: inbox, important, and spam, the accuracy of the ML model is the total number of emails correctly moved into each of the three folders divided by the total number of input emails.

Another formulation of accuracy is based on the Confusion Matrix which is an $N \times N$ matrix and whose definition is

$$\text{Predicted}$$

$$\bar{C} = \text{Actual} \begin{bmatrix} c_{11} & \cdots & c_{1N} \\ \vdots & \ddots & \vdots \\ c_{N1} & \cdots & c_{NN} \end{bmatrix} \qquad (7.2)$$

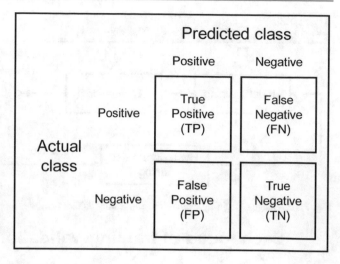

Fig. 7.12 Confusion matrix of binary classification problem

As shown in Eq. (7.2), left axis of the confusion matrix $\bar{C}$ shows the actual class and its top axis shows the predicted classes. In this way, element c_{ij} is the number of predictions whose predicted class is j and actual class is i. The confusion matrix summarizes the performance of a classification model. To illustrate this, consider the confusion matrix of a binary classification problem in which the possible classes are positive and negative, as shown in Fig. 7.12. The confusion matrix shows 4 types of predictions:

- **True Positive (TP)** is correct prediction of an actual positive class.
- **True Negative (TN)** is correct prediction of an actual negative class.
- **False Positive (FP)** is incorrect prediction of an actual negative class.
- **False Negative (FN)** is incorrect prediction of an actual positive class.

Recall that the definition of accuracy is the ratio between correct predictions and total predictions. Naturally, with confusion matrix, accuracy can be formulated as

$$\text{Accuracy} = \frac{\text{TP} + \text{TN}}{\text{TP} + \text{TN} + \text{FP} + \text{FN}}. \qquad (7.3)$$

While an N-class classification problem intuitively has an $N \times N$ confusion matrix, a 2×2 confusion matrix showing (TP, TN, FP, FN) can be built for each specific class, using the *one-vs-rest* strategy. Specifically, for any given class i, it is treated as positive class and all other classes together are treated as negative class. In this way, based on original confusion matrix $\bar{C}$, the 4 types of predictions of class i (denoted by TP_i, TN_i, FP_i, FN_i) are

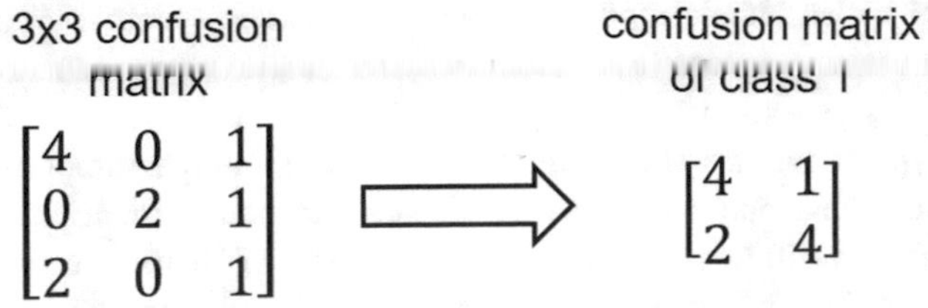

3x3 confusion matrix

$$\begin{bmatrix} 4 & 0 & 1 \\ 0 & 2 & 1 \\ 2 & 0 & 1 \end{bmatrix} \implies$$

confusion matrix of class 1

$$\begin{bmatrix} 4 & 1 \\ 2 & 4 \end{bmatrix}$$

Fig. 7.13 Example of one-vs.-all transformation

$$TP_i = c_{ii}$$

$$FP_i = \sum_{j=1}^{N} c_{ji} - TP_i$$

$$FN_i = \sum_{j=1}^{N} c_{ij} - TP_i \qquad (7.4)$$

$$TN_i = \sum_{j=1}^{N} \sum_{k=1}^{N} c_{jk} - TP_i - FP_i - FN_i$$

Figure 7.13 shows an example of such *one-vs.-all* transformation. In this example, given the 3×3 confusion matrix of a 3-class classification problem, the binary confusion matrix of class 1 is generated using Eq. (7.4).

Besides accuracy, there are other evaluation metrics based on confusion matrix. Two of them are True Positive Rate (TPR) and False Positive Rate (FPR). TPR is the proportion of actual positive input that are correctly predicted. It equals to

$$TPR = Recall = \frac{TP}{TP + FN}. \qquad (7.5)$$

TPR is also known as *recall* or *sensitivity*. Whereas, FPR is the proportion of actual negative input that are mistakenly predicted. It equals to

$$FPR = \frac{FP}{FP + TN}. \qquad (7.6)$$

FPR is also known as *specificity*. Note that both TPR and FPR have a value range of [0,1].

Receiver Operating Characteristics (ROC) curve and Area Under the Curve (AUC) are critical metrics derived from TPR and FPR. The ROC curve is obtained by plotting TPR values against FPR values that are computed using classification thresholds within [0,1], and AUC is the area under ROC curve. A good classification model should have a highly positive gradient when threshold is small, which indicates high TPR with small FPR. While the ROC curve is good for visual analysis, AUC provides a quantified performance measurement. It is the probability that a classification model outputs a higher probability of being positive to an actual positive input than an actual negative input. AUC essentially

measures separability of a classification model. It has a range of [0,1]. The larger the value is, the better separability a classifier has. Note that AUC is one of the most widely used metrics for evaluation.

$F1$ score is another accuracy metric for classification problem. It is the harmonic mean of *recall* (i.e., TPR) and *precision* whose formulation is

$$F1 = 2 * \frac{\text{Precision} \cdot \text{Recall}}{\text{Precision} + \text{Recall}} \qquad (7.7)$$

in which *recall* is the TPR whose definition is shown in Eq. (7.5) and *precision* is the proportion of all positive predictions that are correct. Its formulation is

$$\text{Precision} = \frac{TP}{TP + FP}. \qquad (7.8)$$

As shown in the above equation, while *precision* measures how precise a classifier is, *recall* measures how robust it is The tradeoff between *recall* and *precision* commonly exists when tuning a classification model. This is because the only difference between *recall* and *precision* is using FN or FP in the denominator and a decrease of one of them usually causes an increase of the other one. For example, to miss less spam emails (i.e., higher *recall*), a spam email classifier may generally predict more emails as spam. However, such behavior causes more normal emails incorrectly predicted as spam (i.e., lower *precision*). The $F1$ score provides a measurement on the balance between *recall* and *precision* with a value range of [0,1]. The higher the $F1$ score, the less FN and FP. An $F1$ score with value of 1 means that the classifier correctly predicts all inputs.

While all classification metrics that have been introduced so far are based on counts of the four types of predictions, Logarithmic Loss or Log Loss is an evaluation metric based on the ground truth and the probability assigned by the classifier to each class for each input sample. Assume that there are M input samples and N classes and define y_{ij} as a binary variable whose value is 1 if ith sample's actual class is j, otherwise it is 0. Further denote p_{ij} as the probability that the ith sample's class is class j, i.e., the output of the classifier, Log loss is then formulated as

$$\text{Log Loss} = -\frac{1}{M} \sum_{i=1}^{M} \sum_{j=1}^{N} y_{ij} \times \log(p_{ij}). \qquad (7.9)$$

Log Loss have a value range of $[0, \infty]$. The closer it is to 0, the higher accuracy a classifier has.

Regression Performance Metrics

The conventional evaluation metrics of the regression problem are Mean Absolute Error (MAE) and Mean Squared

Error (MSE). MAE is the mean of the absolute differences between predicted values and their corresponding ground truth. Assume that there are M input data samples and denote the ground truth value and predicted value of the ith sample as y_i and $\hat{y}_i$, respectively. MAE is represented as

$$\text{MAE} = \frac{1}{M} \sum_{i=1}^{M} |y_i - \hat{y}_i|. \qquad (7.10)$$

Note that MAE does not contain any information about the direction of the error, i.e., whether the regression model is over-predicting the input samples or under-predicting those samples.

MSE is the mean of the squared differences between predicated values and their corresponding ground truth, which can be formulated as

$$\text{MSE} = \frac{1}{M} \sum_{i=1}^{M} (y_i - \hat{y}_i)^2. \qquad (7.11)$$

While MSE is very similar to MAE, its gradient is easier to be obtained. Besides, MSE penalizes large errors more heavily than MAE.

Some other common evaluation metrics include Mean Absolute Predication Error (MAPE), Root MSE (RMSE), and Normalized RMSE (NRMSE). MAPE calculates the mean of percentage differences between predicted values and the ground truth. It is defined as

$$\text{MAPE} = \frac{100\%}{M} \sum_{i=1}^{M} |\frac{y_i - \hat{y}_i}{y_i}|. \qquad (7.12)$$

RMSE is squared root of MSE, which essentially is the standard deviation of the prediction errors. Its formula is

$$\text{RMSE} = \sqrt{\frac{1}{M} \sum_{i=1}^{M} (y_i - \hat{y}_i)^2}. \qquad (7.13)$$

NRMSE normalizes RMSE using the observed value range. It can be expressed as

$$\text{NRMSE} = \frac{\text{RMSE}}{O_{\max} - O_{\min}}, \qquad (7.14)$$

in which $O_{\max}$ and $O_{\min}$ are the max and min observed value, respectively.

Clustering Performance Metrics

Evaluating the performance of a clustering model is more complicated than summarizing the prediction errors of a regression model or calculating the precision and the recall of a classification model. Specially, evaluation metrics of a clustering model should measure the quality of the separations generated by a clustering model. To achieve this, one type of metrics checks the satisfaction of some assumption such that data in the same cluster are more similar than data in different clusters. The other type of metrics examines the similarity between data clusters and some ground truth set of classes. Note that the latter type of metrics requires the knowledge of the ground truth classes, which is actually available in many clustering problems. In the following, some typical evaluation metrics of each type are introduced.

Contingency Table A contingency table describes and defines the relationship between two partitions. It is used in many ground truth based evaluation metrics. Let us define the set of output clusters as $\mathcal{C} = \{C_1, C_2, \ldots, C_N\}$ in which C_i is the ith output cluster, and define the set of ground truth classes as $\mathcal{G} = \{G_1, G_2, \ldots, G_M\}$ in which G_j is the jth actual class. The contingency matrix of output clusters and actual classes is an $N \times M$ matrix in which each element e_{ij} is the number of data points that are in both cluster C_i and actual class G_j. Furthermore, denote $e_{i.}$ and $e_{.j}$ as the number of data points in cluster C_i and class G_j, respectively. Intuitively, they are the sum of elements on row i and column j. An example of contingency matrix is given in Fig. 7.14.

Rand Index and Adjusted Rand Index The Rand Index (RI), proposed by Rand [298], is an evaluation metric that measures the similarity of cluster based on agreement and disagreement between data point pairs in clusters. Specifically, given a pair of data points denoted by (d_i, d_j), an agreement is that these two data points are assigned to the same output cluster and the same actual classes or they are assigned to different clusters and different classes. If the

$$\mathcal{G} = \{G_1 = \{a, b\}, G_2 = \{c, d, e, f\}\}$$
$$\mathcal{C} = \{C_1 = \{a, b, f\}, C_2 = \{c, e\}, C_3 = \{d\}\}$$

$N \times M$ contingency table

	G_1	G_2	
C_1	$e_{11} = 2$	$e_{12} = 1$	$e_{1.} = 3$
C_2	$e_{21} = 0$	$e_{22} = 2$	$e_{2.} = 2$
C_3	$e_{31} = 0$	$e_{32} = 1$	$e_{3.} = 1$
	$e_{.1} = 2$	$e_{.2} = 4$	

Fig. 7.14 Example of contingency table

criteria of an agreement is not fulfilled, it is a disagreement instead. Let us define $A(d_i, d_j)$ as a binary variable whose value is 1 if there is an agreement on (d_i, d_j); otherwise, it is 0. We further define $C(d_i)$ and $G(d_i)$ as the cluster and actual classes to which d_i is assigned, respectively. $A(d_i, d_j)$ can be then formulated as

$$A(d_i, d_j) = \begin{cases} 1, & \text{if } C(d_i) == C(d_j) \text{ and } G(d_i) == G(d_j), \\ 1, & \text{if } C(d_i) \neq C(d_j) \text{ and } G(d_i) \neq G(d_j), \\ 0, & \text{otherwise.} \end{cases}$$
(7.15)

As shown in the above equation, an agreement on the data point pair (d_i, d_j) is that d_i and d_j are in the same actual class and are in the same cluster, or they are in the different actual classes and are in the different cluster. With binary variable $A(d_i, d_j)$, rand index (denoted by $Rand(C, G)$) is formulated as

$$Rand(C, G) = \frac{\sum_{i<j}^{k} A(d_i, d_j)}{\binom{k}{2}},$$
(7.16)

in which the denominator is the total number of data point pairs, given k data points. While RI uses an intuitive way to measure the similarity between C and G, one concern is that RI does not guarantee that random assignment gets a value close to zero, especially when the number of clusters is on the same level of magnitude as the number of data points.

Hubert and Arabie [284] proposed a corrected-for-chance version of the Rand index that is known as Adjusted Rand Index (ARI). In ARI, a baseline is established by calculating the similarity between ground truth classes G and clusters generated by a random assignment; the RI is then compared with this baseline. The formulation of ARI is

$$ARI = \frac{Index - E[Index]}{\max(Index) - E[Index]}.$$
(7.17)

In the above equation, $Index$ is the total number of data point pairs that are assigned to the same cluster and the same actual class, i.e., the first case in Eq. (7.15). Its value is given by

$$Index = \sum_{i=1}^{N} \sum_{j=1}^{M} \binom{e_{ij}}{2},$$
(7.18)

where e_{ij} is the number of data points that are in both cluster C_i and actual class G_j. Here $E[index]$, the expected value of $Index$, is the sum of expected number of data point pairs

that are assigned to the same partition for each combination of cluster C_i and actual class G_j, i.e.,

$$E[Index] = \sum_{i=1}^{N} \sum_{j=1}^{M} E\left[\binom{e_{ij}}{2}\right] = \sum_{i=1}^{N} \sum_{j=1}^{M} \frac{\binom{e_{i.}}{2}\binom{e_{.j}}{2}}{\binom{K}{2}},$$
(7.19)

where $e_{i.}$ and $e_{.j}$ are the number of data points in cluster C_i and class G_j, respectively. Also $\max(Index)$ is the maximum value of $Index$ which is the average number of data point pairs in clusters C and actual classes G. It can be presented as

$$\max(Index) = \frac{1}{2}\left(\sum_{i=1}^{N} \binom{e_{i.}}{2} + \sum_{j=1}^{M} \binom{e_{.j}}{2}\right).$$
(7.20)

Putting Eqs. (7.17)–(7.20) together, we have completed ARI formulation as

$$ARI = \frac{\sum_{i,j} \binom{e_{ij}}{2} - \sum_{i,j} \frac{\binom{e_{i.}}{2}\binom{e_{.j}}{2}}{\binom{K}{2}}}{\frac{1}{2}\left(\sum_i \binom{e_{i.}}{2} + \sum_j \binom{e_{.j}}{2}\right) - \sum_{i,j} \frac{\binom{e_{i.}}{2}\binom{e_{.j}}{2}}{\binom{K}{2}}}.$$
(7.21)

The value range of ARI is [0, 1], with only extreme cases below zero meaning that their index value is even worse than that of random assignment.

Fowlkes–Mallows Index Fowlkes–Mallows Index (FMI) is another evaluation metric that measures the similarity between output clusters C and actual classes G [276]. Given an $N \times M$ contingency table, the FMI, denoted by $B(N, M)$, is defined as

$$B(N, M) = \frac{T}{\sqrt{P \cdot Q}},$$
(7.22)

in which

$$T = \sum_{i,j} e_{ij}^2 - k,$$
(7.23)

$$P = \sum_i e_{i.}^2 - k,$$
(7.24)

$$Q = \sum_j e_{.j}^2 - k.$$
(7.25)

FMI can also be defined based on agreements and disagreement of data point pairs between C and G. Specifically, let us define

$$TP = |S| \text{ where } S = \{(d_i, d_j) \mid C(d_i) = C(d_j) \text{ and } G(d_i) = G(d_j)\}, \tag{7.26}$$

$$FP = |S| \text{ where } S = \{(d_i, d_j) \mid C(d_i) = C(d_j) \text{ and } G(d_i) \neq G(d_j)\}, \tag{7.27}$$

$$TN = |S| \text{ where } S = \{(d_i, d_j) \mid C(d_i) \neq C(d_j) \text{ and } G(d_i) \neq G(d_j)\}, \tag{7.28}$$

$$FN = |S| \text{ where } S = \{(d_i, d_j) \mid C(d_i) \neq C(d_j) \text{ and } G(d_i) = G(d_j)\}. \tag{7.29}$$

FMI then can be defined as

$$FMI = \sqrt{\frac{TP}{TP+FP} \cdot \frac{TP}{TP+FN}}. \qquad (7.30)$$

Silhouette Coefficient The Silhouette Coefficient is an evaluation metric that does not require ground truth classes [300]. Given a data point d_i, its Silhouette Coefficient $s(d_i)$ is derived from the mean distance between d_i and all of the other points in the same cluster and the smallest mean distance between d_i and all of the other points in any other cluster, of which d_i is not a member. Let us denote the earlier one by $a(d_i)$ and the latter one by $b(d_i)$, we then have

$$a(d_i) = \frac{1}{|C(d_i)|-1} \sum_{d_j \in C(d_i),\, d_j \neq d_i} d(d_i, d_j), \qquad (7.31)$$

in which $d(d_i, d_j)$ is the distance between data points d_i and d_j. This distance function can be any distance metric, such as the Euclidean distance or the Manhattan distance. We also have

$$b(d_i) = \min_{C_q \neq C(d_i)} \frac{1}{|C_q|} \sum_{d_j \in C_q} d(d_i, d_j). \qquad (7.32)$$

Based on $a(d_i)$ and $b(d_i)$, the Silhouette Coefficient $s(d_i)$ is defined as

$$s(d_i) = \frac{b(d_i) - a(d_i)}{\max\{a(d_i), b(d_i)\}}. \qquad (7.33)$$

From the above definition, it is straightforward that $s(d_i)$ has a value range of $[-1, 1]$. The closer $s(d_i)$ is to 1, the better performance the clustering model has, because it requires a smaller $a(d_i)$ and larger $b(d_i)$ to make $s(d_i)$ become larger. A smaller $a(d_i)$ indicates d_i is more like other data points in the same cluster, while a larger $b(d_i)$ means d_i is more dissimilar to data points in its nearest cluster. Overall, these indicates a better separation between clusters. However, $s(d_i)$ only measures the quality of separations of a single data point.

To evaluate the quality of separations of individual cluster, we define the average silhouette width $s(C_q)$ as the average silhouette value of all data points in cluster C_q, which can be presented as

$$s(C_q) = \frac{1}{|C_q|} \sum_{d_i \in C_q} s(d_i). \qquad (7.34)$$

Based on the average silhouette width of individual cluster, we can further obtain the average silhouette width of the entire clustering $s(C)$ by using the following equation.

$$s(C) = \frac{1}{N} \sum_j s(C_j). \qquad (7.35)$$

7.3 Traffic Classification

Traffic classification is the problem of assigning network flows to a set of predefined classes based on the behaviors or properties of those flows. It is a fundamental part of network analysis and management. Traffic classification has been applied in many networking areas, such as Quality of Service (QoS), network security and intrusion detection, performance monitoring, flow planning, resource provisioning, and etc. For instance, an enterprise intranet solution usually includes prioritizing network flows of business critical applications, characterizing behaviors of different network flows for better planning and designing, and identifying unknown network traffic for anomaly detection. All these functionality require accurate traffic classification.

The traditional traffic classification approach is the port-based approach which essentially identifies traffic class based on the mapping between applications and their port numbers registered with Internet Assigned Numbers Authority (IANA). While there are many applications, such as WWW and email, that use (or have used) fixed port numbers, more and more applications do not have fixed port number anymore. Examples include P2P applications, online games and multimedia applications. With the increasing usage of dynamic port numbers, the port-based approach is less and less efficient. Whereas, with the maturity of ML techniques, ML-based traffic classification solutions have been more popular and have shown the superior performance in the much more complicated network environment of today.

The following sections introduce the goals, the features, and the state-of-the-art ML solutions of the traffic classification problem.

7.3.1 Classification Goals

The predefined classes to which network traffic is assigned are the fundamental part of the traffic classification problem. Three general types of classes have been used in the existing traffic classification solutions: traffic categories, application protocols, and end-user applications.

Traffic categories are common groups of applications [294, 321]. For example, the category "interactive" usually refers to network traffic of those applications having interactive communications such as ssh, klogin, telnet, and etc. Table 7.1 shows some common traffic categories and applications belonging to each of these categories. At first glance, using traffic categories as predefined classes may reduce the complexity of the traffic classification problem, because similar applications are clustered. Such a clustering may increase the separation between the traffic behaviors of different

Table 7.1 Example of traffic categories and applications of each category

Categories	Applications
Interactive	ssh, telnet, rlogin
Mail	IMAP, SMTP, POP3
Bursty	FTP, software updates, video downloads
Games	Overwatch, League of Legends, World of Warcraft
Streaming	YouTube, Netflix, Hulu
P2P	BitTorrent, Gnutella, KaZaA
Service	DNS, NTP, Ident

categories, therefore make the classification easier. However, such a grouping is artificial and may lack of the support from sophisticated data analysis. This leads to the fact that the traffic behaviors of the applications within the same category may not necessarily be similar with each other, while the traffic behaviors of the applications across categories may not be separated clearly. For example, both streaming and P2P applications may have short-lived bidirectional control flows at the beginning of data transmission. In addition, in modern networking solutions, more sophisticated network control at the application level is commonly required, which will not be fulfilled by using traffic categories. As a result, the usage of traffic categories is less popular today.

Application protocols and end-user applications are the other two popular choices of the predefined classes. Application protocols are the protocols used by network applications, for example, HTTP, DNS, NTP, SMTP, SSH, BitTorrent. Note that many of the applications listed in Table 7.1 are essentially application protocols. Using application protocols to define a network traffic is very straightforward. Meanwhile, end-user applications are the actual applications of network traffic. For example, YouTube, Netflix, and Hulu (listed in Table 7.1 under streaming category) are end-user applications. Both application protocols and end-user applications provide a fine-grained classification of network traffic. In many cases, they are mixed together as predefined classes [275].

There are studies focusing on even more fine-grained classification [267], in which given a set of end user applications, network flows of each application are further clustered into groups and the classification goal is to assign a given network flow to one of the pre-clustered groups of a specific application. For example, network flows of an FTP application may be clustered into control flows and data flows. In this way, the classification model is required to distinguish network flows of a single application with different characteristics and therefore enables more sophisticated network control and planning.

7.3.2 Features

As introduced in the previous sections, the quality of the features used in an ML model is critical to the success of the model. In the traffic classification problem, there are many choices of the features which can be broadly divided into four categories: payload-based features, flow statistics, connection statistics, and multi-flow features [299]. Among those categories, the most commonly used features are payload-based features and flow statistics.

Payload-Based Features

Payload-based features are the features derived from payload of flow packets. Depending on the application protocol used, there are two types of payload: text-based and binary-based. In the text-based protocol, packet payload is plain text, part of which may be human readable. Whereas, in the binary-based protocol, packet payload is composed of binary values with certain rules on the packet header. Figure 7.15 shows an example of payload of the SMTP and RTCP protocols which are text-based and binary-based, respectively. The key insight of using payload as features is that for network flows of a single application, there are certain invariant parts across packets and also unique information carried by a sequence

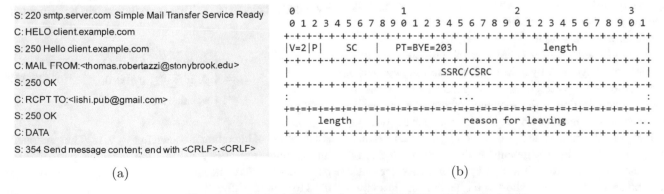

Fig. 7.15 Examples of SMTP and RTCP packet payload. (**a**) SMTP session. (**b**) RTCP BYE packet header definition

Fig. 7.16 Discrete byte encoding

Fig. 7.17 Example of byte segments

Fig. 7.18 Example of Bitcoding

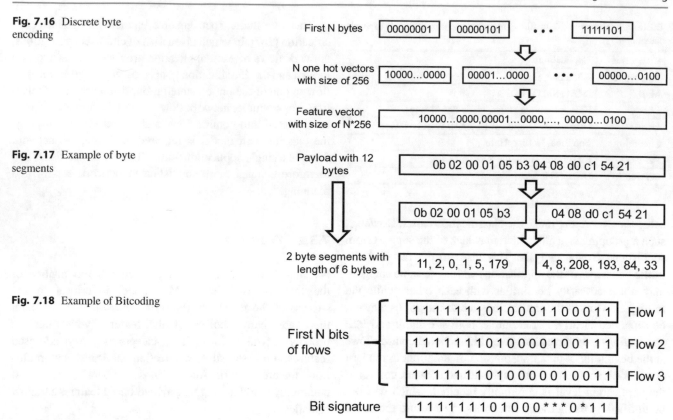

of packets. In the following, several methods of deriving ML features from the packet payload are introduced.

Discrete byte encoding [278] is an approach that transforms the first N bytes of a network flow into a feature vector with size of $N \times 256$. The procedure comprises two steps, as shown in Fig. 7.16. The first step is transforming the first N bytes into N one-hot vectors, each of which has a length of 256. Let B_i denote the ith byte with value of c_i, then v_i is the one-hot vector derived from B_i whose c_ith element is 1 while all other elements are 0, i.e.,

$$v_i[j] = \begin{cases} 1, & \text{if } j = c_i, \\ 0, & \text{otherwise.} \end{cases} \quad (7.36)$$

Note that the length of v_i is 256, because the value range of a byte (8 bits) is [0, 255]. After generating N vectors, the second step is simply concatenating them together to form the feature vector with length of $N \times 256$.

The byte segments [287] are vectors composed of a fixed number of integers. A byte segment represents a segment of byte values of the packet payload. To generate byte segments, one can simply divide the sequence of bytes in the packet payload into segments with length of L and then convert byte values into decimal values. Figure 7.17 shows an example of generating 2 byte segments with length of 6 from a payload with 12 bytes. Note that byte values are presented as hexadecimal values to shorten the length for illustration

purpose. Such a set of byte segments of a packet payload is then used as the ML feature.

While the discrete byte encoding and the byte segments are relatively intuitive ways to generate feature vectors from payload, there are more advanced approaches proposed recently.

The Bitcoding [283] is a method that generates a bit signature of a network application from the first N bits of flows of that application. Let us define $\mathcal{F} = \{F_1, F_2, \ldots, F_N\}$ as the set of flows in which F_i is the ith flow and c_{ij} is the value of jth bit of F_i. Further define the bit signature as vector $\bar{S}$ in which s_i is the ith element with three possible values: 1, 0, and $*$. Specifically, if all flows in $\mathcal{F}$ have a value of 1 on ith bit, then s_i is set to 1; similarly, if all flows in $\mathcal{F}$ have value 0 on the ith bit, then s_i is set to 0; otherwise, s_i is represented by wildcard character $*$. This formulation can be presented as

$$s_i = \begin{cases} 1, & \text{if } \sum_{j=1}^{N} c_{ji} = N, \\ 0, & \text{if } \sum_{j=1}^{N} c_{ji} = 0, \\ *, & \text{otherwise.} \end{cases} \quad (7.37)$$

An example of bitcoding is shown in Fig. 7.18. The generated bit signature essentially catches the invariant part of flow payload of a network application. Such signatures can be used to identify different network applications.

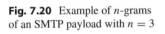
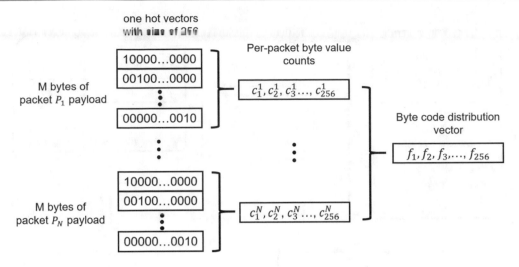

Fig. 7.19 Byte value frequency vector

Fig. 7.20 Example of n-grams of an SMTP payload with $n = 3$

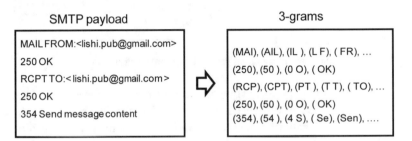

The byte code distribution vector [316] is a vector that records the frequency of each possible value of a byte appearing in the first N packets of a network flow. Assume that the first N packets are denoted by $\mathcal{P} = \{P_1, P_2, \ldots, P_N\}$ and each of them has M bytes. The procedure of generating the byte code distribution vector consists of two steps, as shown in Fig. 7.19. The first step is generating byte value counts of each packet. For packet P_i, the byte value counts are defined as a vector $\bar{C} = \{c_1^i, c_2^i, \ldots, c_{256}^i\}$ in which c_j^i is the total number of bytes in the P_i's payload with value of j. When presenting the value of each byte in the packet payload as a one-hot vector with the length of 256, such a vector $\bar{C}$ can be obtained by summing up those one-hot vectors. After obtaining the byte value count vector for all packets in $\mathcal{P}$, the next step is constructing the byte code distribution vector which is defined as $\mathcal{F} = \{f_1^i, f_2^i, \ldots, f_{256}^i\}$. Specifically, f_j^i is the frequency that byte value j appears in all bytes of all packets in $\mathcal{P}$. It can be calculated by

$$f_j^i = \frac{\sum_{i=1}^{N} c_j^i}{N * M}. \tag{7.38}$$

The byte code distribution vector essentially records statistical information of a network flow's payload.

The protocol keywords distribution [320] is a method that derives a set of keywords to describe a given protocol from the packet payload and generates a keywords distribution vector for any new packet as ML features. This method contains two steps. The first step is to infer a set of protocol keywords from the packet payload of a specific protocol. This problem is treated as a topic modeling problem in the Natural Language Processing (NLP) area, in which a small set of topics are generated from a set of observed documents composed of words with the assumption that each word is attributed to one of the topics. In the protocol keywords inference problem, each historical packet payload is a document and the n-grams of that payload are the words contained in the given document. An n-gram is a subsequence of n elements in a given sequence of elements (longer than n). An example of n-grams of an SMTP payload with $n = 3$ is shown in Fig. 7.20. Assume that the set of n-grams of all observed payload is $\mathcal{W}$ with $|\mathcal{W}| = V$. Further assume that there exist N keywords for the given protocol, then the ith protocol keyword is defined as vector $\vec{\varphi}_i = \{\varphi_1^i, \varphi_2^i, \ldots, \varphi_V^i\}$ in which φ_j^i is the probability that the jth n-gram in $\mathcal{W}$ belongs to the keyword i. The Latent Dirichlet Allocation (LDA) method [270] is used to infer such protocol keywords. In the NLP area, LDA is a generative statistical model that generates a set of unobserved groups from sets of observations to explain why some parts of those observations are similar. The detailed procedure of LDA is omitted here, as it is beyond the scope of this book.

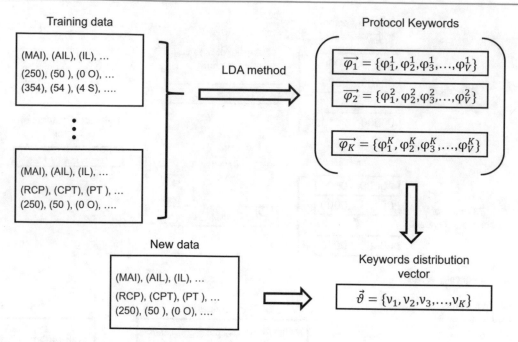

Fig. 7.21 Procedure of generating protocol keywords distribution vector

After obtaining protocol keyword vectors, the second step is extracting the keyword distribution vector from the packet payload as the feature vector. A keyword distribution vector for a given packet payload is defined as $\vec{\vartheta} = \{v_1, v_2, \ldots, v_N\}$ in which v_i represents the likelihood of the given payload being attributed to keyword i and is defined as

$$v_i = \frac{c_i + \alpha}{\sum_{j=1}^{N} c_j + N\alpha}, \tag{7.39}$$

where c_i is the number of times keyword i is observed in n-grams of the given payload. Figure 7.21 shows the overall procedure of generating the keyword distribution vector.

Flow Statistics

Flow statistics are the statistics of network flow metrics that can be used to define and describe a network flow. They usually include unique properties of a network flow, like the source and destination IP addresses and ports, and a set of characteristics parameterizing the flow behavior such as flow duration, mean and variance of packet payload, mean and variance of packet inter-arrival time, and etc. In practice, there exist hundreds of available flow statistics [292]. How to select or extract a set of features from these choices is a critical problem when proposing an ML solution for the traffic classification problem. Both the feature selection methods and the feature extraction methods have been applied to solve this problem. In the following, several existing approaches are introduced as examples.

The consistency-based feature selection algorithm [274] is an algorithm that evaluates the goodness of a set of features

via its inconsistency level which is defined as the ratio of the number of samples with the same value of these features but different class labels (i.e., inconsistent samples) to the number of all samples. The algorithm has been used in traffic classification problem [273]. Specifically, the inconsistent level of each pair of features is measured and the top k pairs of features with the least inconsistency level are selected as final features.

While the consistency-based feature selection algorithm focuses on the relevance between features, there are other methods that consider both the relevance between a given feature and the output class and the redundancy between those selected features [294, 304].

The Fast Correlation-Based Filter (FCBF) [319] is a feature selection algorithm that measures the goodness of a feature by its correlation with the output class and other good features. In the FCBF, the correlations are measured via *symmetrical uncertainty* which is based on the entropy of random variables. Assume that a random variable X takes values from $\{x_1, x_2, \ldots, x_N\}$, then the entropy of this random variable is

$$H(X) = -\sum_{x_i \in X} p(x_i) \log_2 p(x_i), \tag{7.40}$$

in which $p(x_i)$ is the probability that variable X takes value x_i. The entropy of random variable X given another random variable Y is

$$H(X|Y) = -\sum_{y_j \in Y} p(y_j) \sum_{x_i \in X} p(x_i|y_j) \log_2 p(x_i|y_j). \tag{7.41}$$

Based on $H(X)$ and $H(X|Y)$, the information gain is defined as

$$IG(X|Y) = H(X) - H(X|Y) = H(Y) - H(Y|X) = IG(Y|X), \tag{7.42}$$

which is a measure of the correlation between X and Y in practice. With Eqs. (7.40) and (7.42), the symmetrical uncertainty is defined as

$$SU(X, Y) = 2\left[\frac{IG(X|Y)}{H(X) + H(Y)}\right]. \tag{7.43}$$

The $SU(X, Y)$ has a value range of $[0, 1]$, where value 0 indicates that X and Y are independent and value 1 indicates that X and Y are strongly correlated. The FCBF algorithm calculates the symmetrical uncertainty between each feature and the output (denoted by $SU_{i,c}$) and between each pair of features (denoted by $SU_{i,j}$). All features that satisfy $SU_{i,c} > \theta$ are selected into a set $\mathcal{F}$. After constructing the set $\mathcal{F}$, the features in $\mathcal{F}$ are sorted in the descending order of $SU_{i,c}$. Subsequently, each selected feature F_p in $\mathcal{F}$ is considered iteratively. For any other feature F_q, it is identified as redundant and gets removed from $\mathcal{F}$, if $SU_{p,q} > SU_{p,c}$. After iterating over all features in $\mathcal{F}$, the remaining features are the final features. The FCBF algorithm is used in an ML solution of traffic classification problem [294]. In this solution, a wrapper method is used to determine the best value of the threshold θ.

The *WMI_AUC* algorithm [304] is a feature selection algorithm for traffic classification that considers the relevance between features and the output class. Similar to the feature selection algorithm proposed in [294], the WMI_AUC algorithm selects a set of features based on their relevance with the output class and then refines the selected set of features based on the Area Under Curve (AUC) metric. To begin with, the relevance between a given feature and the output class is measured via *Mutual Information* which is defined as

$$H(X; Y) = -\sum_{y_j \in Y} \sum_{x_i \in X} p(x_i|y_j) \log_2 \left(\frac{p(x_i, y_j)}{p(x_i)p(y_j)}\right), \tag{7.44}$$

in which X and Y are two random variables. Any feature F_i with $H(F_i, c) > \theta$, where θ is a pre-defined threshold, is selected into a feature set $\mathcal{F}$. In the following, the WMI_AUC algorithm removes the features that cannot increase AUC metric of the proposed classifier from $\mathcal{F}$. The remaining features in $\mathcal{F}$ are the final features.

Autoencoder [312] is a feature extraction method that has been used in traffic classification [291] to reduce the dimensionality of selected features. It aims at learning a representation of the original input features with reduced dimensionality (i.e., encoding). To achieve this, the autoen-coder leverages a Neural Network (NN) with encoding layers and decoding layers, both are full connected layers. Passing through encoding layers, the input features are gradually reduced to a set of new features with smaller dimensionality, represented as neurons in the NN. This reduced set of features is further sent through decoding layers with the goal of reconstructing the input features, so that this reduced set of features is trained to contain critical information of the original features. In short, the autoencoder is an NN which approximates its input in its output and has multiple hidden layers whose number of neurons is decreasing at the beginning and is increasing thereafter. The output value of the hidden layer with the smallest number of neurons is the encoded representation of the input features, i.e., the reduced set of features. Figure 7.22 shows an example of autoencoder with 3 hidden layers. An autoencoder-based ML solution has been proposed [291], in which several autoencoders are stacked together to form a deep neural network (DNN). More details of this DNN classifier are introduced in the following section.

7.3.3 State-of-the-Art Solutions

Traditional models used in the traffic classification problem include Random Forest [321], Naive Bayesian [294], SVM [275], C5.0 [268] and many others. Much recent research that use these traditional methods focus more on new feature engineering methods, transfer learning, and incremental learning. As an example, a modified version of the incremental SVM (ISVM) model named AISVM has been proposed to solve the traffic classification problem [307]. The AISVM model is initially trained using the original training dataset. After the initial training, the Support Vectors (SVs) and the calculated coefficients are retained while all training data are dropped. In the following incremental training process, the model is trained using the combination of the retained SVs and the new training data. After each incremental training process, some of the SVs in the previously retained set may not be valid support vectors anymore. In the classic ISVM model, these SVs are dropped immediately. However, differing from the classic ISVM model, the AISVM model assigns a weight to each of such SVs and still keeps them in the set of retained SVs. The weights decrease along with more incremental training process. Once the weight of a specific SV becomes zero, that SV is dropped. In this way, some potentially useful information contained in those invalid SVs are utilized by the AISVM model for a longer time and therefore improves the model performance.

While there is still active research on applying/optimizing those traditional methods on traffic classification problem, deep learning methods are becoming much more popular, attributed to the huge improvement on the performance and

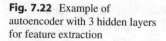

Fig. 7.22 Example of autoencoder with 3 hidden layers for feature extraction

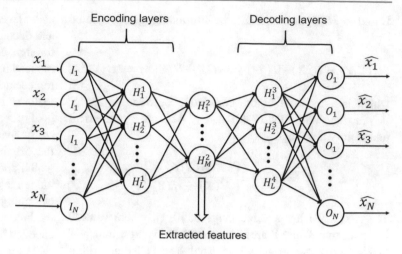

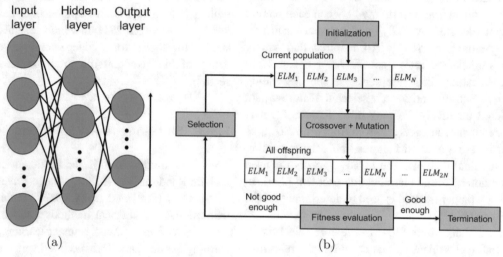

Fig. 7.23 Examples of the GA-WK-ELM model. (**a**) Structure of the ELM model. (**b**) Flowchart of the GA-WK-ELM model

scalability of those methods. Many novel deep learning models have been proposed to solve the traffic classification problem. Some representative ones are introduced in the following.

The GA-WK-ELM model [89] is a combination of the Genetic Algorithm (GA) and the Wavelet Kernel based Extreme Learning Machine (WK-ELM) model. The ELM model [281],whose structure is shown in Fig. 7.23, is a neural network with a single hidden layer. The training process of an ELM model is very different from that of a conventional neural network. Specifically, the weights of the hidden layer are randomly generated and the weights of the output layer are directly calculated. Without backpropagation, the training process of the ELM model is fast and the performance of the ELM model has been proved to be promising. In addition, kernel functions can be used in the ELM model. However, the performance of an ELM model is influenced by the random initialization. To overcome this drawback, the GA-WK-ELM model combines the ELM model with the GA algorithm. The GA algorithm is a meta-heuristic which optimizes a cer-

tain objective function by simulating the process of natural selection. Its general procedure is shown in Fig. 7.23b. The GA algorithm maintains a population of candidate solutions and performs crossover and mutation on these candidate solutions to get more candidate solutions. Subsequently, only the better solutions are kept in the population. Note that the quality of a candidate solution is measured via the value of the objective function which is also known as the fitness value. By repeating this process, the quality of the candidate solutions keeps improving and eventually it converges to the global optimum or stops when a certain termination criteria is met. In the case of GA-WK-ELM algorithm, the candidate solutions are the ELM models with different weights of the hidden layer.

As one of the most popular deep learning methods, Convolutional Neural Network (CNN) [286] has been used in solving the traffic classification problem [291]. The basic block of a CNN model is the convolution layer whose input is a three-dimensional matrix with the size of $N \times M \times K$, denoted by $\bar{I}$. In the convolutional layer, W filters are applied

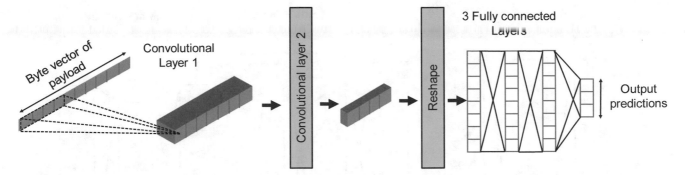

Fig. 7.24 Example of one-dimensional CNN

on the input matrix. Each filter w is a three-dimensional matrix with the size of $P \times Q \times K$. It is applied on the input matrix in a sliding manner to generate an output matrix with the size of $(N - P + 1) \times (M - Q + 1)$.

Combining W such two-dimensional matrix together forms the final output matrix $\bar{O}$ with size of $(N - P + 1) \times (M - Q + 1) \times W$. The element o_{ijk} of $\bar{O}$ is computed as

$$o_{ijl} = f\left(\sum_{a=1}^{P} \sum_{b=1}^{Q} \sum_{c=1}^{K} w_{abc}^l \cdot I_{(i+a)(j+b)c} \right), \quad (7.45)$$

where w^l is the lth filter and function f is usually a nonlinear function such as Rectified Linear Unit (ReLU). A one-dimensional CNN (1D-CNN) model is proposed to solve the traffic classification problem [291]. The input of this 1D-CNN model is a one-dimensional feature vector that contains byte values of packet payload. Its overall structure is shown in Fig. 7.24. This 1D-CNN model contains two convolutional layers. The output of these convolutional layers is a two-dimensional matrix which is then reshaped into a one-dimensional vector. In the following, three fully-connected layers are applied sequentially on this one-dimensional vector to obtain the output per-class predictions. Using the convolutional layer, the correlations between consecutive bytes of packet payload is considered when classifying network traffic.

A stacked autoencoder (SAE) model is proposed in [291] also. While the autoencoder is an NN that performs dimensionality reduction on the input features as introduced in Sect. 7.3.2, a stacked autoencoder is a deep learning model that stacks multiple autoencoders together. The output of one autoencoder is the input of the successive autoencoder, as shown in Fig. 7.25. The training process of the SAE model includes training each autoencoder separately and then training the whole NN. When training an individual autoencoder, it essentially trains the weights of the encoding layer. After the weights of all encoding layers are determined, the final round training on the whole NN is to fine-tune the weights of all layers to achieve the best performance.

Another popular deep learning method is Recurrent Neural Network (RNN). It is a type of neural network that focuses on discovering data patterns from a sequence of data samples. It has memory to carry over auto-learned critical information from previous samples. When processing an input training sample, an RNN combines this input sample with information carried by the memory and generates an output. Meanwhile, it updates its memory with new information learned from this latest training sample. RNN has been widely used in the area of Natural Language Processing (NLP), such as handwriting recognition and speech recognition, where the correlation among sequential data can be critical.

The Byte Segment Neural Network (BSNN) [287] is an RNN-based neural network that solves the traffic classification problem. The structure of the BSNN is shown in Fig. 7.26. Its input is the packet payload of network traffic. The payload is formatted as byte segments whose definition has been introduced in Sect. 7.3.2. Each byte segment is processed by an individual RNN encoder in the following manner: Each byte in the segment is represented as a one-hot vector with a length of 255; All such one-hot vectors are processed by the RNN encoder sequentially. The output of each byte is a vector and all output vectors are concatenated together to form the output vector of this byte segment. The output vectors of all byte segments are processed by another RNN encoder to get a vector that is the final representation of the whole traffic. At last, this vector is sent through a fully connected layer with softmax function to get the final predictions. The BSNN model essentially explores the correlation between bytes in each individual segment and then explores the correlation between the representations of byte segment to achieve a high-quality classification. Note that popular RNN encoder includes Long Short-Term Memory (LSTM) and Gated Recurrent Unit (GRU).

While RNN explores the temporal behaviors and CNN explores the correlated local behaviors, an intuitive thought is to combine them together. Such a combined model has been proposed to solve the traffic classification problem [290]. As shown in Fig. 7.27, the proposed model comprises

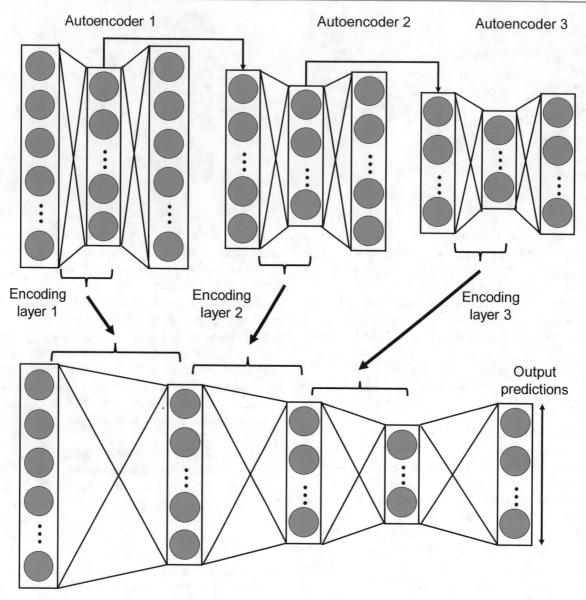

Fig. 7.25 Example of stacked autoencoder

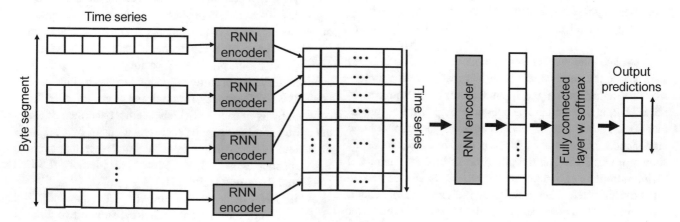

Fig. 7.26 General structure of the byte segment neural network

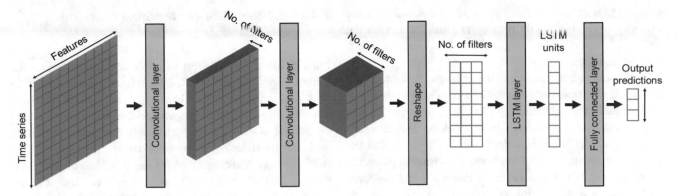

Fig. 7.27 Combination of CNN and RNN

Fig. 7.28 Recurrent neural network with input of byte segments

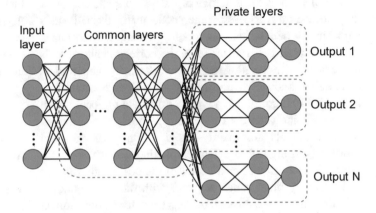

two consecutive convolutional layers followed by an LSTM layer. The input of this model is a matrix composed of the time-series of feature vectors representing all packets of a given network flow. The feature set includes source port, destination port, packet direction, bytes in payload, packet inter-arrival time, and window size. After applying two convolutional layers on the input matrix, a three-dimensional (3D) matrix is generated as the output. Subsequently, this 3D matrix is reshaped into a 2D matrix to fit the following LSTM layer. The output of the LSTM layer is a 1D vector. At last, a fully connected layer is used to transform this 1D LSTM vector into the final prediction vector which is the output of the whole model.

A multi-output Deep Neural Network (DNN) is proposed to accommodate multiple traffic classification tasks into one solution [308]. For example, for network traffic in a data center, it may be required to distinguish elephant flows and mice flows as well as classifying flows based on pre-determined application protocols. These classification results are then applied at different level of network planning and resource management. Instead of training multiple classifiers separately, the proposed multi-output DNN can complete all these tasks in one model. As shown in Fig. 7.28, the model starts with a set of common layers followed by N private layers, one for each classification task. With such setup, the common layers are responsible for capturing the common

knowledge that are useful for all tasks, while each private layer is responsible for capturing the knowledge that is uniquely critical for a specific task. The training process can be either Round-Robin Backpropagation (RRB) or Parallel-Private and Sum-Common Backpropagation (PPSCB). In the RRB training, each task trains its private layer and the common layer in turn. In the PPSCB training, all tasks are trained at the same time. While the private layers are updated independently, the common layers are updated as a weighted combine of the updates from all tasks. With such a multi-output DNN, enabling a new classification task, i.e., transfer learning, is very straightforward. It only requires adding a new private layer for the new task and training the newly added private layer and the common layer.

7.4 Traffic Routing

Traffic routing is the problem of finding the path along which the packets of a network flow are forwarded from its source to its destination. It is the most fundamental problem in the networking area, as the basic usage of a computer network is to transmit data from one place to another. The most common goal of traffic routing is to find the shortest path between the source and destination of a given network flow. Meanwhile, unique requirements may exist when it is performed in some specific types of network. For example, in a wireless sensor network, traffic routing is usually required to extend the lifetime of sensors and to be able to handle dynamic network topology changes.

Traditionally, the traffic routing problem is treated as a distributed decision making problem in which each router is responsible for determining a neighboring router to forward a received network packet to. The decision is made based on each router's knowledge of the network and the incoming packets. Most of the traditional routing protocols belong to one of the two families: link-state and distance-vector. The link-state protocols are mainly based on the Dijkstra's algorithm. When employing a link-state protocol, each router stores the topology of the network and computes the shortest path to every other node using the Dijkstra's algorithm. When the network topology changes, for instance, a link between two routers fails, the corresponding routers broadcast the information to the network and all other routers update their knowledge accordingly. The Open Shortest Path First (OSPF) algorithm is a typical link-state protocol which has been widely used in modern networks. The distance-vector protocols are mainly based on the Bellman-Ford shortest path algorithm When employing a distance-vector protocol, each router has a vector of the cost of the shortest path to every other router together with the corresponding neighboring node to forward packets to each of those shortest paths. The routers periodically broadcast their cost vector to their neighbors. When a router receives a cost vector from one of its neighbors, it updates its own cost vector accordingly.

In Software Defined Networks (SDNs), which have been widely deployed recently, the traffic routing problem becomes a centralized/hierarchical decision making problem. Instead of letting the routers make routing decisions, the centralized controllers in an SDN are responsible for determining a route for each network flow in the network. With more comprehensive knowledge about the network, the controllers are able to perform routing with more complicated objectives, for example, meeting QoS requirements, allocating bandwidth to guarantee flow completion time, or saving energy consumption.

In recent years, the deployment scenarios of networks have been expanded greatly, while the traffic pattern has been much more varied and the traffic load has increased significantly. All these facts have imposed novel challenges on existing traffic routing solutions. Whereas, the ML techniques have been more and more used in solving the traffic routing problem, as they are capable to scale with the complicated and dynamic network topology, to learn the relationship between the selected routes and the required QoS metrics, and to foresee the impact of routing decisions. In the following, several state-of-the-art ML based traffic routing solutions are introduced.

7.4.1 State-of-the-Art Machine Learning Solutions

Most of the state-of-the-art ML solutions are based on the Reinforcement Learning (RL) technique. As introduced in Sect. 7.2.1, reinforcement learning is a type of machine learning solution that builds an agent to take actions based on current state in an interactive environment with the goal of maximizing long-term return. Mathematically, reinforcement learning is based on the Markov Decision Process (MDP) model [309] which can be described by $(\mathcal{S}, \mathcal{A}, \mathcal{P}, \mathcal{R})$. Here $\mathcal{S}$ is the set of states in which s_t is the state at time t observed by the agent. $\mathcal{A}$ is the set of actions in which a_t is the action taken by the agent corresponding to the observed state s_t. $\mathcal{P}$ is the set of transition probabilities in which $P_{s_t s_{t+1}}^{a_t}$ is the probability of transitioning the state s_t to the state s_{t+1} when taking the action a_t. Naturally, the sum of the transition probabilities of all possible state s_{t+1} is 1, i.e.,

$$\sum_{s_{t+1} \in \mathcal{S}} P_{s_t s_{t+1}}^{a_t} = 1. \tag{7.46}$$

At last, $\mathcal{R}$ is the set of rewards in which r_t is the expected reward obtained by taking the action a_t on the state s_t. It can be formulated as

$$
\begin{aligned}
r_t &= r(s, a)|_{s=s_t, a=a_t} \\
&= \sum_{s_{t+1} \in \mathcal{S}} P_{s_t s_{t+1}}^{a_t} R_{s_t s_{t+1}}^{a_t},
\end{aligned}
\tag{7.47}
$$

where $R_{s_t s_{t+1}}^{a_t}$ is the reward of transitioning to the state s_{t+1} from the state s_t with taking the action a_t. In addition, the strategy used by the agent to determine the action to take on an observed state is modeled as policy $\pi(s_t, a_t)$ which is a function that returns the probability of taking the action a_t when observing the state s_t. Figure 7.29 shows the procedure of such a decision making chain.

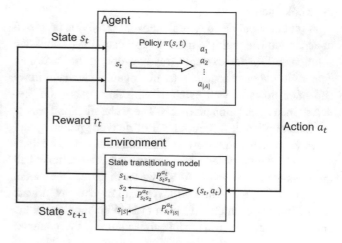

Fig. 7.29 General procedure of a decision making chain

The goal of an RL solution is to learn a parameterized policy $\pi^\theta(s_t, a_t)$ which maximizes the expectation of the accumulative overall rewards when starting from an initial state s_0, i.e., find θ^* such that

$$\theta^* = \arg\max_\theta \mathbb{E}\left[\sum_t \gamma^t r_t\right], \tag{7.48}$$

in which γ is the discount factor that determines the importance of the future rewards. Based on the method employed to find the optimal θ^*, there exist several types of RL solutions, including value-based RL solutions, policy gradients RL solutions, and actor-critic RL solutions. In the following, existing RL solutions of the traffic routing problem from each of these types are introduced.

Value-Based Reinforcement Learning Solutions

In value-based RL solutions, the policy $\pi^\theta(s_t, a_t)$ is calculated based on the value function $V^\pi(s)$, which measures the value of a given state s, and the Q function $Q^\pi(s, t)$, which measures the value of the state-action pair (s, t). The value function $V^\pi(s)$ is defined as the expected cumulative rewards that the agent can obtain when starting from the given state s following the policy. It can be formulated as

$$V^\pi(s) = \mathbb{E}_\pi\left[\sum_{t \geq 0} \gamma^t r_t \mid s_0 = s\right]. \tag{7.49}$$

Whereas, the Q function $Q^\pi(s, t)$ is defined as the expected cumulative rewards that the agent can obtain when starting from the state s and taking the action t. It can be formulated as

$$Q^\pi(s, a) = \mathbb{E}_\pi\left[\sum_{t \geq 0} \gamma^t r_t \mid s_0 = s, a_0 = a\right]. \tag{7.50}$$

Taking some transformation on Eq. (7.50), one can obtain the following expression:

$$Q^\pi(s, a) = \left(r_t + \gamma \sum_{s_{t+1} \in \mathcal{S}} P_{s_t s_{t+1}}^{a_t} V^\pi(s_{t+1})\right)\bigg|_{s_t = s, a_t = a}, \tag{7.51}$$

which describes the relationship between the value function and the Q function.

With the Q function, the agent can employ several different strategies to determine the action a_t to take given the input state s_t. The most commonly used three strategies are greedy, $1 - \epsilon$ greedy, and softmax. The greedy policy simply takes the action with the largest Q value, i.e.,

$$\pi^\theta(s_t, a_t) = \begin{cases} 1, & \text{if } a_t = \arg\max_{a \in \mathcal{A}} Q^\pi(s_t, a), \\ 0, & \text{otherwise.} \end{cases} \tag{7.52}$$

The $1 - \epsilon$ greedy policy follows the greedy policy with the probability of $1 - \epsilon$ and takes a random action with the probability of ϵ, i.e.,

$$\pi^\theta(s_t, a_t) = \begin{cases} 1 - \epsilon, & \text{if } a_t = \arg\max_{a \in \mathcal{A}} Q^\pi(s_t, a), \\ \frac{\epsilon}{|A|-1}, & \text{otherwise.} \end{cases} \tag{7.53}$$

The softmax policy employs the following equation:

$$\pi^\theta(s_t, a_t) = \frac{\exp(Q^\pi(s_t, a_t)/\tau)}{\sum_{a_i \in \mathcal{A}} \exp(Q^\pi(s_t, a_i)/\tau)}, \tag{7.54}$$

in which τ is the temperature parameter that controls the behavior of the softmax function: the larger it is, the probability is more evenly distributed across all actions, i.e., the policy behaviors more like a random policy; the smaller it is, the probability of the action with the largest Q value is closer to 1, i.e., the policy behaviors more like the greedy policy.

In the value-based policy, the most critical task is to find the optimal Q-value function $Q^*(s, a)$ that represents the maximum expected cumulative reward that the agent can obtain starting from the pair (s_t, a_t). Following the Bellman Equation [309], one has the following condition:

$$Q^*(s_t, a_t) = r_t + \gamma \sum_{s_{t+1} \in \mathcal{S}} P_{s_t s_{t+1}}^{a_t} \max_a Q^*(s_{t+1}, a). \tag{7.55}$$

Correspondingly, the optimal value function $V^*(s_t)$ is

$$V^*(s_t) = \max_a Q^*(s_t, a). \tag{7.56}$$

The optimality condition described in Eq. (7.55) can be approximated via reinforcement iterations using the following equation:

$$Q_{t+1}^\pi(s_t, a_t) = Q_t^\pi(s_t, a_t) + \alpha\left[r_t + \gamma \max_a Q_t^\pi(s_{t+1}, a) - Q_t^\pi(s_t, a_t)\right], \tag{7.57}$$

which essentially is the well-known Q-learning algorithm [309].

With the development of deep learning technique, more and more Deep Reinforcement Learning (DRL) is proposed to solve the traffic routing problem. In DRL solutions, the Q-value function is approximated by a deep neural network, denoted by $Q^\pi(s_t, a_t; \theta)$ in which θ is the neural network weights.

The most popular value-based reinforcement algorithm used in those DRL solutions is the Deep Q-learning algorithm in which the reinforcement iteration is composed of the feedforward and the backpropagation of the neural network. In the feedforward, current loss is calculated based on the loss function

$$L_t(\theta_t) = (r_t + \gamma \max_a Q_t^\pi(s_{t+1}, a; \theta_t) - Q_t^\pi(s_t, a_t; \theta_t))^2. \tag{7.58}$$

In the backpropagation, the weight variable θ is updated based on the gradient derived from the loss function, i.e.,

$$\theta_{t+1} = \theta_t + \nabla_{\theta_t} L_t(\theta_t). \tag{7.59}$$

Note that the details of the procedure of calculating the gradient are beyond the scope of this chapter and therefore is omitted.

In the following, examples of both regular RL solutions and DRL solutions of the traffic routing problem are introduced.

The QoS-aware Adaptive Routing (QAR) [289] is a reinforcement learning based approach that performs routing in a multi-layer hierarchical Software Defined Networking

(SDN) environment in which the control plane is composed of three levels of controllers: super, domain, and slave controllers. Both slave controllers and domain controllers control the underlying switches. The domain controllers have full access to those switches and are responsible for making routing decisions, whereas the slave controllers mainly serve for more basic tasks like providing some simple control functions, sharing control workloads with the domain controllers, and etc. While there exist one slave controller and one domain controller for each domain of switches, there is only one super controller which controls all domain controllers and regulates the entire network functionality.

As shown in Fig. 7.30, the proposed QAR algorithm routes each network traffic in two steps: In the first step, the super controller determines the domain-level routing path, i.e., the sets of domains through which the given network traffic will pass through, and the corresponding border switches of each of those domains; in the second step, the domain controller of each of those selected domains calculates the intra-domain path from the given source border switch to the destination border switch.

Both the super controller and the domain controllers employ a reinforcement learning approach to learn the best routing policy with a QoS-aware reward function. The proposed RL approach is a value-based approach and uses a modified Q-learning algorithm [301] in which the Q-value function is updated using the following reinforcement equation:

$$Q_{t+1}^\pi(s_t, a_t) = Q_t^\pi(s_t, a_t) + \alpha \left[r_t + \gamma Q_t^\pi(s_{t+1}, a_{t+1}) - Q_t^\pi(s_t, a_t) \right]. \tag{7.60}$$

While the conventional Q-learning (Eq. (7.57)) adapts the Q-value function based on the available actions, this modified Q-learning simply adopts the future rewards that it actually obtains. The QoS-aware reward function used in the QAR algorithm is defined as the reward received for forwarding packets from node i to node j when the system is at state s_t and takes action a_t. It can be formulated as

$$\begin{aligned} r_t &= r(i \rightarrow j|_{s_t, a_t}) \\ &= -g(a_t) + \beta_1(\theta_1 delay_{ij} + \theta_2 queue_{ij}) + \beta_2 loss_j + \beta_3(\phi_1 B1_{ij} + \phi_2 B2_{ij}), \end{aligned} \tag{7.61}$$

Fig. 7.30 Multi-layer Hierarchical routing framework

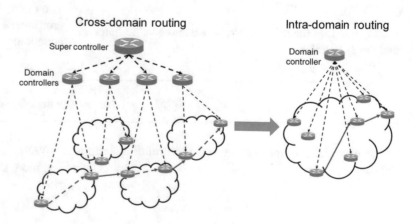

where $\varrho(a_t)$ is a constant indicating the fixed cost of forwarding a packet and $(\beta_1, \beta_2, \theta_1, \theta_2, \phi_1, \phi_2)$ are all weight variables with the value range of $[0, 1)$. All the QoS metrics in the equation are defined as

$$delay_{ij} = \frac{2}{\pi} \arctan\left(d_{ij}^l - \frac{1}{|\mathcal{N}(i)|} \sum_{k \in \mathcal{N}(i)} d_{ik}^l\right), \quad (7.62)$$

$$queue_{ij} = \frac{2}{\pi} \arctan\left(d_{ij}^q - \frac{1}{|\mathcal{E}|} \sum_{(m,n) \in \mathcal{E}} d_{mn}^q\right), \quad (7.63)$$

$$loss_{ij} = 1 - 2 \cdot \%loss_{ij}, \quad (7.64)$$

$$B1_{ij} = \frac{2BW_{ij}^A}{BW_{ij}^T} - 1, \quad (7.65)$$

$$B2_{ij} = \frac{2}{\pi} \arctan\left(0.01(BW_{ij}^A - \frac{1}{|\mathcal{E}|} \sum_{(m,n) \in \mathcal{E}} BW_{mn}^A)\right), \quad (7.66)$$

where d_{ij}^l and d_{ij}^q are the link transmission delay and packet queuing delay from node i to node j, respectively; $\mathcal{N}(i)$ is the set of neighbor nodes of node i and $\mathcal{E}$ is the set of links; $\%loss_{ij}$ is the percentage of packet loss on link ij; BW_{ij}^A and BW_{ij}^T are the available bandwidth and total bandwidth of link ij, respectively. Equation (7.62) considers the comparative difference between the link transmission delay of link ij and the average delay of all links starting from node i. Equation (7.63) considers the comparative difference between the packet queuing delay of link ij and the average delay of all links in the domain. Equation (7.64) considers the loss rate. Equations (7.65) and (7.66) consider the characteristics of the available bandwidth on the link ij. Combining all these QoS metrics together, Eq. (7.61) gives an overall evaluation on the goodness of the action of forwarding packet from node i to node j.

The QELAR [280] is a reinforcement learning based adaptive routing protocol for energy-aware underwater sensor networks. The proposed protocol dynamically routes the network traffic from one sensor to another in the underwater environment with the goal of making the residual system energy more evenly distributed across all sensors and therefore prolonging the lifetime of the network. The core part of QELAR is a value-based reinforcement learning model that runs on each sensor node and decides the next hop of sensor to which each received packet is forwarded. Different from the conventional Q-learning algorithm, the learning process used in the QELAR protocol is based on Eqs. (7.55) and (7.56), which requires the knowledge of the underlying state transitioning probabilistic model.

In the QELAR protocol, the state s_i is the sensor node s_i and the action a_j is forwarding the received packet to the sensor node s_j. With taking the action a_j, there exist two

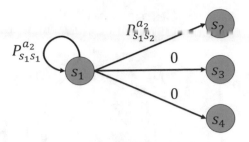

Fig. 7.31 State transitioning model of taking action a_2 in the state s_1 in the QELAR protocol

consequences: the packet is successfully forwarded to the node s_j, i.e., the state s_i is transited to the state s_j; or the packet forwarding fails and the system stays on the state s_i. The probability of the former case is denoted by $P_{s_i s_j}^{a_j}$ and that of the latter case is denoted by $P_{s_i s_i}^{a_j}$. The sum of these two probabilities is 1. Figure 7.31 shows an example of this state transitioning probabilistic model. With this model, Eq. (7.55) becomes

$$Q^t(s_i, a_j) = r_t + \gamma\left(P_{s_i s_j}^{a_j} V^t(s_j) + P_{s_i s_i}^{a_j} V^t(s_i)\right). \quad (7.67)$$

In the above equation, the reward function is defined as

$$r_t = P_{s_i s_j}^{a_j} R_{s_i s_j}^{a_j} + P_{s_i s_i}^{a_j} R_{s_i s_i}^{a_j}, \quad (7.68)$$

where $R_{s_i s_j}^{a_j}$ and $R_{s_i s_i}^{a_j}$ are the reward received when the packet forward succeeds and fails, respectively. Their formulation is

$$R_{s_i s_j}^{a_j} = -g - \alpha_1[c(s_i) + c(s_j)] + \alpha_2[d(s_i) + d(s_j)], \quad (7.69)$$

$$R_{s_i s_i}^{a_j} = -g - \alpha_1 c(s_i) + \alpha_2 d(s_i), \quad (7.70)$$

in which g is a constant cost, $c(s)$ and $d(s)$ are residual energy functions, and α_1 and α_2 are their weights. Specifically, $c(s_i)$ is the cost function of the residual energy of sensor node s_i. It is defined as

$$c(s_i) = 1 - \frac{E_{res}(s_i)}{E_{init}(s_i)}, \quad (7.71)$$

where $E_{res}(s_i)$ and $E_{init}(s_i)$ are the residual and initial energy of node s_i, respectively. Meanwhile, $d(s_i)$ is the comparative difference between the residual energy of node s_i and the average residual energy of all its neighbor nodes. It is formulated as

$$d(s_i) = \frac{2}{\pi} \arctan\left(E_{res}(s_i) - \frac{1}{|\mathcal{N}(s_i)|} \sum_{s_k \in \mathcal{N}(s_i)} E_{res}(s_k)\right), \quad (7.72)$$

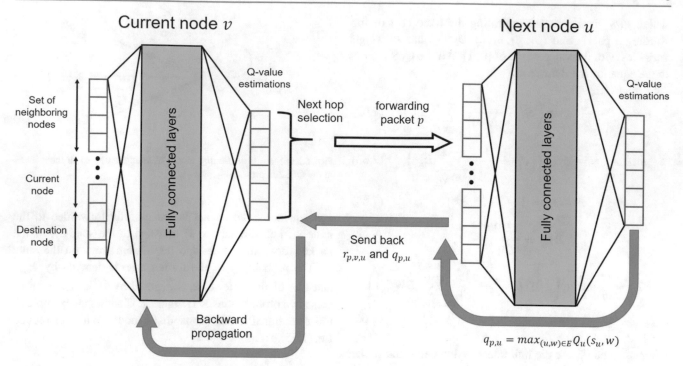

Fig. 7.32 The DQN-routing algorithm with the DNN architecture

where $\mathcal{N}(s_i)$ is the set of neighbor nodes of node s_i. With cost functions $c(s)$ and $d(s)$, the QELAR protocol is naturally inclined to distribute the energy consumption of forwarding packets more evenly across all nodes.

The DQN-Routing algorithm [295] is a multi-agent deep reinforcement learning routing algorithm. In this algorithm, each node/router in the network is an agent running its own deep reinforcement learning model. The proposed DRL model is largely based on the conventional Q-learning algorithm. The critical difference is that the Q-value function $Q(s, a)$ is approximated by a Neural Network.

In the DQN-routing algorithm, a reinforcement iteration of the model running on node v includes forwarding a received packet p to another node u and updating the Neural Network based on the feedback sent back from node u. Specifically, when node v receives the packet p, it estimates the Q-value of each available next hop and uses the softmax strategy to determine the next hop to forward the packet p over. Assume that the selected next hop is node u with a Q-value of $Q_v(s_v, u)$. After node u receives the packet p, it also observes the link latency of forwarding the packet p from node u to node v, which is the actual reward of this action and is denoted by $r_{p,v,u}$. Meanwhile, node u estimates the Q-value of each of its available next hops and obtains its own value $q_{p,u}$ based on Eq. (7.56). It can be formulated as

$$q_{p,u} = \max_{(u,w)\in\mathcal{E}} Q_u(s_u, w), \qquad (7.73)$$

in which w is a neighbor node of node u and $Q_u(s_u, w)$ is the Q-value of forwarding the packet p from node u to node w. Subsequently, node u sends $r_{p,v,u}$ and $q_{p,u}$ back to node v. Node v then updates its Neural Network based on Eqs. (7.58) and (7.59).

Three Neural Network architectures are proposed in the DQN-routing algorithm: DNN architecture, RNN with router memory architecture, and RNN with packet memory architecture. Figure 7.32 shows the DNN architecture in which the Neural Network is simply a set of fully connected layers. Figure 7.33 shows the RNN with router memory architecture in which the Long Short Term Memory (LSTM) layer is used. With the hidden state of the LSTM layer from previous reinforcement iterations, the model essentially has memory (useful information) of those packets previously forwarded by node v. Figure 7.34 shows the RNN with packet memory architecture. While this architecture uses the LSTM layer also, the hidden state of the LSTM layer is forwarded to node u together with the packet and is used by the LSTM layer of node u as input. In this way, node u essentially obtains some memory of previously visited nodes of packet p and uses this information to make a decision.

Policy Gradients and Actor-Critic Reinforcement Learning Solutions

Recall that the goal of an RL solution is to learn a parameterized policy $\pi^\theta(s, t)$ that maximizes the expectation of the total rewards which is defined as

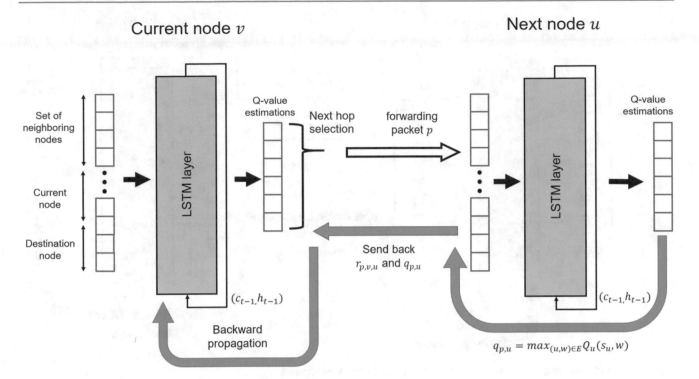

Fig. 7.33 The DQN-routing algorithm with the RNN with router memory architecture.

$$\mathbb{J}(\theta) = \mathbb{E}\left[\sum_t \gamma^t r_t\right] = \mathbb{E}\left[r(\tau)\right], \qquad (7.74)$$

where τ is a sequence of states, actions, and rewards known as a trajectory. In the value-based RL solutions, the policy $\pi^\theta(s,t)$ is based on the value of the state $V^\pi(s,\theta)$ or the value of the state-action pair $Q^\pi(s,t,\theta)$, and the optimal policy is searched in the reinforcement iterations on those values. The policy gradients RL solutions are another type of RL solutions in which the objective function is directly differentiated and the parameter θ is updated using the gradient of the objective function. This update operation can be formulated as

$$\theta_{t+1} = \theta_t + \alpha \nabla_\theta \mathbb{J}(\theta_t). \qquad (7.75)$$

in which the gradient of the objective function $\nabla_\theta \mathbb{J}(\theta_t)$ is calculated by

$$\nabla_\theta \mathbb{J}(\theta_t) = \mathbb{E}\left[\sum_{t \geq 0} r(\tau) \nabla_\theta \log \pi^\theta(s_t, a_t)\right]. \qquad (7.76)$$

The mathematical derivation process of the above equation is omitted as it is out of the scope of this book. In Eq. (7.53), the expectation can be further approximated by sampling a large number of trajectories and average them out which is known as Markov Chain Monte-Carlo (MCMC). As a result, for a given sampled trajectory τ, the gradient of the objective is approximated as

$$\nabla_\theta \mathbb{J}(\theta_t) = \sum_{t \geq 0} r(\tau) \nabla_\theta \log \pi^\theta(s_t, a_t)$$
$$= \sum_{t \geq 0}\left(\sum_{0 \leq t' \leq t} \gamma^{t'} r_t'\right) \nabla_\theta \log \pi^\theta(s_t, a_t). \qquad (7.77)$$

This is the classic policy gradient algorithm called REINFORCE. The REINFORCE algorithm usually suffers from high variance because the reward estimation is very hard.

The Actor-Critic method [285] improves the REINFORCE algorithm on the variance reduction by modeling the reward function $r(\tau)$ as

$$r(\tau) = \sum_{t' \geq t} \gamma^{t'-t} r_t' - V^\omega(s_t), \qquad (7.78)$$

in which $\sum_{t' \geq t} \gamma^{t'-t} r_t'$ is the future total reward obtained by taking action a_t in state s_t and $V^\omega s_t$ is the expected value of the state s_t that is learned by another policy ω using the Q-learning algorithm. The intuition here is that if an action a_t in the state s_t is a good action, one should obtain more future rewards than what is expected to be obtained from the state s_t. Essentially, the Actor-Critic method trains two policies, one is the agent's policy (the actor) while the other provides feedback on the goodness of the action taken by the agent's policy, i.e., the critic.

The DRL-TE algorithm [317] is a DRL-based algorithm for a specific traffic engineering problem in which K

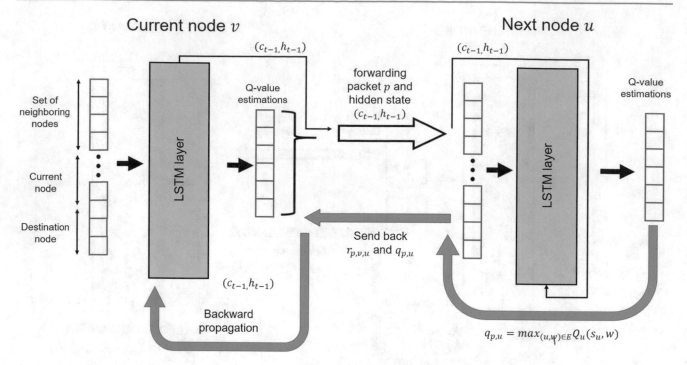

Fig. 7.34 The DQN-routing algorithm with the RNN with packet memory architecture

communication sessions exist in a general communication network while each communication session k has a set of candidate paths $\mathcal{P}_k$. The goal is to find a rate allocation solution for those communication sessions which specifies the amount of traffic load f_{kj} flowing through the jth path of $\mathcal{P}_k$. With such a rate allocation solution, each packet of the session k is routed via a specific path j with a probability of $w_{kj} = f_{kj}/\sum_j f_{kj}$.

In this problem, the state is the throughput and delay of each communication sessions which is defined as

$$s = (x1, z1, x2, z2, \ldots, x_K, z_k), \qquad (7.79)$$

where x_i and z_i are the throughput and delay of the session i. The action is the rate allocation solution of each communication sessions which is defined as

$$a = (w_{11}, w_{12}, \ldots, w_{1|P_1|}, \ldots, w_{K|P_K|}). \qquad (7.80)$$

The reward is the total utility of all the communication sessions which is defined as

$$r = \sum_{k=1}^{K} U(x_k, z_k), \qquad (7.81)$$

in which $U(x_k, z_k)$ is the utility function of the session k. Its definition is

$$U(x_k, z_k) = \frac{(x_k)^{1-\alpha_1}}{1-\alpha_1} - \sigma \cdot \frac{(z_k)^{1-\alpha_2}}{1-\alpha_2} \qquad (7.82)$$

To solve this problem, the DRL-TE algorithm utilizes the Deep Deterministic Policy Gradient (DDPG) method [288] which is the state-of-the-art Actor-Critic method for continuous control, because the rate allocation solution, as the output action, is a vector of continuous variables.

An Actor-critic DRL solution is proposed to solve the problem of QoS-aware routing in SDN [296]. In the proposed solution, the state is the traffic demands of network flows that is defined as a matrix $\bar{M}$ with a size of $N \times N$. The entry M_{ij} of this matrix is the traffic demand between source node i and destination node j. The action is the link-weight vector $\bar{L}$ in which L_i is the weight of link i. This link-weight vector is then used to generate routing configurations using the Dijkstra's algorithm. Two reward functions are proposed: (1) the mean of the number of flows that fulfill their QoS requirement and (2) the mean of QoS metrics including latency and packet loss. A CNN whose structure is showed in Fig. 7.35 is used in the solution to generate the link-weight vector from the input traffic demand matrix. The learning algorithm used in this solution is the DDPG algorithm as the link-weight vector is composed of continuous variables.

7.5   Resource Management

Resource management is the problem of allocating available network resources to the incoming requests in a manner that

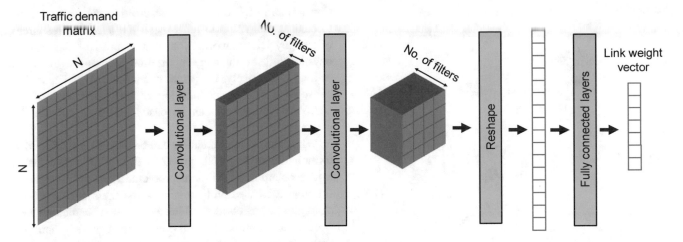

Fig. 7.35 The CNN structure of the QoS-aware routing framework proposed in [296]

maximizes a certain goal. There are many types of network resources. While typical ones include link bandwidth and computing resources of nodes in the network, the modern network techniques have enabled many new types of resources management. For example, in the SDN architecture, there may exist multiple controllers that manage the same network. All incoming packet routing requests need to be distributed over the fleet of controllers in a way that the response time is minimized. Another example is the usage of an optical switch which can activate only a limited number of links at any given time. The resource management problem in this case is how to selectively activate the connected links to achieve a certain goal, for instance, minimizing the average network flow completion time.

From these examples, one can see that resource management in networking covers a great many different problems which have their own specific criteria and goals. In the following, two categories of resource management problems are introduced along with some existing solutions. The first category is the Virtual Network Embedding problem which is a classic networking management problem existing in both Wide Area Networks and Data Center Networks. The second category is resource management in Software-Defined Networks which covers several new resource management problems introduced by the usage of the SDN techniques.

7.5.1 The Virtual Network Embedding Problem

The Virtual Network Embedding (VNE) problem is the problem of allocating the resource of a physical network, also called substrate network, to a virtual network request (VNR), that is composed of a set of virtual nodes (virtual machines) and a set of virtual links that interconnect those virtual nodes. In the following, a mathematical formulation of the VNE problem is given.

The substrate network is modeled as a graph $\mathcal{G}^S = (\mathcal{N}^S, \mathcal{E}^S, \mathcal{R}^S, \mathcal{P}^S)$. $\mathcal{N}^S = (n_1, n_2, \ldots, n_N)$ is the set of nodes in which n_i^S is the ith node in the network. $\mathcal{E}^S$ is the set of physical links with $\mathcal{E}^S \subseteq \mathcal{N}^S \times \mathcal{N}^S$ and e_{ij}^S denotes the physical link between node n_i and node n_j. $\mathcal{R}^S$ is a set of vectors in which a vector $R^S(o)$ represents the available resources of the network entity $o \in \mathcal{N}^S \cup \mathcal{E}^S$. The resource vector $R^S(o)$ is defined as $(r_{o1}^S, r_{o2}^S, \ldots, r_{ol}^S)$ in which each element is a type of resources. For example, the resource vector of a node can contain CPU, disk, and memory. $\mathcal{P}^S$ is the set of loop free paths in the physical network, defined as $\mathcal{P}^S = (p_1^S, p_2^S, \ldots, p_P^S)$. Furthermore, $\mathcal{P}_{ij}^S \in \mathcal{P}^S$ is the subset of paths between node n_i and node n_j.

The virtual network request is modeled as a graph $\mathcal{G}^V = (\mathcal{N}^V, \mathcal{E}^V, \mathcal{R}^V)$. $\mathcal{N}^V$ is the set of virtual nodes in which n_i^V is the ith virtual node in the request. $\mathcal{E}^V$ is the set of virtual links with $\mathcal{E}^V \subseteq \mathcal{N}^V \times \mathcal{N}^V$ and e_{ij}^V denotes the virtual link between virtual node n_i and virtual node n_j. $\mathcal{R}^V$ is a set of vectors in which a vector $R^V(o)$ represents the requested resources of the network entity $o \in \mathcal{N}^V \cup \mathcal{E}^V$. The vector $R^V(o)$ is defined as $(r_{o1}^V, r_{o2}^V, \ldots, r_{ol}^V)$ in which each element is a type of resources.

Given the definition of the substrate network G^S and the virtual network request G^V, and further define variable $x_{n_i^V n_j^S}$ as a binary variable whose value is 1 if the virtual node n_i^V is allocated to physical node n_j^S, otherwise it is value is 0, and define variable $y_{e_{ij}^V p_k^S}$ as a binary variable whose value is 1 if the virtual link e_{ij}^V is allocated to the path p_k^S. The VNE problem can be formulated as a Mixed Integer Programming (MIP) problem as shown in the following

VNE

$$\text{Minimize } f\left(x_{n_i^V n_j^S},\ y_{e_{ij}^V p_k^S}\right) \qquad (7.83)$$

Subject to

$$\sum_{n_j^S \in \mathcal{N}^S} x_{n_i^V n_j^S} = 1, \ \forall n_i^V \in \mathcal{N}^V, \qquad (7.84)$$

$$\sum_{p_k^S \in \mathcal{P}^S} y_{e_{ij}^V p_k^S} = 1, \ \forall e_{ij}^V \in \mathcal{E}^V, \qquad (7.85)$$

$$\sum_{n_p^s \in \mathcal{N}^s} \sum_{n_q^s \in \mathcal{N}^s} x_{n_i^V n_p^S} \cdot x_{n_j^V n_q^S} \left(\sum_{p_k^S \in P_{pq}^S} y_{e_{ij}^V p_k^S} \right) = 1, \ \forall e_{ij}^V \in \mathcal{E}^V, \qquad (7.86)$$

$$\sum_{n_i^V \in \mathcal{N}^V} x_{n_i^V n_j^S} \cdot r_{n_i^V l}^V \le R_{n_j^S l}^S, \ \forall l, \ \forall n_j^S \in \mathcal{N}^S, \qquad (7.87)$$

$$\sum_{e_{ij}^V \in \mathcal{E}^V} \sum_{p_k^S | e_{pq}^S \in p_k^S} y_{e_{ij}^V p_k^S} \cdot r_{e_{ij}^V l}^V \le R_{e_{pq}^S l}^S, \ \forall l, \ \forall e_{pq}^S \in \mathcal{E}^S. \qquad (7.88)$$

Equation (7.83) is the objective function which can be defined based on the goal of the problem to be solved. For example, if the goal is to minimize the network cost, the objective function can be the total number of hops of all physical paths allocated to the request. Constraints (7.84) and (7.85) ensure that each virtual node is allocated to only one physical node and each virtual link is allocated to one physical path, respectively. Constraints (7.86) together ensure that the physical path allocated for any virtual link e_{ij}^V interconnects the physical nodes allocated for the source and destination virtual nodes of that virtual link.

Constraints (7.87) and (7.88) ensure that the resources allocated to the VNR do not exceed the available resources of the physical network. Note that in most of the cases a physical node or link can support more than one virtual node or link, respectively. In special cases that a 1-to-1 mapping between physical nodes/links to virtual nodes/link is needed, additional constraints can be easily added to enforce the mapping.

While the optimal solution can be derived from the MIP formulation of the VNE problem, it is well-known that a MIP problem is an NP-hard problem. Most of the existing solutions are based on heuristics or metaheuristics. In cases of large scale networks, which is common in modern networks, the solution space is huge and it is very hard for those solutions to find the optimal allocation or even a near-optimal allocation. Whereas, ML techniques have been used to improve the efficiency of those heuristic solutions. In the following, some examples of the ML-based solutions are introduced.

State-of-the-Art Solutions

A Recurrent Neural Network (RNN) based admission control mechanism is proposed to facilitate the virtual network request embedding process in [271]. As shown in Fig. 7.36, the incoming VNR is sent through an RNN together with the current status of the substrate network. The RNN model predicts whether the incoming VNR should be accepted. Only when the model predicts that the request should be accepted, does the embedding system proceed to the next step in which the VNE algorithm tries to embed the VNR

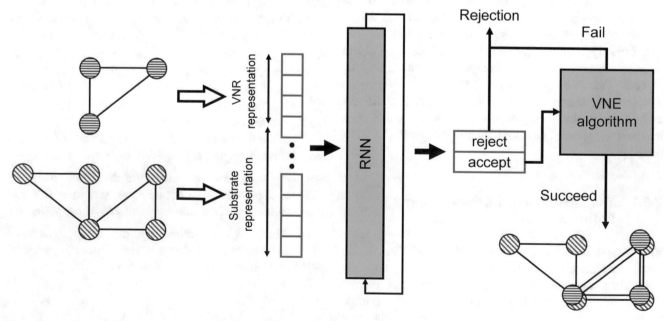

Fig. 7.36 RNN-based admission control mechanism in the VNE problem

into the substrate network. If successful, the system allocates resources to the VNR according to the output of the algorithm; otherwise, it rejects the request. With such an admission control mechanism, the system is able to filter out those infeasible requests and therefore saves the cost of running the VNE algorithm which can be huge as searching the solution space of the VNE problem may be very costly.

The features of the RNN model include the substrate network representation and the virtual network request representation. To make these representations independent of each specific entity in the network for generalization, 21 graph features are proposed. Examples of those graph features include average/standard deviation degree, average path length, number of nodes, number of edges, spectral radium. The substrate network representation is a vector containing all these 21 graph features and another 8 VNE-Features including three CPU features, three bandwidth features, and two embedding features. The virtual network request representation contains just a subset of those 21 graph features. To select this subset of graph features, a set of graphs are generated according to different models and a Principal Component Analysis (PCA) is performed on the set of graph feature vectors of those graphs; 7 graph features with the highest load factor—the coefficient in the linear transformation of the input data obtained via the PCA—are selected.

The VNE-TD algorithm [315] is a DRL-based VNE solution in which the value function $V^\pi(s)$ is approximated by a Deep Neural Network (DNN). The overall structure of the VNE-TD algorithm is shown in Fig. 7.37. The state s_t contains the current status of the substrate network. It is a vector composed of the normalized residual node capacities and link bandwidths which can be presented as

$$s_t = \left(\frac{c^t_{n^S_1}}{c^t_{\max}}, \frac{c^t_{n^S_2}}{c^t_{\max}}, \ldots, \frac{c^t_{n^S_N}}{c^t_{\max}}, \frac{b^t_{e^S_1}}{b^t_{\max}}, \frac{b^t_{e^S_2}}{b^t_{\max}}, \ldots, \frac{b^t_{e^S_E}}{b^t_{\max}} \right),$$
(7.89)

in which $c^t_{n^S_i}$ is the residual node capacity of node n^S_i and $c^t_{\max}$ is the maximum residual node capacity, while $b^t_{e^S_j}$ is the residual link bandwidth of link e^S_j and $b^t_{\max}$ is the maximum residual link bandwidth. The set of possible actions is a set of embedding plans generated by three existing heuristics. The reward is the revenue of embedding the incoming VNR which is defined as

$$r(t) = \begin{cases} \alpha \sum_{n^V_i \in \mathcal{N}^V} c_{n^V_i} \\ \quad + \beta \sum_{e^V_j \in \mathcal{E}^V} b_{e^V_j}, & \text{if the VNR is embedded,} \\ 0, & \text{otherwise.} \end{cases}$$
(7.90)

where $c_{n^V_i}$ and $b_{e^V_j}$ are the requested computing resource and the request link bandwidth, respectively, while α and β are the unit prices charged for the computing resource and the link bandwidth.

When making decisions, the policy calculates the next state $s^{a_i}_{t+1}$ for each embedding plan a_i and uses the DNN model to calculate the value of that state. In the following, the embedding plan a_m with the highest value is chosen and executed. Meanwhile, the reward $r(t)$ and the value $V^\pi(s^{a_m}_{t+1})$ is back propagated to update the weights of the DNN model.

The NeuroViNE preprocessor [272] is an RNN-based preprocessor that extracts a subgraph containing "good" substrate nodes to which the incoming VNR is allocated. It leverages the Hopfield Network [279] which computes a probability for each substrate node to be selected as part of the output subgraph. The input of the Hopfield Network

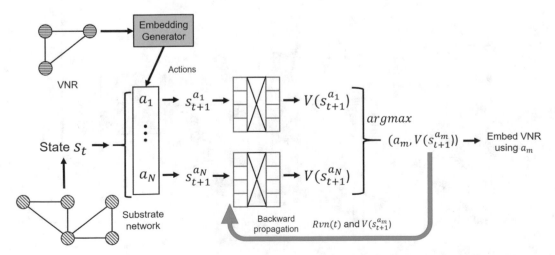

Fig. 7.37 The VNE-TD algorithm

contains the size of the subgraph, the node ranking vector Ξ, and the path ranking vector Ψ.

The size of the output subgraph is determined by a selection function ζ. The selection functions are proposed: ζ_{const}, $\zeta_{factor}(\mathcal{G}^V)$, and $\zeta_{inverse}(\mathcal{G}^V)$. Function ζ_{const} simply uses a constant as the size of subgraph, i.e.,

$$\zeta_{const} = K. \tag{7.91}$$

Function $\zeta_{factor}(\mathcal{G}^V)$ sets the size of the subgraph to be proportional to the size of the VNR, i.e.,

$$\zeta_{factor}(\mathcal{G}^V) = \alpha |\mathcal{N}^V|. \tag{7.92}$$

Function $\zeta_{inverse}(\mathcal{G}^V)$ sets the size of the subgraph to be inversely proportional to the size of the VNR, i.e.,

$$\zeta_{factor}(\mathcal{G}^V) = \frac{\alpha}{|\mathcal{N}^V|}. \tag{7.93}$$

The node ranking vector Ξ contains a rating of each substrate node. For substrate node n_i^S, its rating $\Xi_{n_i^S}$ is defined

$$\Xi_{n_i^S} = \beta \cdot \frac{c_{\max} - c_{n_i^S}}{c_{\max}}, \tag{7.94}$$

in which $c_{n_i^S}$ is the residual node capacity of node n_i^S, $c_{\max}$ is the maximum residual node capacity, and β is a constant.

The path ranking vector Ψ contains ratings of the shortest paths in the substrate network. The procedure of determining the ratings contains three steps. First, a weight $w_{e_{ij}^S}$ is assigned to each link e_{ij}^S. The weight $w_{e_{ij}^S}$ is defined as

$$w_{e_{ij}^S} = b_{\max} - \frac{b_{e_{ij}^S}}{b_{\max}}, \tag{7.95}$$

in which $b_{e_{ij}^S}$ is the residual link bandwidth of link e_{ij}^S and $b_{\max}$ is the maximum residual link bandwidth. Second, the shortest paths of all pairs of nodes are calculated based on the weights obtained in the first step. Denote the shortest path between node n_i^S and node n_j^S by p_{ij}^S and denote its cost by D_{ij}. At last, the rating of the shortest path p_{ij}^S, denoted by $\Psi_{p_{ij}^S}$, is calculated as follows:

$$\Psi_{p_{ij}^S} = \gamma \frac{D_{ij}}{D_{\max}}, \tag{7.96}$$

in which $D_{\max}$ is the maximum cost of all the shortest paths and γ is a constant.

After obtaining the three inputs introduced above, they are sent into the Hopfield Network and a subgraph $\mathcal{G}^{S,subgraph}$ is output. Subsequently, a VNE algorithm is called to embed the incoming VNR into this subgraph. Figure 7.38 shows the procedure of embedding the VNR with the NeuroViNE preprocessor. Note that the implementation and working mechanism of the Hopfield Network is omitted here due to being out of scope.

7.5.2 Resource Management in Software-Defined Networks

The Software-Defined Network (SDN) technique greatly improves the flexibility of managing a large amount of network resources by decoupling the control plane from the forwarding plane. It provides more dynamic provisioning on link bandwidth, traffic routing, and network topology. Meanwhile, it also introduces new type of resource management, for instance, how to distribute the incoming packet routing requests over a fleet of SDN controllers. Several recent

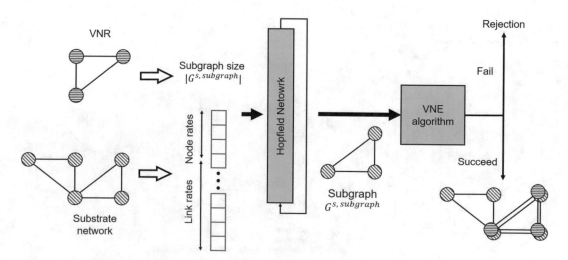

Fig. 7.38 The procedure of embedding the VNR with the NeuroViNE preprocessor

studies on resource management in SDNs are introduced in the following.

DeepConf [302] is a Reinforcement Learning (RL) based general resource management framework for software-defined data center network. In DeepConf framework, an ML layer is added on the top of the SDN controller layer, in which multiple DeepConf agents run at the same time, each one for a specific network management problem, examples include traffic engineering, bandwidth allocation, energy-saving scheduling, and network topology augmentation. A network simulator is used for the offline training of those RL models. The policies of DeepConf agents share the same general RL model in which a CNN is utilized to compute decisions from the input state. The state space consists of two components. The first one is the general network state required by all DeepConf agents which is the information of the flows in the network. This information is represented as a matrix named Traffic Matrix. The other component is the DeepConf agent specific state. Depending on the problem with which the DeepConf agent is dealing, this component of the state includes the corresponding critical information.

A use case of the DeepConf agent is to solve the network topology augmentation problem in which the data center network is composed of electrical packet switches and an optical switch. While the optical switch is physically connected to all electrical packet switches, at any given time, the optical switch can only activate K number of links. The DeepConf agent is required to periodically determine the K links to be activated for the following time period. The structure of the DeepConf agent policy for this problem is shown in Fig. 7.39. The state of the policy includes the traffic matrix and the network topology which is presented as matrix also. The state is sent through two convolutional layers followed by three fully connected layers with the softmax selection. The output is the probability distribution describing the prob-

ability that each link between the optical switch and those electrical switches is selected to be activated. At last, the K links with the largest probabilities are selected to output as the action and will be activated in the next time period. The goal of the agent is to maximize the link utilization while minimizing the average flow-completion time. Based on this goal, the reward function of the RL model is defined as

$$r(t) = \sum_{f \in \mathcal{F}} |\mathcal{L}_f| \frac{b_f}{d_f}, \qquad (7.97)$$

in which $\mathcal{F}$ is the set of all active and completed flows during the time period starting from timestamp t and f represents a given flow f; $\mathcal{L}_f$ is the set of links used by flow f and $|\mathcal{L}_f|$ is its cardinality; b_f is the number of bytes transferred during the time period and d_f is the total duration of the flow f. Essentially, this reward function is to reward for short path and link utilization and to penalize for long running flows.

Distributed controller architectures have been widely used to handle the increasing demand on the scalability of the SDN. In such an architecture, a critical problem to solve is properly dispatching requests issued from every switch to suitable controllers so that each request can be addressed in a timely manner. The logical part that handles this problem is usually called Scheduling Function (SF). An RL-based scheduling function design in [282] is shown in Fig. 7.40. The policy (i.e., the scheduling function) takes the entire network information as the state and outputs the controller to which a request is sent as the action. The policy is trained using the Proximal Policy Optimization [303] algorithm which is an Actor-Critic algorithm proposed recently with certain advantages over the conventional Actor-Critic algorithms. The details of the Proximal Policy Optimization are omitted since it is outside of the scope of this book.

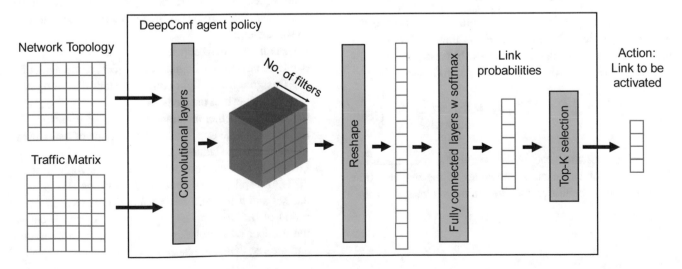

Fig. 7.39 Using the DeepConf agent policy for the topology augmentation

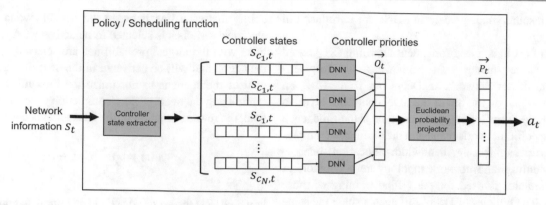

Fig. 7.40 The RL-based scheduling function design in SDN

The scheduling function contains three parts. The first part is a controller state extractor. Assume that there exist N_c controller and N_s switches in the network, and assume that c_i is the ith controller and s_j is the jth switch. When receiving a new request t from switch s_j, the extractor extracts the controller state $s_{c_i,t}$ for each control c_i from the entire network information s_t. The controller state $s_{c_i,t}$ includes the responding time τ_{c_i}, the utilization u_{c_i}, the processing capacity α_{c_i}, the request arrival rate α_{s_j}, and the communication delay $D_{s_j}^{c_i}$ between the controller c_i and the switch s_j.

The second part is a DNN model that takes each controller state individually and sequentially and outputs the controller priorities $\vec{O}_t = (o_{c_1,t}, o_{c_2,t}, \ldots, o_{c_{N_c},t})$ in which $o_{c_i,t}$ is the priority of the controller c_i.

The last part is a Euclidean probability projector that transforms the controller priority vector $\vec{O}_t$ into a probability vector $\vec{P}_t = (p_{1,t}, p_{2,t}, \ldots, p_{N_c,t})$ which specifies the probabilities that each controller to be chosen. Specifically, the scheduling function is designed to select a controller from the M controllers with the highest priority. To achieve this design, the Euclidean probability projector generates a new vector $\widetilde{O}_t = (\widetilde{o_{1,t}}, \widetilde{o_{2,t}}, \ldots, \widetilde{o_{N_c,t}})$ by normalizing the elements in the priority vector $\vec{O}_t$ and sorting them in the descending order. The element $\widetilde{o_{i,t}}$ is then the normalized priority of the controller with the ith highest priority. Subsequently, the projector calculates $p_{i,t}$ as follows:

$$p_{i,t} = \begin{cases} \widetilde{o_{i,t}} + \frac{1}{M}\left(1 - \sum_{j=1}^{M} \widetilde{o_{j,t}}\right), & \text{if } 1 \le i \le M, \\ 0, & \text{otherwise.} \end{cases} \quad (7.98)$$

After obtaining the probability vector $\vec{P}_t$, the scheduling function randomly selects a controller as the output action based on the probabilities specified in the vector.

7.6 Conclusion

In this chapter state-of-the-art machine learning applications in the networking area have been reviewed. Machine learning techniques have been well leveraged to solve some networking problems like traffic classification, but their applications on some other networking problems are still in the relatively early stage. Most of the cutting-edge machine learning techniques are initially applied to computer vision, natural language processing, and pattern recognition problems. Their applications to the network problems have either lagged behind or not even started yet. Therefore, there is still long way to go towards realizing knowledge-driven networking and there are many opportunities for researchers and developers in the networking area in machine learning.

7.7 Problems

1. For the problems described in the following, choose one of the three categories of the machine learning techniques (i.e., supervised learning, unsupervised learning, reinforcement learning) as the most suitable techniques to solve them:
   (a) Predict the market price of a stock.
   (b) Given a photo, recognize the objects contained in it.
   (c) Design a traffic light control system that mitigates the congestion problem.
   (d) Predict whether a patient has a certain disease.
   (e) For a mobile app, find several groups of users with different use behaviors.
2. Given a dataset containing 12,000 samples, if a 70%/30% split is used between the training and the test dataset and a fourfold cross validation is used in each round of training and hyperparameter tuning. What is the number of samples in the training, the validation, and the test datasets, respectively?

Sample id	1	2	3	4	5	6	7	8	9	10
Prediction	Yes	Yes	Yes	No	No	Yes	No	Yes	No	Yes
Truth	Yes	Yes	No	Yes	Yes	No	No	Yes	Yes	Yes

Fig. 7.41 Predictions of 10 patients on whether they get a certain disease

Sample id	1	2	3	4	5	6	7	8	9	10
Truth	Yes	Yes	No	Yes	Yes	No	No	Yes	Yes	Yes
Probability of Yes	0.8	0.6	0.3	0.5	0.7	0.2	0.5	0.8	0.9	0.9
Probability of No	0.2	0.4	0.7	0.5	0.3	0.8	0.5	0.2	0.1	0.1

Fig. 7.42 Probability predictions of 10 patients on whether they get a certain disease

house id	1	2	3	4	5	6	7	8	9	10
Sold price (thousands)	120	90	190	260	420	360	210	130	280	310
Prediction (thousands)	100	95	180	255	380	390	240	120	290	320

Fig. 7.43 Predictions of the value of 10 houses

Sample id	1	2	3	4	5	6
Actual class C	1	1	2	1	3	2
Cluster G	1	1	1	3	3	2

Fig. 7.44 Clustering results of 10 samples into 3 clusters together with samples' actual class

3. Given a system that predicts whether a patient has a certain disease, Fig. 7.41 shows the predictions of ten patients. Given these predictions, draw the confusion matrix.

4. Based on the predictions shown in Fig. 7.41, calculate the accuracy, the recall, the precision, and the $F1$ score.

5. Figure 7.42 shows the probability predictions of 10 patients on whether they get a certain disease or not. Based on these data, calculate the log loss of the output.

6. Assume that a regression model is used to predict the value of 10 houses. Figure 7.43 shows the sold price (i.e., the ground truth) and the predicted value of these 10 houses. Based on these data, calculate the MAE, MSE, MAPE, and RMSE.

7. Assume that 10 samples whose actual class is available have been clustered into 3 clusters, as shown in Fig. 7.41, draw the corresponding contingency table.

8. Given the clustering results shown in Fig. 7.44, calculate the Rand Index.

9. Given the clustering results shown in Fig. 7.44, calculate the Fowlkes–Mallows Index.

10. A neural network with 1 hidden layer is shown in Fig. 7.45. The weight matrix between each pair of layers is given in the figure also. Given an input matrix including 2 samples, calculate the output.

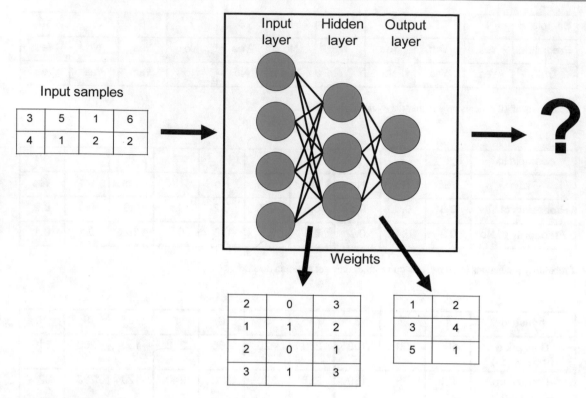

Fig. 7.45 Forward propagation of a neural network with 1 hidden layer

A.1 Some Summation Formulas

The following summation formulas are used in the text:

$$\sum_{n=0}^{\infty} x^n = \frac{1}{1-x} \qquad 0 \le x < 1 \qquad \text{(A.1)}$$

$$\sum_{n=0}^{N} x^n = \frac{1-x^{N+1}}{1-x} \qquad \text{(A.2)}$$

$$\sum_{n=0}^{\infty} n x^n = \frac{x}{(1-x)^2} \qquad 0 \le x < 1 \qquad \text{(A.3)}$$

$$\sum_{n=0}^{\infty} n^2 x^n = \frac{x(1+x)}{(1-x)^3} \qquad 0 \le x < 1 \qquad \text{(A.4)}$$

$$\sum_{n=0}^{\infty} \frac{1}{n!} x^n = e^x \qquad \text{(A.5)}$$

© Springer Nature Switzerland AG 2020
T. G. Robertazzi, L. Shi, *Networking and Computation*, https://doi.org/10.1007/978-3-030-36704-6

Bibliography

1. "8b/10b encoding", *www.knowledgetransfer.net/dictionary*.
2. B. Abernathy and T.G. Robertazzi, "Loading and Spatial Location in Wire and Radio Communication Networks," *IEEE MILCOM '91*, 1991, pp. 391–395.
3. N. Abramson, "The ALOHA System - Another Alternative for Computer Communications," *Proceedings of the Fall Joint Computer Conference*, 1970.
4. N. Abramson, "The Development of the ALOHANET," *IEEE Transactions on Information Theory*, vol. 31, 1985, pp. 119–123.
5. H. Adler, Y. Gong and A.L. Rosenberg, "Optimal Sharing of Bags of Tasks in Heterogeneous Clusters," *Proc. of ACM SPAA'03*, June 2003.
6. R. Agrawal and H.V. Jagadish, "Partitioning Techniques for Large Grained Parallelism," *IEEE Transactions on Computers*, vol. 37, no. 12, Dec. 1988, pp. 1627–1634.
7. R. Ahlswede, N. Cai, S.-Y. R. Li and R.W. Yeung, "Network Information Flow", IEEE Transactions on Information Theory, vol. 46, no. 4, July 2000, pp. 1204–1216.
8. H. Ahmadi and W.E. Denzel, "A Survey of Modern High-Performance Switching Techniques," *IEEE Journal on Selected Areas in Communications*, vol. 7, no. 7, Sept. 1989, pp. 1091–1103.
9. R. Alléaume, C. Brassard, et al., "Using Quantum Key Distribution for Cryptographic Purposes: A Survey," *Theoretical Computer Science*, vol. 560, 2014, pp. 62–81.
10. G.M. Amdahl, "Validity of the Single Processor Approach to Achieving Large Scale Computing Capabilities," *Proceedings of the AFIPS Conference*, AFIPS Press, 1967, pp. 483–485.
11. G.M. Amdahl, "Computer Architecture and Amdahl's Law," *Computer*, 2013, pp. 38–46.
12. E.J. Anderson, "A New Continuous Model for Job Shop Scheduling," *International Journal of System Science*, vol. 12, no. 12, 1981, pp. 1469–1475.
13. G. Apostolopoulos, R. Guérin, S. Kamat and S.K. Tripathi, "Quality of Service Based Routing: A Performance Perspective," *Proceedings of SIGCOMM'98*, 1998, pp. 17–28.
14. M. Arafat, S. Bataineh and I. Khalil, "Probabilistic Approach to Scheduling Divisible Load on Network of Processors," *Sensor Network Data Communication*, vol. 4, no. 130, 2015.
15. J.R. Artalejo, "G-networks: A Versatile Approach for Work Removal in Queueing Networks," *European Journal of Operations Research*, vol. 126, issue 2, Oct. 2000, pp. 233–249.
16. M. Baker, A. Apon, C. Feiner and J. Brown, "Emerging Grid Standards," *Computer*, vol. 38, no. 4, 2005, pp. 43–50.
17. G.D. Barlas, "VoD on Steroids: Optimized Content Delivery using Distributed Video Servers over Best-Effort Internet," *Journal of Parallel and Distributed Computing*, vol. 65, no. 9, 2005, pp. 1057–1071.
18. G.D. Barlas and K. El-Fakih, "A GA-Based Movie-on-Demand Platform using Multiple Distributed Servers," *Multimedia Tools and Applications*, vol. 40, no. 3, Dec. 2008, pp. 361–383.
19. G.D. Barlas, "An Analytical Approach to Optimizing Parallel Image Registration," *IEEE Transactions on Parallel and Distributed Systems*, vol. 21, no. 8, 2010, pp. 1074–1088.
20. G.D. Barlas, "Cluster-based Optimized Parallel Video Transcoding," *Parallel Computing*, vol. 38, March 2012, pp. 226–244.
21. S.T. Başaran, G.K. Kurt, M. Uysal and İ. Altunbaş, "A Tutorial on Network Coded Cooperation", IEEE Communication Surveys and Tutorials, vol. 18, no. 4, Fourth Quarter 2016, pp. 2970–2990.
22. F. Baskett, K.M. Chandy, R.R. Muntz and F. Palacios, "Open, Closed and Mixed Networks of Queues with Different Classes of Customers," *Journal of the ACM*, vol. 22, no. 2, April 1975, pp. 248–260.
23. S. Bataineh and T.G. Robertazzi, "Bus Oriented Load Sharing for a Network of Sensor Driven Processors," *special issue on Distributed Sensor Networks of the IEEE Transactions on Systems, Man and Cybernetics*, vol. 21, no. 5, Sept. 1991, pp. 1202–1205.
24. S. Bataineh, T. Hsiung and T.G. Robertazzi, "Closed Form Solutions for Bus and Tree Networks of Processors Load Sharing a Divisible Job", *IEEE Transactions on Computers*, vol. 43, no. 10, Oct. 1994, pp. 1184–1196.
25. S. Bataineh and M. Al-Ibrahim, "Effect of Fault Tolerance and Communications Delay on Response Time in a Multiprocessor System with a Bus Topology," *Computer Communications*, 1994.
26. S. Bataineh and T.G. Robertazzi, "Performance Limits for Processor Networks with Divisible Jobs," *IEEE Transactions on Aerospace and Electronic Systems*, vol. 33, no. 4, Oct. 1997, pp. 1189–1198.
27. O. Beaumont, L. Carter, J. Ferrante, A. Legrand and Y. Robert, "Bandwidth-Centric Allocation of Independent Tasks on Heterogeneous Platforms," *Proceedings of the International Parallel and Distributed Processing Symposium (IPDPS'02)*, June 2002.
28. O. Beaumont, A. Legrand and Y. Robert, "The Master-Slave Paradigm with Heterogeneous Processors," *IEEE Transactions on Parallel and Distributed Systems*, vol. 14, no. 9, 2003, pp. 897–910.
29. C.H. Bennett and G. Brassard, "Quantum Cryptography: Public Key Distribution and Coin Tossing," *Proc. of IEEE International Conference on Computer Systems and Signal Processing*, vol. 175, pg. 8, 1984.
30. J. Berlińska and M. Drozdowski, "Heuristics for Multi-round Divisible Load Scheduling with Limited Memory," *Parallel Computing*, vol. 36, 2010, pp. 199–211.
31. J. Berlińska and M. Drozdowski, "Scheduling Divisible MapReduce Computations," *Journal of Parallel and Distributed Computing*, vol. 71, no. 3, 2011, pp. 450–459.
32. J. Berlińska and M. Drozdowksi, "Mitigating Partitioning Skew in MapReduce Computations," *Proceedings of the 6th Multidisciplinary International Scheduling Conference: Theory and Applications*, Ghent, 2013, 80–90.
33. J. Berlińska and M. Drozdowski, "Scheduling Multilayer Divisible Computation, RAIRO Operations Research, vol. 49, no. 2, 2015, pp. 339–368.

34. J. Berlińska and M. Drozdowski, "Scheduling Data Gathering with Variable Communication Speed," *Proc. First International Workshop on Dynamic Scheduling Problems*, 2016.

35. J. Berlińska and M. Drozdowski, "Comparing Load-Balalncing Algorithms for MapReduce under Zipfian Data Skews," *Parallel Computing* vol. 72, 2018.

36. D. Bertsekas and R. Gallager, *Data Networks*, 2nd ed., Prentice-Hall, 1991.

37. V. Bharadwaj, D. Ghose and V. Mani, "Optimal Sequencing and Arrangement in Distributed Single-Level Tree Networks with Communication Delays," *IEEE Transactions on Parallel and Distributed Systems*, vol. 5, no. 9, pp. Sept. 1994, pp. 968–976.

38. V. Bharadwaj, D. Ghose and V. Mani, "Multi-installment Load Distribution in Tree Networks with Delays," *IEEE Transactions on Aerospace and Electronic Systems*, vol. 31, no. 2, April 1995, pp. 555–567.

39. (a) V. Bharadwaj, H.F. Li and T. Radhakrishnan, "Scheduling Divisible Loads in Bus Networks with Arbitrary Processor Release Times," *Computers and Mathematics with Applications*, Pergamon Press, vol. 32, no. 7, 1996, pp. 57–77.

40. (b) V. Bharadwaj, D. Ghose, V. Mani, and T.G. Robertazzi, *Scheduling Divisible Loads in Parallel and Distributed Systems*, IEEE Computer Society Press, Los Alamitos CA, Sept. 1996, 292 pages.

41. (a) V. Bharadwaj, L. Xiaolin, and C.C. Ko, "Design and Analysis of Load Distribution Strategies with Start-up Costs in Scheduling Divisible Loads on Distributed Networks," *Mathematical and Computer Modelling*, Pergamon Press, vol. 32, 2000, pp. 901–932.

42. (b) V. Bharadwaj, X. Li and C.C. Ko, "Efficient Partitioning and Scheduling of Computer Vision and Image Processing Data on Bus Networks using Divisible Load Analysis," *Image and Vision Computing*, vol. 18 no. 11, Aug. 2000, pg. 919.

43. (c) V. Bharadwaj and N. Viswanadham, "Sub-Optimal Solutions Using Integer Approximation Techniques for Scheduling Divisible Loads on Distributed Bus Networks," *IEEE Transactions on Systems, Man, and Cybernetics: Part A*, vol. 30, no. 6, pp. 680–691, November 2000.

44. (d) V. Bharadwaj, X. Li and K.C. Chuang, "On the Influence of Start-Up Costs in Scheduling Divisible Loads in Bus Networks", *IEEE Transactions on Parallel and Distributed Systems*, vol. 11, no. 12, Dec. 2000, pp. 1288–1305.

45. V. Bharadwaj and S. Ranganath, "Theoretical and Experimental Study of Large Size Image Processing Applications using Divisible Load Paradigm on Distributed Bus Networks," *Image and Vision Computing*, Elsevier Publishers, vol. 20, issues 13–14, Dec. 2002, pp. 917–936.

46. (a) V. Bharadwaj, D. Ghose, T. G. Robertazzi, "A New Paradigm for Load Scheduling in Distributed Systems," in special issue of *Cluster Computing* on Divisible Load Scheduling, Kluwer Academic Publishers, vol. 6, no. 1, Jan. 2003, pp. 7–18.

47. (b) V. Bharadwaj and G. Barlas, "Scheduling Divisible Loads with Processor Release Times and Finite Size Buffer Capacity Constraints," in special issue of *Cluster Computing* on Divisible Load Scheduling, Kluwer Academic Publishers, vol. 6, no. 1, Jan. 2003, pp. 63–74.

48. J. Blazewicz and M. Drozdowski, "Scheduling Divisible Jobs on Hypercubes," *Parallel Computing*, vol. 21, 1995, 1945–1956.

49. J. Blazewicz and M. Drozdowski, "The Performance Limits of a Two-Dimensional Network of Load Sharing Processors," *Foundations of Computing and Decision Sciences*, vol. 21, no. 1, 1996, pp. 3–15.

50. J. Blazewicz and M. Drozdowski, "Distributed Processing of Divisible Jobs with Communication Startup Costs," *Discrete Applied Mathematics*, vol. 76, issue 1–3, 1997, pp. 21–41.

51. J. Blazewicz, M. Drozdowski, F. Guinand and D. Trystram, "Scheduling a Divisible Task in a 2-Dimensional Mesh," *Discrete Applied Mathematics*, May 1999, pg. 35.

52. S.H. Bokhari, *Assignment Problems in Parallel and Distributed Computing*, Springer, 2012.

53. S. Borkar, "Getting Gigascale Chips: Challenges and Opportunities in Continuing Moore's Law," *ACM Queue*, vol. 1, Oct. 2003, pp. pp. 26–33.

54. S.C. Bruell and G. Balbo, *Computational Algorithms for Closed Queueing Networks*, North-Holland, N.Y. 1980.

55. J.P. Buzen, "Computational Algorithms for Closed Queueing Networks with Exponential Servers," *Communications of the ACM*, vol. 16, no. 9, Sept. 1973, pp. 527–531.

56. Y. Cao, *Algorithms and Performance Evaluation for Scheduling in Parallel Systems*, PhD Thesis, Stony Brook University, 2019.

57. H. Casanova, A. Legrand and Y. Robert, *Parallel Algorithms*, CRC Press, Florida, 2009.

58. A.S. Cassidy and A.G. Andreou, "Analytic Methods for the Design and Optimization of Chip-Multiprocessor Architectures," *Proceedings of the 43rd Annual Conference on Information Sciences and Systems*, 2009.

59. A.S. Cassidy and A.G. Andreou, "Beyond Amdahl's Law: An Objective Function that Links Multiprocessor Performance Gains to Delay and Energy," IEEE Transactions on Computers, vol. 61, no. 8, Aug. 2012, pp. 1110–1126.

60. Y.K. Chang, J.-H. Wu, C.-Y. Chen and C.P. Chu, "Improved Methods for Divisible Load Distribution on k-Dimensional Meshes using Multi-Installment," *IEEE Transactions on Parallel and Distributed Systems*, vol. 18, no. 11, Nov. 2007.

61. X. Chao and M. Pinedo, "On G-Networks: Queues with Positive and Negative Arrivals," *Probability in the Engineering and Information Sciences*, 1993.

62. S. Charcranoon, T.G. Robertazzi and S. Luryi, "Parallel Processor Configuration Design with Processing/Transmission Costs," *IEEE Transactions on Computers*, vol. 49, no. 9, Sept. 2000, pp. 987–991.

63. S. Charcranoon, T.G. Robertazzi, and S. Luryi, "Load Sequencing for a Parallel Processing Utility," *Journal of Parallel and Distributed Computing*, 2004.

64. S. Chen and K. Nahrstedt, "An Overview of Quality of Service Routing for Next-Generation High-Speed Networks: Problems and Solutions," *IEEE Network*, Nov./Dec. 1998, pp. 64–79.

65. C.Y. Chen and C.P. Chu, "Novel Methods for Divisible Load Distribution with Start-Up Costs on a Complete b-Ary Tree," *IEEE Transcations on Parallel and Distributed Systems*, vol. 26, no. 10, 2015, pp. 2836–2848.

66. Y.C. Cheng and T.G. Robertazzi, "Distributed Computation with Communication Delays," *IEEE Transactions on Aerospace and Electronic Systems*, vol. 24, no. 6, Nov. 1988, pp. 700–712.

67. Y.C. Cheng, and T.G. Robertazzi, "Distributed Computation for a Tree Network with Communication Delays," *IEEE Transactions on Aerospace and Electronic Systems*, vol. 26, no. 3, May 1990, pp. 511–516.

68. Y.C. Cheng and T.G. Robertazzi, "A New Spatial Point Process for Multihop Radio Network Modeling," *Proceedings of the International Conference on Communications*, ICC '90, 1990, pp. 1241–1245.

69. Benoit Claise. Specification of the ip flow information export (ipfix) protocol for the exchange of ip traffic flow information. Technical report, 2008.

70. N. Comino and V.L. Narasimhan, "A Novel Data Distribution Technique for Host-Client Type Parallel Applications," *IEEE Transactions on Parallel and Distributed Systems*, vol. 13, no. 2, 2002, pp. 97–110.

71. S. Deb, M. Effros, T. Ho, et al., "Network Coding for Wireless Applications: A Brief Tutorial", see Google Scholar.

72. F. Díaz-del-Río, J. Salmerón-García and J. Luis Sevillano, "Extending Amdahl's Law for the Cloud Computing Era," *Computer*, Feb. 2016, pp. 14–22.

73. P. Diggle, *Statistical Analysis of Spatial Point Patterns*, Academic Press, London, 1983.

74. E. W. Dijkstra, "Cooperating Sequential Processes," in F. Genuys, editor, Academic Press, 1983.

75. R. Disney, *Traffic Processes in Queueing Networks: A Markov Renewal Approach*, The Johns Hopkins University Press, Baltimore MD, 1987.

76. Divisble Load Theory publications web page at Robertazzi's web page at www.ece.stonybrook.edu.

77. M. Drozdowski, *Selected Problems of Scheduling Tasks in Multiprocessor Computer Systems*, Politechnika Poznanska, Book No. 321, Poznan, Poland, 1997.

78. M. Drozdowski and W. Glazek, "Scheduling Divisible Loads in a Three Dimensional Mesh of Processors," *Parallel Computing*, vol. 25, 1999, pp. 381–404.

79. M. Drozdowski and P. Wolniewicz, "Experiments with Scheduling Divisible Tasks in Clusters of Workstations," in A. Bode, T. Ludwig, W. Karl and R. Wismüler, editors, *EURO-Par 2000*, Lecture Notes in Computer Science 1900, Springer-Verlag, 2000, pp. 311–319.

80. (a) M. Drozdowski and P. Wolniewicz, "Divisible Load Scheduling in Systems with Limited Memory," in special issue of *Cluster Computing* on Divisible Load Scheduling, Kluwer Academic Publishers, vol. 6, no. 1, Jan. 2003, pp. 19–30.

81. (b) M. Drozdowksi and P. Wolniewicz, "Out of Core Divisible Load Processing," IEEE Transactions on Parallel and Distributed Systems, vol. 14, no. 10, Oct. 2003, pp. 1048–1056.

82. M. Drozdowski and M. Lawenda, "Multi-installment Divisible Load Processing in Heterogeneous Systems with Limited Memory," Lecture Notes in Computer Science 3911, 2006, pp. 847–854, in R. Wyrzykowski et al. (eds.), *Proceeding of PPAM 2005*.

83. (a) M. Drozdowski, "Energy Considerations for Divisible Load Processing," Lecture Notes in Computer Science 6068, 2010, pp. 92–101, in R. Wyrzykowski, J. Dongarra, K. Karczewski, J. Wasniewski (eds.), *Proceedings of the 8th International Conference PPAM 2009*, part II.

84. (b) M. Drozdowski, *Scheduling for Parallel Processing*, Springer, New York, 2009.

85. M. Drozdowski and J.M. Marszalkowski, "Divisible Load Scheduling in Hierarchical Memory Systems with Time and Energy Constraints," *Parallel Processing and Applied Math in LNCS vol. 9574*, Springer, 2016, pp. 111–120.

86. P.-F. Dutot, "Divisible Load on a Heterogeneous Arrays," *Proc. of the International Parallel and Distributed Processing Symposium (IPDS'03)*, Nice, France, April 2003.

87. C. Eklund, R.B. Marks, K.L. Stanwood and S. Wang, "IEEE Standard 802.16: A Technical Overview of the WirelessMAN Air Interface for Broadband Wireless Access," *IEEE Communications Magazine*, June 2002, pp. 98–107.

88. D. England, B. Veeravalli and J.B. Weissman, "A Robust Spanning Tree Topology for Data Collection and Dissemination in Distributed Environments, *IEEE Transactions on Parallel and Distributed Systems*, vol. 18, 2007, pp. 608–620.

89. Fatih Ertam and Engin Avcı. A new approach for internet traffic classification: Ga-wk-elm. *Measurement*, 95:135–142, 2017.

90. I. Foster and C. Kesselman, *The Grid 2: Blueprint for a New Computing Infrastructure*, Morgan-Kaufmann, 2003.

91. P.A. Franaszek and A. X. Widmer, "Byte Oriented DC Balanced (0.4) 8B/10B Partitioned Block Transmission Code," *US Patent 4486739*, December 4, 1984.

92. M.A. Franklin, "A VLSI Performance Comparison on Banyan and Crossbar Communication Networks," *IEEE Transactions on Computers*, vol. C-30, no. 4, April 1981, 283–290.

93. M. W. Garrett and S.-Q. Li, "A Study of Slot Reuse in Dual Bus Multiple Access Networks," *Proceedings of INFOCOM '90*, San Francisco, CA, June 1990.

94. T. Garritano, "Globus: An Infrastructure for Resource Sharing," *Clusterworld*, vol. 1, no. 1, pp. 30–31, 50.

95. (a) E. Gelenbe, "Product Form Networks with Negative and Positive Customers," *Journal of Applied Probability*, vol. 28, 1991, pp. 656–663.

96. (b) E. Gelenbe, P. Glynn and K. Sigman, "Queues with Negative Arrivals," *Journal of Applied Probability*, vol. 28, 1991, pp. 245–250.

97. S. Ghanbari, M. Othman and W.J. Leong, "Multi-criteria Based Algorithm for Scheduling Divisible Loads," *Proceedings of First International Conference on Advanced Data and Information Engineering*, 2013, pp. 547–554.

98. D. Ghose and V. Mani, "Distributed Computation with Communication Delays: Asymptotic Performance Analysis," *Journal of Parallel and Distributed Computing*, vol. 23, 1994, pp. 293–305.

99. D. Ghose and H.J. Kim, "Load Partitioning and Trade-Off Study for Large Matrix Vector Computations in Multicast Bus Networks with Communication Delays," *Journal of Parallel and Distributed Computing*, vol. 54, 1998, pp. 32–59.

100. D. Ghose, "A Feedback Strategy for Load Allocation in Workstation Clusters with Unknown Network Resource Capabilities using the DLT Paradigm," *Proceedings of the International Conference on Parallel and Distributed Processing Techniques and Applications (PDPTA'02)*, Las Vegas, Nevada, vol. 1, June 2002, pp. 425–428.

101. D. Ghose and T.G. Robertazzi, editors, special issue on Divisible Load Scheduling, *Cluster Computing*, vol. 6, 2003.

102. W. Glazek, "A Multistage Load Distribution Strategy for Three Dimensional Meshes," in special issue of *Cluster Computing* on Divisible Load Scheduling, Kluwer Academic Publishers, vol. 6, no. 1, Jan. 2003, pp. 31–40.

103. L. Goldberg, "802.11 Wireless LANs: A Blueprint for the Future?," *Electronic Design*, Aug. 4, 1997, pp. 44–52.

104. D. Goodman and R. Yates, *Probability and Stochastic Processes*, 2nd ed., Wiley, 2004.

105. W.J. Gordon and G.F. Newell, "Closed Queueing Systems with Exponential Servers," *Operations Research*, vol. 15, 1967, pp. 254–265.

106. J.M. Griffiths, "Binary Code Suitable for Line Transmission," *Electronics Letters*, vol. 5, 1969, pp. 79–81.

107. D. Gross and C.M. Harris, *Fundamentals of Queueing Theory*, Wiley, New York, 1974, 1985.

108. R.A. Guérin, A. Orda and D. Williams, "QoS Routing Mechanisms and OSPF Extensions, *Proceedings of IEEE Globecomm'97*, 1997, pp. 1903–1908.

109. J.J. Gustafson, "Reevaluating Amdahl's Law," *Communications of the ACM*, vol. 31, no. 5, May 1988, pp. 532–533.

110. J.C. Haartsen and S. Mattisson, "Bluetooth - A New Low-Power Radio Interface Providing Short-Range Connectivity," *Proceedings of the IEEE*, vol. 88, no. 10, Oct. 2000, pp. 1651–1661.

111. J.L. Hammond and P.J.P. O'Reilly, *Performance Analysis of Local Computer Networks*, Addison-Wesley, Reading, Mass. 1986.

112. U. Herzog, L. Woo and K.M. Chandy, "Solution of Queueing Problems by a Recursive Technique," *IBM Journal of Research and Development*, May 1975, pp. 295–300.

113. M.D. Hill and M.R. Marty, "Amdahl's Law in the Multicore Era," *Computer*, July 2008, pp. 33–38.

114. M.D. Hill and M.R. Marty, "Retrospective on Amdahl's Law in the Multicore Era," *Computer*, June 2017, pp. 12–14.

115. F.S. Hillier and G.J. Lieberman, *Introduction to Operations Research*, 8th edition, McGraw-Hill, 2005.

116. G.J. Holzmann, "The Model Checker Spin," *IEEE Transactions on Software Engineering*, vol. 23, no. 5, May 1997, pp. 279–295.

117. G.J. Holzmann, *The SPIN Model Checker: Primer and Reference Manual*, Addison-Wesley, Boston, 2004.

118. C.H. Hsu, T.L. Chen and J.H. Park, "On Improving Resource Utilization and System Throughput of Master Slave Job Scheduling in Heterogeneous Systems," *Journal of Supercomputing*, vol. 45, no. 1, 2008, pp. 129–150.

119. J.Y. Hui and E. Arthurs, "A Broadband Packet Switch for Integrated Transport," *IEEE Journal on Selected Areas in Communications*, vol. SAC-5, no. 8, Oct. 1987, pp. 1264–1273.

120. T.S. Humble, "Quantum Security for the Physical Layer", *IEEE Communications Magazine*, vol. 51, Aug. 2013, pp. 56–62.

121. J.T. Hung, H.J. Kim and T.G. Robertazzi, "Scalable Scheduling in Parallel Processors," *Proceedings of the 2002 Conference on Information Sciences and Systems*, Princeton University, Princeton NJ, March 2002.

122. J.T. Hung, *Scalable Scheduling in Parallel, Distributed and Grid Systems*, Ph.D Thesis, Dept. of Electrical and Computer Engineering, Stony Brook University, Stony Brook NY, Aug. 2003.

123. (a) J.T. Hung and T.G. Robertazzi, "Scalable Scheduling for Clusters and Grids using Cut Through Switching," *International Journal of Computers and their Applications*, ACTA Press, vol. 26, no. 3, 2004, pp. 147–156.

124. (b) J.T. Hung and T.G. Robertazzi, "Divisible Load Cut Through Switching in Sequential Tree Networks," *IEEE Transactions on Aerospace and Electronic Systems*, vol. 40, no. 3, July 2004, pp. 968–982.

125. J.T. Hung and T.G. Robertazzi, "Scheduling Nonlinear Computational Loads," *IEEE Transactions on Aerospace and Electronic Systems*, vol. 44, no. 3, July 2008, pp. 1169–1182.

126. J.R. Jackson, "Networks of Waiting Lines," *Operations Research*, vol. 5, 1957, pp. 518–521.

127. B.H.H. Juurlink and C.H. Meenderinck, "Amdahl's Law for Predicting the Future of Multicores Considered Harmful," *ACM SIGARCH Computer Architecture News*, vol. 40, no. 2, May 2012, 9 pages.

128. J.M. Kahn, R.H. Katz and K.S.J. Pister, "Emerging Challenges: Mobile Networking for 'Smart Dust'," *Journal of Communications Networks*, vol. 2, no. 3, Sept. 2000.

129. S. Kang, B. Veeravalli and K.M.M. Aung, "Dynamic Scheduling Strategy with Efficient Node Availability Prediction for Handling Divisible Loads in Multi-Cloud Systems," *Journal of Parallel and Distributed Computing*, vol. 113, 2018, pp. 1–16.

130. S. Kapp, "802.11: Leaving the Wire Behind," *IEEE Internet Computing*, vol. 6, no. 1, Jan-Feb 2002, pp. 82–85.

131. M.J. Karol, M.G. Hluchyj and S.P. Morgan, "Input vs. Output Queueing on a Space Division Packet Switch," *IEEE Transactions on Communications*, vol. COM-35, no. 12, Dec. 1987, pp. 1345–1356.

132. B. Karp and H.T. Kung, "GPSR: Greedy Perimeter Stateless Routing for Wireless Networks," *Proceedings of the 6th Annual ACM/IEEE International Conference on Mobile Computing and Networking (MobiCom)*, 2000.

133. J.S. Kaufman, "Blocking in a Shared Resource Environment," *IEEE Transactions on Communications*, vol. COM-29, no. 10, Oct. 1981, pp. 1474–1481.

134. D.G. Kendall, "Some Problems in the Theory of Queues," *Journal of the Royal Statistical Society*, series B, vol. 13, no. 2, 1951, pp. 151–185.

135. P. Kermani and L. Kleinrock, "Virtual Cut-Through: A New Computer Communications Switching Technique," *Computer Networks*, vol. 3, 1979, pp. 267–286.

136. A.Y. Khinchin, "Mathematisches uber die Erwatung vor einem offentlichen Schalter," in English "Mathematical Theory of Stationary Queues," *Matem. Sbornik*, vol. 39, 1932, pp. 73–84.

137. H.J. Kim, G.-I. Jee and J.G. Lee, "Optimal Load Distribution for Tree Network Processors," *IEEE Transactions on Aerospace and Electronic Systems*, vol. 32, no. 2, April 1996, pp. 607–612.

138. H.J. Kim, "A Novel Optimal Load Distribution Algorithm for Divisible Loads," in special issue of *Cluster Computing* on Divisible Load Scheduling, Kluwer Academic Publishers, vol. 6, no. 1, Jan. 2003, pp. 41–46.

139. S.-H. Kim and T.G. Robertazzi, "Spatial Network Traffic Intensity," *Proceedings of the 2000 Conference on Information Sciences and Systems*, Princeton University, Princeton NJ, 2000, pp. TP2–24.

140. L. Kleinrock, *Queueing Systems*, vol. I and II, Wiley, New York, 1975.

141. (a) K. Ko, *Scheduling Data Intensive Parallel Processing in Distributed and Networked Environments*, Ph.D Thesis, Dept. of Electrical and Computer Engineering, University at Stony Brook, Aug. 2000.

142. (b) K. Ko and T.G. Robertazzi, "Record Search Time Evaluation," *Proceedings of the Conference on Information Sciences and Systems*, Princeton University, Princeton, N.J., March 2000.

143. K. Ko and T.G. Robertazzi, "Scheduling in an Environment of Multiple Job Submission," *Proceedings of the 2002 Conference on Information Sciences and Systems*, Princeton University, Princeton NJ, March 2002.

144. K. Ko and T.G. Robertazzi, "Signature Search Time Evaluation in Flat File Databases," *IEEE Transactions on Aerospace and Electronic Systems*, vol. 44, no. 2, April 2008, pp. 493–502.

145. H. Kobayashi, *Modeling and Analysis: An Introduction to System Performance Evaluation*, Addison-Wesley, Reading, Mass. 1978.

146. C.S. Kong, V. Bharadwaj and D. Ghose, "Large Matrix-Vector Products on Distributed Bus Networks with Communication Delays using the Divisible Load Paradigm: performance and simulation," *Computers and Mathematics in Simulation*, Elsevier Press, vol. 58, 2001, pp. 71–92.

147. B. Kreaseck, L. Carter, H. Casanova, and J. Ferrante, "Autonomous Protocols for Bandwidth-Centric Scheduling of Independent-task Applications," *Proceedings of the International Parallel and Distributed Processing Symposium (IPDPS'03)*, Nice, France, April 2003.

148. Y. Kyong and T.G. Robertazzi, "Greedy Signature Processing with Arbitrary Location Distributions: A Divisible Load Framework," *IEEE Transactions on Aerospace and Electronic Systems*, vol. 48, no. 4, 2012, pp. 3027–3041.

149. R.O. LaMaire, A. Krishnan, P. Bhagwat and J. Panian, "Wireless LANs and Mobile Networking: Standards and Future Directions," *IEEE Communications Magazine*, vol. 34, no. 8, Aug. 1996, pp. 86–94.

150. A.A. Lazar and T.G. Robertazzi, "Markovian Petri Net Protocols with Product Form Solution," *Performance Evaluation*, vol. 12, 1991, pp. 67–77.

151. C.E. Leiserson, "Fat-Trees: Universal Networks for Hardware-Efficient Supercomputing," *IEEE Transactions on Computers*, vol. C-34, no. 10, 1985, pp. 892–901.

152. X. Li, V. Bharadwaj and C.C. Ko, "Optimal Divisible Task Scheduling on Single-Level Tree Networks with Finite Size Buffers," *IEEE Transactions on Aerospace and Electronic Systems*, vol. 36, no. 4, Oct. 2000, pp. 1298–1308.

153. S.-Y. R. Li, R.W, Yeung and N. Cai, "Linear Network Coding", IEEE Transactions on Information Theory, vol. 49, no.2, Feb. 2003, pp. 371–381.

154. W. Lin and J.Z. Wang, "Bandwidth-aware Divisible Task Scheduling for Cloud Computing," *Journal of Software: Practice and Experience*, vol. 44, no. 2, 2012, pp. 163–174.

155. X. Lin, Y. Lu, J. Deogun and S. Goddard, "Real-Time Divisible Load Scheduling with Different Processor Available Times," *2007*

International Conference on Parallel Processing (ICPP 2007), 2007.

156. X. Lin, A. Mamat, Y. Lu, J. Deogun and S. Goddard, "Real Time Scheduling of Divisible Loads in Cluster Computing Environments," *Journal of Parallel and Distributed Computing*, vol. 70, no. 3, 2010, pp. 296–308.

157. M. Littlewood and I.D. Gallagher, "Evolution Toward an ATD Multi-service Network," *British Telecom Technology Journal*, vol. 5, no. 2, April 1987.

158. D. Lun, M. Médard, R. Koetter and M. Effros, "On Coding for Reliable Communication over Packet Networks," Physical Communications, vol. 1, no. 1, 2008, pp 3–20.

159. L.O. Mailloux, M.R. Grimaila et al., "Performance Evaluations of Quantum Key Distribution System Architectures," *IEEE Security and Privacy*, vol. 13, Jan/Feb 2015, pp. 30–40.

160. (a) A. Mamat, Y. Lu, J. Deogun and S. Goddard, "Efficient Real-Time Divisible Load Scheduling," *Journal of Parallel and Distributed Computing*, vol. 72, no. 12, 2012, pp. 1603–1616.

161. (b) A. Mamat, Y. Lu, J. Deogun and S. Goddard, "Scheduling Real-Time Divisible Loads with Advance Reservations," *Real-Time Systems*, 2012, vol. 48, issue 3, pp. 264–293.

162. V. Mani and D. Ghose, "Distributed Computation in a Linear Network: Closed-Form Solutions and Computational Techniques," *IEEE Transactions on Aerospace and Electronic Systems*, vol. 30, no. 2, April 1994.

163. A. Marowka, "Extending Amdahl's Law for Heterogeneous Computing," *2012 10th IEEE Symposium on Parallel and Distributed Processing with Applications*, 2012, pp. 309–316.

164. A. Marowka, "Analytical Modeling of Energy Efficiency in Heterogeneous Processors," *Computers in Electrical Engineering*, vol. 39, 2013, pp. 2566–2578.

165. M.A. Marsan, G. Balbo, G. Conte and F. Gregoretti, "Modeling Bus Contention and Memory Interference in a Multiprocessor System," *IEEE Transactions on Computers*, vol. C-32, no. 1, 1983, pp. 60–72.

166. M.A. Marsan, G. Balbo and G. Conte, *Performance Models of Multiprocessor Systems*, The MIT Press, 1986.

167. M. Mauve, H. Hastenstein and A. Widmer, "A Survey on Position-Based Routing in Mobile Ad Hoc Networks," *IEEE Network*, vol. 15, no. 3, Nov./Dec. 2001, pp. 30–39.

168. H. Michiel and K. Laevens, "Teletraffic Engineering in a Broad-Band Era," *Proc. of the IEEE*, vol. 85, no. 12, Dec. 1997, pp. 2007–2033.

169. L.E. Miller, "Distributional Properties of Inhibited Random Positions of Mobile Radio Terminals," *Proceedings of the 2002 Conference on Information Sciences and Systems*, Princeton University, March 2002.

170. Tom M. Mitchell. *Machine learning.* McGraw Hill series in computer science. McGraw-Hill, 1997. ISBN 978-0-07-042807-2. URL http://www.worldcat.org/oclc/61321007.

171. M. Moges and T.G. Robertazzi, "Divisible Load Scheduling and Markov Chain Models, *Computers and Mathematics with Applications*, vol. 52, 2006, pp. 1529–1542.

172. C.S.R. Murthy and B.S. Manoj, *Ad Hoc Wireless Networks: Architectures and Protocols*, Prentice-Hall, 2004.

173. Y. Oie, M. Murata, K. Kubota and H. Miyahara, "Effect of Speedup in Nonblocking Packet Switches," *Proceedings of IEEE International Conference on Communications*, 1989, pp. 410–414.

174. R.O. Onvural, *Asynchronous Transfer Mode Networks: Performance Issues*, 2nd edition, Artech House, 1995.

175. N. Papanikolaou, "An Introduction to Quantum Cryptography," *Crossroads*, ACM, vol. 11, no. 3, March 2005.

176. A. Papoulis and S.U. Pillai, *Probability, Random Variables and Stochastic Processes*, McGraw-Hill, 2002.

177. B. Parhami, "Amdahl's Reliability Law: A Simple Quantification of the Weakest-Link Phenomena," *Computer*, July 2015, pp. 55–58.

178. J.H. Patel, "Performance of Processor-Memory Interconnection for Multiprocessors, *IEEE Transactions on Computers*, vol. C-30, no. 10, Oct. 1981, pp. 771–780.

179. C.E. Perkins, *Ad Hoc Networking*, Addison-Wesley, 2000.

180. W.W. Peterson and D.T. Brown, "Cyclic Codes for Error Detection," *Proceedings of the IRE (Institute of Radio Engineers)*, 1961, pp. 228–235.

181. C.A. Petri, *Kommunikation mit Automaten*, Ph.D Thesis, University of Bonn, Germany, 1962.

182. D.L.H. Ping, B. Veeravallia and D. Bader, "On the Design of High-Performance Algorithms for Aligning Multiple Protein Sequences in Mesh-Based Multiprocessor Architectures," *Journal of Parallel and Distributed Computing*, vol. 67, no. 9, 2007, pp. 1007–1017.

183. D.A.L. Piriyakumar and C.S.R. Murthy, "Distributed Computation for a Hypercube Network of Sensor-Driven Processors with Communication Delays Including Setup Time," *IEEE Transactions on Systems, Man and Cybernetics-Part A: Systems and Humans*, vol. 28, no. 2, March 1998, pp. 245–251.

184. F.J. Pollack, "New Microarchitecture Challenges in the Coming Generations of CMOS Process Technologies," keynote address, abstract only, *Proceedings of the of the 32nd Annual ACM/IEEE International Symposium on Microarchitecture*, MICRO 32, 1999.

185. F. Pollaczek, "Uber eine Aufgabe dev Wahrscheinlichkeitstheorie," *Math. Zeitschrift*, vol. 32, 1930, pp. 64–100, 729–750.

186. J. Quereshi, C.H. Foh and J. Cai, "Optimal Solution for the Index Coding Problem using Network Coding over GF(2)", IEEE SECON 2012, pp. 134–142.

187. J.M. Rabaey, M.J. Ammer and J.L. da Silva Jr., et al., "PicoRadio Supports Ad Hoc Ultra-Low Power Wireless Networking," *Computer*, vol. 33, no. 7, July 2000, pp. 42–48.

188. M. Reiser and H. Kobayashi, "Recursive Algorithms for General Queueing Networks with Exponential Servers," *IBM Research Report RC 4254*, Yorktown Heights NY, 1973.

189. M. Reiser and S.S. Lavenberg, "Mean-value Analysis of Closed Multichain Queueing Networks," *Journal of the ACM*, vol. 27, no. 2, April 1980, pp. 313–322.

190. S. Ristov, R. Prodan et al., "Superlinear Speedup in HPC Systems: Why and When?," *Proceedings of the Federated Conference on Computer Science and Information Systems*, 2016, pp. 889–898.

191. (a) T.G. Robertazzi, *Performance Evaluation of High Speed Switching Fabrics and Networking: ATM, Broadband ISDN and MAN Technology*, IEEE Press, 1993 (now distributed by Wiley).

192. (b) T.G. Robertazzi, "Processor Equivalence for a Linear Daisy Chain of Load Sharing Processors," *IEEE Transactions on Aerospace and Electronic Systems*, vol. 29, no. 4, Oct. 1993, pp. 1216–1221.

193. T.G. Robertazzi, *Planning Telecommunication Networks*, Wiley and IEEE Press, 1999.

194. T.G. Robertazzi, *Computer Networks and Systems: Queueing Theory and Performance Evaluation*, 3rd edition., Springer-Verlag, NY, NY, 2000.

195. T.G. Robertazzi, "Ten Reasons to Use Divisible Load Theory," *Computer*, vol. 36, no. 5, May 2003, pp. 63–68.

196. T.G. Robertazzi, *Networks and Grids: Technology and Theory*, Springer, NY, 2007.

197. T.G. Robertazzi, "A Product Form Solution for Tree Networks with Divisible Loads," *Parallel Processing Letters*, vol. 21, no.1, March 2011, pp. 13–20.

198. T.G. Robertazzi, *Introduction to Computer Networking*, Springer, 2017.

199. L. Roberts, "Extensions of Packet Communication Technology to a Hand Held Personal Terminal," *Proceedings of the Spring Joint Computer Conference*, AFIPS, 1972, pp. 295–298.

200. M. A. Rodrigues, "Erasure Node: Performance Improvements for the IEEE 802.6 MAN," *Proceedings of IEEE INFOCOM '90*, San Francisco CA, June 1990, pp. 636–643.

201. R. Rom and M. Sidi, *Multiple Access Protocols: Performance and Analysis*, Springer-Verlag, New York, 1990.

202. K.W. Ross, *Multiservice Loss Models for Broadband Telecommunication Networks*, Spring Verlag, NY, NY, 1995.

203. J.P. Ryan, "WDM: North American Deployment Trends," *IEEE Communications Magazine*, vol. 36, no. 2, Feb. 1998, pp. 40–44.

204. T.N. Saadawi, M.H. Ammar and A. El Hakeem, *Fundamentals of Telecommunication Networks*, Wiley, NY, NY, 1994.

205. A. L. Samuel. Some Studies in Machine Learning Using the Game of Checkers. *IBM Journal of Research and Development*, 3(3):210–229, jul 1959. ISSN 0018-8646. doi: 10.1147/rd.33.0210. URL http://ieeexplore.ieee.org/document/5392560/.

206. V. Scarani and C. Kurtsiefer, "The Black Paper of Quantum Cryptography: Real Implementation Problems," *Theoretical Computer Science*, vol. 560, 2014, pp. 27–32.

207. J.M. Schopf and B. Nitzberg, "Grids: The Top Ten Questions," *Scientific Programming*, IOS Press, vol. 10, no. 2, 2002, pp. 103–111.

208. M. Schwartz, *Telecommunication Networks: Protocols, Modeling and Analysis*, Addison-Wesley, Reading, Mass., 1987.

209. L. Schwiebert, S.K.S. Gupta and J. Weinmann, "Research Challenges in Wireless Networks of Biomedical Sensors," *ACM Sigmobile*, 2001, pp. 151–165.

210. R.C. Shah and J.M. Rabaey, "Energy Aware Routing for Low Energy Ad Hoc Sensor Networks," *Proc. of the 3rd IEEE Wireless, Communications and Networking Conference*, 2002, pp. 350–355.

211. L. Shi, *Task Scheduling in Modern Data Center: Task Placement and Resource Allocation*, PhD Thesis, Dept. of Electrical and Computer Engineering, Stony Brook University, 2016.

212. B.A. Shirazi and A.R. Hurson, *Scheduling and Load Balancing in Parallel and Distributed Systems*, Wiley -IEEE Computer Society Press, 1995.

213. A. Shokripour, M. Othman, H. Ibrahim and S. Subramaniam, "A Method for Scheduling Heterogeneous Multi-Installment Systems," *Lecture Notes in Artificial Intelligence 6592*, Springer, 2011, pp. 31–41. In N.T. Nquyen et al. (eds.).

214. A. Shokripour, M. Othman, H. Ibrahim and S. Subramaniam, "New Method for Scheduling Heterogeneous Multi-Installment Systems, *Future Generation Computer Systems*, vol. 28, no. 8, 2012, pp. 1205–1216.

215. C.E. Siller, editor, *SONET/SDH: A Sourcebook of Synchronous Networking*, Wiley - IEEE Press, 1996.

216. S. Siwamogsatham, "10 Gigabit Ethernet," *www.cse.wustl.edu/~jain/cis788--99/ftp/10gbe/index.html*, 1999.

217. J. Sohn and T.G. Robertazzi, "Optimal Load Sharing for a Divisible Job on a Bus Network," *IEEE Transactions on Aerospace and Electronic Systems*, vol. 32, no. 1, Jan. 1996, pp. 34–40.

218. (a) J. Sohn, T.G. Robertazzi and S. Luryi, "Optimizing Computing Costs using Divisible Load Analysis," *IEEE Transactions on Parallel and Distributed Systems*, vol. 9, March 1998, pp. 225–234. Also related: T.G. Robertazzi, S. Luryi and J. Sohn, Load Sharing Controller for Optimizing Monetary Cost, US Patent 5,889,989, March 30, 1999. T.G. Robertazzi, S. Luryi and S. Charcranoon, Load Sharing Controller for Optimizing Resource Utilization Cost, US Patent 6,370,560, April 9, 2002.

219. (b) J. Sohn and T.G. Robertazzi, "Optimal Time Varying Load Sharing for Divisible Loads," *IEEE Transactions on Aerospace and Electronic Systems*, vol. 34, no. 3, July 1998, pp. 907–924.

220. W. Stallings, *High-Speed Networks and Internets: Performance and Quality of Service*, Prentice-Hall, Upper Saddle River NJ, 2002.

221. S. Suresh, V. Mani and S.N. Omkar, "The Effect of Start-Up Delays in Scheduling Divisible Load on Bus Networks: An Alternate Approach," *Computers and Mathematics with Applications*, vol. 46, 2003, 1545–1557.

222. (a) S. Suresh, C. Run, H.J. Kim, T.G. Robertazzi and Y.-I. Kim, "Scheduling Second Order Computational Load in Master-Slave Paradigm," *IEEE Transactions on Aerospace and Electronic Systems*, vol. 48, no. 1, 2012, pp. 380–393.

223. (b) S. Suresh, H.J. Kim, C. Run and T.G. Robertazzi, "Scheduling Nonlinear Divisible Loads in a Single Level Tree Network," *Journal of Supercomputing*, vol. 61, 2012, pp. 1068–1088.

224. S. Suresh, H. Huang and H.J. Kim, "Scheduling in Computer Cloud with Multiple Data Banks using Divisible Load Paradigm", *IEEE Transactions on Aerospace and Electronic Systems*, vol. 51, no. 2, 2015, pp. 1288–1297.

225. T.H. Szymanski, "A VLSI Comparison between Crossbar and Switch Recursive Banyan Interconnection Networks," *Proc. of the International Conference on Parallel Processing*, Aug. 1986, pp. 192–199.

226. A.S. Tanenbaum, *Computer Networks*, 3rd edition, Prentice-Hall, 1996.

227. A.S. Tanenbaum, *Computer Networks*, 4th edition, Prentice-Hall, 2002.

228. C.-K. Toh, *Ad Hoc Mobile Wireless Networks: Protocols and Systems*, Prentice-Hall, 2002.

229. S.J. Vaughan-Nichols, "Will 10-Gigabit Ethernet Have a Bright Future?," *Computer*, June 2002, pp. 22–24.

230. B. Veeravalli, X. Li, and, K.C. Chung, "On the Influence of Start-up Costs in Scheduling Divisible Loads on Bus Networks," *IEEE Transactions on Parallel and Distributed Systems*, vol. 11, no. 12, pp. Dec. 2000, pp. 1288–1305.

231. B. Veeravalli and W.H. Min, "Scheduling Divisible Loads on Heterogenous Linear Daisy Chain Networks with Arbitrary Processor Release Times," *IEEE Transactions on Parallel and Distributed Systems*, vol, 15, no. 3, 2004, pp. 273–288.

232. R.C. Walker, B. Amrutur and R.W. Dugan, "Decoding Method and Decoder for 64b/66b Coded Packetized Serial Data," *US Patent 6650638 B1*, November 18, 2003.

233. R.C. Walker, B. Amrutur and R.W. Dugan, "Coding Method and Coder for Coding Packetized Serial Data with Low Overhead," *US Patent 6718491 B1*, April 6, 2004.

234. (a) I.Y. Wang and T.G. Robertazzi, "Recursive Computation of Steady State Probabilities of Non-Product Form Queueing Networks Associated with Computer Network Models," *IEEE Transactions on Communications*, vol. 38, no. 1, Jan. 1990, pp. 115–117.

235. (b) I.Y. Wang and T.G. Robertazzi, "Service Stage Petri Net Models with Product Form Solution," *Queueing Systems*, vol. 7, no. 3, 1990, pp. 355–374.

236. Z. Wang and J. Crowcroft, "Quality-of-Service Routing for Supporting Multimedia Applications," *IEEE Journal of Selected Areas in Communications*, vol. 14, no. 7, Sept. 1996, pp. 1228–1234.

237. K. Wang and T.G. Robertazzi, "Scheduling Divisible Loads with Nonlinear Communication Time," *IEEE Transactions on Aerospace and Electronic Systems*, vol. 51, no. 3, 2015, pp. 2479–2485.

238. X. Wang, and B. Veeravalli, "Performance Characterization on Handling Large-scale Partitionable Workloads on Heterogeneous Networked Compute Platforms," *IEEE Transactions on Parallel and Distributed Computing*, accepted 2019.

239. A.X. Widmer and P.A. Franaszek, "A DC-Balanced, Partitioned Block, 8b/10b Transmission Code," *IBM Journal of Research and Development*, vol. 27, no. 5, Sept. 1985, pp. 440–451.

240. J. Williams, "The 802.11b Security Problem - Part I," *IEEE ITPro (Information Technology Professional)*, Nov./Dec. 2001, pp. 91–96.

241. D.H. Woo and H.-H. S. Lee, "Extending Amdahl's Law for Energy-Efficient Computing in the Many-Core Era," *Computer*, Dec. 2008, pp. 24–31.

242. M.E. Woodward, *Communication and Computer Networks: Modelling with Discrete-Time Queues*, Wiley and IEEE Computer Society Press, 1983.

243. C.-I. Wu and T.-Y. Feng, "Tutorial: Interconnection Networks for Parallel and Distributed Processing," IEEE Computer Society Press, Los Alamitos CA, 1984.

244. F. Wu, *Scheduling Load Computation and Communication in Virtualized Networks*, PhD Thesis, Stony Brook University, Dec. 2018.

245. L. Xiaolin, V. Bharadwaj and C.C. Ko, "Experimental Study on Processing Divisible Loads for Large Size Image Processing Applications using PVM Clusters," *International Journal of Computers and Applications*, ACTA Press, July 2001.

246. H. Xuan, S. Wei, W. Tong, et al., "Fault-Tolerant Scheduling Algorithm with Re-allocation for Divisible Task," *IEEE Access*, 2018.

247. H. Yamamoto, M. Tsuru and Y. Oie, "Parallel Transferable Uniform Multi-round Algorithm for Achieving Minimum Application Turnaround Times for Divisible Workload," Lecture Notes in Computer Science 3726, 2005, pp. 817–828, in L.T. Yang et al. (eds.), *Proceedings of HPCC 2005*.

248. H. Yamamoto, M. Tsuru and Y. Oie, "Parallel Transferable Uniform Multi-round Algorithm in Heterogeneous Distributed Computing Environment," Lecture Notes in Computer Science 4208, 2006, pp, 51–60, in M. Gerndt and D. Kranzlmuller (eds.), *Proceedings of HPCC 2006*.

249. Y. Yang, K. van der Raadt and H. Casanova, "Multiround Algorithms for Scheduling Divisible Loads," *IEEE Transactions on Parallel and Distributed Systems*, vol. 16, no. 11, 2005, pp. 1092–1102.

250. Y.-S. Yeh, M.G. Hluchyj and A.S. Acampora, "The Knockout Switch: A Simple, Modular Architecture for High Performance Packet Switching," *Journal on Selected Areas in Communication*, vol. SAC-5, no. 8, Oct. 1987, pp. 1274–1283.

251. (a) Z. Ying, *Signature Search and Computing Cost Optimization in Distributed Networks*, Ph.D Thesis, Stony Brook University, 2014.

252. (b) Z. Ying and T.G. Robertazzi, "Signature Searching in a Networked Collection of Files," *IEEE Transactions on Parallel and Distributed Systems*, vol. 25, no. 5, 2014, pp. 1339–1348.

253. H. Yoon, K.Y. Lee and M.T. Liu, "Performance Analysis of Multi-buffered Packet-Switching Networks in Multiprocessor Systems," *IEEE Transactions on Computers*, vol. 39, no. 3, March 1990, pp. 319–327.

254. M.C. Yuang, "Survey of Protocol Verification Techniques Based on Finite State Machine Models," *Proceedings of Computer Networking Symposium*, Washington D.C., 1988, pp. 164–172.

255. Z. Zhang, *A Novel Analysis Method for Parallel Processing*, Ph.D Thesis, Stony Brook University, Dec. 2014.

256. Z. Zhang and T.G. Robertazzi, "Scheduling Divisible Loads in Gaussian, Mesh and Torus Networks of Processors," *IEEE Transactions on Computers*, vol. 64, no. 11, 2015, pp. 3249–3264.

257. J. Zhang, *Data Distribution Equivalence for Data Intensive Interconnection Networks*, Ph.D Thesis, Stony Brook University, Aug. 2018.

258. J. Zheng and M.J. Lee, "Will 802.15.4 Make Ubiquitous Networking A Reality? A Discussion on a Potential Low Power, Low Bit Rate Standard," *IEEE Communications Magazine*, June 2004, pp. 140–146.

259. "Application Layer Packet Classifier for Linux", L7-filter.sourceforge.net, 2019. [Online]. Available: *http://l7-filter.sourceforge.net/*. [Accessed: 23-Sep-2019].

260. "Datasets of Canadian Institute for Cybersecurity", unb.ca, 2019. [Online]. Available: *https://www.unb.ca/cic/datasets/index.html*. [Accessed: 23-Sep-2019].

261. "MAWI Working Group Traffic Archive", mawi.wide.ad.jp, 2019. [Online]. Available: *http://mawi.wide.ad.jp/mawi/*. [Accessed: 23-Sep-2019].

262. "WAND Group", wand.net.nz, 2019. [Online]. Available: *https://wand.net.nz/wits/*. [Accessed: 23-Sep-2019].

263. SNAP: Web data: Amazon reviews", Snap.stanford.edu, 2019. [Online]. Available: *https://snap.stanford.edu/data/web-Amazon.html*. [Accessed: 23-Sep-2019].

264. "Data.gov", Data.gov, 2019. [Online]. Available: *https://www.data.gov/*. [Accessed: 23-Sep-2019].

265. "Database of Eurostat", ec.europa.eu. [Online]. Available: *https://ec.europa.eu/eurostat/data/database*. [Accessed: 23-Sep-2019].

266. "Open Images Dataset V5", storage.googleapis.com. [Online]. Available: *https://storage.googleapis.com/openimages/web/index.html*. [Accessed: 23-Sep-2019].

267. Taimur Bakhshi and Bogdan Ghita. On internet traffic classification: A two-phased machine learning approach. *Journal of Computer Networks and Communications*, 2016.

268. Taimur Bakhshi and Bogdan Ghita. On internet traffic classification: A two-phased machine learning approach. *Journal of Computer Networks and Communications*, 2016.

269. Stephen D Bay, Dennis F Kibler, Michael J Pazzani, and Padhraic Smyth. The uci kdd archive of large data sets for data mining research and experimentation. *SIGKDD explorations*, 2(2):81–85, 2000.

270. David M Blei, Andrew Y Ng, and Michael I Jordan. Latent dirichlet allocation. *Journal of machine Learning research*, 3 (Jan):993–1022, 2003.

271. Andreas Blenk, Patrick Kalmbach, Patrick Van Der Smagt, and Wolfgang Kellerer. Boost online virtual network embedding: Using neural networks for admission control. In 2016 12th International Conference on Network and Service Management (CNSM), pages 10–18. IEEE, 2016.

272. Andreas Blenk, Patrick Kalmbach, Johannes Zerwas, Michael Jarschel, Stefan Schmid, and Wolfgang Kellerer. Neurovine: A neural preprocessor for your virtual network embedding algorithm. In *IEEE INFOCOM* 2018-IEEE Conference on Computer Communications, pages 405–413. IEEE, 2018.

273. Yu-ning Dong, Jia-jie Zhao, and Jiong Jin. Novel feature selection and classification of internet video traffic based on a hierarchical scheme. *Computer Networks*, 119:102–111, 2017.

274. Zeon Trevor Fernando, I Sumaiya Thaseen, and Ch Aswani Kumar. Network attacks identification using consistency based feature selection and self organizing maps. In 2014 First International Conference on Networks & Soft Computing (ICNSC2014), pages 162–166. IEEE, 2014.

275. Alessandro Finamore, Marco Mellia, Michela Meo, and Dario Rossi. Kiss: Stochastic packet inspection classifier for udp traffic. *IEEE/ACM Transactions on Networking (TON)*, 18(5): 1505–1515, 2010.

276. Edward B Fowlkes and Colin L Mallows. A method for comparing two hierarchical clusterings. *Journal of the American statistical association*, 78(383):553–569, 1983.

277. Francesco Gringoli, Luca Salgarelli, Maurizio Dusi, Niccolo Cascarano, Fulvio Risso, et al. Gt: picking up the truth from the ground for internet traffic. *ACM SIGCOMM Computer Communication Review*, 39(5):12–18, 2009.

278. Patrick Haffner, Subhabrata Sen, Oliver Spatscheck, and Dongmei Wang. Acas: automated construction of application signatures. In *Proceedings of the 2005 ACM SIGCOMM workshop on Mining network data*, pages 197–202. ACM, 2005.

279. John J Hopfield. Neural networks and physical systems with emergent collective computational abilities. *Proceedings of the national academy of sciences*, 79(8):2554–2558, 1982.

280. Tiansi Hu and Yunsi Fei. Qelar: A machine-learning-based adaptive routing protocol for energy-efficient and lifetime-extended underwater sensor networks. *IEEE Transactions on Mobile Computing*, 9(6):796–809, 2010.

281. Guang-Bin Huang, Qin-Yu Zhu, and Chee-Kheong Siew. Extreme learning machine: theory and applications. *Neurocomputing*, 70 (1–3):489–501, 2006.

282. Victoria Huang, Gang Chen, and Qiang Fu. Effective scheduling function design in sdn through deep reinforcement learning. In *ICC 2019–2019 IEEE International Conference on Communications (ICC)*, pages 1–7. IEEE, 2019.

283. Neminath Hubballi and Mayank Swarnkar. Bitcoding: Network traffic classification through encoded bit level signatures. *IEEE/ACM Transactions on Networking*, (99):1–13, 2018.

284. Lawrence Hubert and Phipps Arabie. Comparing partitions. *Journal of classification*, 2(1):193–218, 1985.

285. Vijay R Konda and John N Tsitsiklis. Actor-critic algorithms. In *Advances in neural information processing systems*, pages 1008–1014, 2000.

286. Yann LeCun, Léon Bottou, Yoshua Bengio, Patrick Haffner, et al. Gradient-based learning applied to document recognition. *Proceedings of the IEEE*, 86(11):2278–2324, 1998.

287. Rui Li, Xi Xiao, Shiguang Ni, Haitao Zheng, and Shutao Xia. Byte segment neural network for network traffic classification. In *2018 IEEE/ACM 26th International Symposium on Quality of Service (IWQoS)*, pages 1–10. IEEE, 2018.

288. Timothy P Lillicrap, Jonathan J Hunt, Alexander Pritzel, Nicolas Heess, Tom Erez, Yuval Tassa, David Silver, and Daan Wierstra. Continuous control with deep reinforcement learning. *arXiv preprint arXiv:1509.02971*, 2015.

289. Shih-Chun Lin, Ian F Akyildiz, Pu Wang, and Min Luo. Qos-aware adaptive routing in multi-layer hierarchical software defined networks: A reinforcement learning approach. In *2016 IEEE International Conference on Services Computing (SCC)*, pages 25–33. IEEE, 2016.

290. Manuel Lopez-Martin, Belen Carro, Antonio Sanchez-Esguevillas, and Jaime Lloret. Network traffic classifier with convolutional and recurrent neural networks for internet of things. *IEEE Access*, 5:18042–18050, 2017.

291. Mohammad Lotfollahi, Ramin Shirali Hossein Zade, Mahdi Jafari Siavoshani, and Mohammdsadegh Saberian. Deep packet: A novel approach for encrypted traffic classification using deep learning. *arXiv preprint arXiv:1709.02656*, 2017.

292. Andrew Moore, Denis Zuev, and Michael Crogan. Discriminators for use in flow-based classification. Technical report, 2013.

293. Andrew W Moore and Konstantina Papagiannaki. Toward the accurate identification of network applications. In *International Workshop on Passive and Active Network Measurement*, pages 41–54. Springer, 2005.

294. Andrew W Moore and Denis Zuev. Internet traffic classification using bayesian analysis techniques. In *ACM SIGMETRICS Performance Evaluation Review*, volume 33, pages 50–60. ACM, 2005.

295. Dmitry Mukhutdinov, Andrey Filchenkov, Anatoly Shalyto, and Valeriy Vyatkin. Multi-agent deep learning for simultaneous optimization for time and energy in distributed routing system. *Future Generation Computer Systems*, 94:587–600, 2019.

296. Tran Anh Quang Pham, Yassine Hadjadj-Aoul, and Abdelkader Outtagarts. Deep reinforcement learning based qos-aware routing in knowledge-defined networking. In *International Conference on Heterogeneous Networking for Quality, Reliability, Security and Robustness*, pages 14–26. Springer, 2018.

297. Lutz Prechelt. Early stopping-but when? In *Neural Networks: Tricks of the trade*, pages 55–69. Springer, 1998.

298. William M Rand. Objective criteria for the evaluation of clustering methods. *Journal of the American Statistical association*, 66 (336):846–850, 1971.

299. Matthew Roughan, Subhabrata Sen, Oliver Spatscheck, and Nick Duffield. Class-of-service mapping for qos: a statistical signature-based approach to ip traffic classification. In *Proceedings of the 4th ACM SIGCOMM conference on Internet measurement*, pages 135–148. ACM, 2004.

300. Peter J Rousseeuw. Silhouettes: a graphical aid to the interpretation and validation of cluster analysis. *Journal of computational and applied mathematics*, 20:53–65, 1987.

301. Gavin A Rummery and Mahesan Niranjan. *On-line Q-learning using connectionist systems*, volume 37. University of Cambridge, Department of Engineering Cambridge, England, 1994.

302. Saim Salman, Christopher Streiffer, Huan Chen, Theophilus Benson, and Asim Kadav. Deepconf: Automating data center network topologies management with machine learning. In *Proceedings of the 2018 Workshop on Network Meets AI & ML*, pages 8–14. ACM, 2018.

303. John Schulman, Filip Wolski, Prafulla Dhariwal, Alec Radford, and Oleg Klimov. Proximal policy optimization algorithms. *arXiv preprint arXiv:1707.06347*, 2017.

304. Muhammad Shafiq, Xiangzhan Yu, Ali Kashif Bashir, Hassan Nazeer Chaudhry, and Dawei Wang. A machine learning approach for feature selection traffic classification using security analysis. *The Journal of Supercomputing*, 74(10):4867–4892, 2018.

305. David Silver, Julian Schrittwieser, Karen Simonyan, Ioannis Antonoglou, Aja Huang, Arthur Guez, Thomas Hubert, Lucas Baker, Matthew Lai, Adrian Bolton, et al. Mastering the game of go without human knowledge. *Nature*, 550(7676):354, 2017.

306. William Stallings. *SNMP, SNMPv2, SNMPv3, and RMON 1 and 2*. Addison-Wesley Longman Publishing Co., Inc., 1998.

307. Guanglu Sun, Teng Chen, Yangyang Su, and Chenglong Li. Internet traffic classification based on incremental support vector machines. *Mobile Networks and Applications*, 23(4):789–796, 2018.

308. Haifeng Sun, Yunming Xiao, Jing Wang, Jingyu Wang, Qi Qi, Jianxin Liao, and Xiulei Liu. Common knowledge based and one-shot learning enabled multi-task traffic classification. *IEEE Access*, 2019.

309. Richard S Sutton and Andrew G Barto. *Reinforcement learning: An introduction*. MIT Press, 1988.

310. Gerald Tesauro. Reinforcement learning in autonomic computing: A manifesto and case studies. *IEEE Internet Computing*, 11(1): 22–30, 2007.

311. "thomasbhatia/OpenDPI", GitHub. [Online]. Available: *https://github.com/thomasbhatia/OpenDPI*. [Accessed: 23-Sep-2019].

312. Pascal Vincent, Hugo Larochelle, Yoshua Bengio, and Pierre-Antoine Manzagol. Extracting and composing robust features with denoising autoencoders. In *Proceedings of the 25th international conference on Machine learning*, pages 1096–1103. ACM, 2008.

313. Oriol Vinyals, Timo Ewalds, Sergey Bartunov, Petko Georgiev, Alexander Sasha Vezhnevets, Michelle Yeo, Alireza Makhzani, Heinrich Küttler, John Agapiou, Julian Schrittwieser, et al. Starcraft ii: A new challenge for reinforcement learning. *arXiv preprint arXiv:1708.04782*, 2017.

314. Mowei Wang, Yong Cui, Xin Wang, Shihan Xiao, and Junchen Jiang. Machine learning for networking: Workflow, advances and opportunities. *IEEE Network*, 32(2):92–99, 2018.

315. Sen Wang, Jun Bi, Jianping Wu, Athanasios V Vasilakos, and Qilin Fan. Vne-td: a virtual network embedding algorithm based on temporal-difference learning. *Computer Networks,* 2019.

316. Yuxi Xie, Hanbo Deng, Lizhi Peng, and Zhenxiang Chen. Accurate identification of internet video traffic using byte code distribution features. In *International Conference on Algorithms and Architectures for Parallel Processing*, pages 46–58. Springer, 2018.

317. Zhiyuan Xu, Jian Tang, Jingsong Meng, Weiyi Zhang, Yanzhi Wang, Chi Harold Liu, and Dejun Yang. Experience-driven networking: A deep reinforcement learning based approach. In *IEEE INFOCOM* 2018-IEEE Conference on Computer Communications, pages 1871–1879. IEEE, 2018.

318. Jinghua Yan. A survey of traffic classification validation and ground truth collection. In 2018 8th International Conference on Electronics Information and Emergency Communication (ICEIEC), pages 255–259. IEEE, 2018.

319. Lei Yu and Huan Liu. Feature selection for high-dimensional data: A fast correlation-based filter solution. In *Proceedings of the 20th international conference on machine learning (ICML-03)*, pages 856–863, 2003.

320. Xiaochun Yun, Yipeng Wang, Yongzheng Zhang, and Yu Zhou. A semantics-aware approach to the automated network protocol identification. *IEEE/ACM Transactions on Networking (TON)*, 24 (1):583–595, 2016.

321. Jun Zhang, Xiao Chen, Yang Xiang, Wanlei Zhou, and Jie Wu. Robust network traffic classification. *IEEE/ACM Transactions on Networking (TON)*, 23(4):1257–1270, 2015.

Index

© Springer Nature Switzerland AG 2020
T. G. Robertazzi, L. Shi, *Networking and Computation*, https://doi.org/10.1007/978-3-030-36704-6

Printed in the United States
by Baker & Taylor Publisher Services